LONDON
ENGLAND
LEAK
RESEARCH · ENGINEERING · MANUFACTURING

The Leak premises at 470 Uxbridge Road, Shepherds Bush, where the first "Point One" amplifier was developed. (See picture credits overleaf).

firsts in High Fidelity

The Products and History of H. J. Leak & Co. Ltd.

By Stephen Spicer

With Contributions by Former Employees:

Ted Ashley Phil Biggs John Bolingbroke Jackie Cadby
Angus Gilchrist Keshoe Gulhane John Leak Simon Leak Ian Pullen
Norman Rowe Ray Whitcombe Steve Wilkins George Wise

And Also:

Gordon Arbib Ken Barker Sylvia Barlow Tony Coleman
Syd McClory Joe Middleton Peter Stinson

Audio Amateur Press

Published by: Audio Amateur Press
Agents: Old Colony Sound Laboratory
P.O. Box 243, Peterborough, NH 03458, USA.
www.audioXpress.com
e-mail: custserv@audioXpress.com

Text by S. Spicer except where shown otherwise.
Photography and Digital Image Processing by S. Spicer
except where shown otherwise.
Typeset, Graphic Design and Layout by S. Spicer.
Author e-mail: stephenspicer@telstra.com

Printed in the United States of America

First Edition, 2000; Second Revision, 2000.

ISBN: 1-882580-31-1
Library of Congress Catalog Card: No. 00 131 652

Abbreviations for Sources of Reproduced Images:
HFN: Hi-Fi News magazine, Link House Publications, UK.
WW: Wireless World magazine, Reed Business Information Ltd, UK.
AE: Audio Engineering magazine, USA.
T&R Group: Transformer and Rectifier Group, UK.

Note:
The term "Leak" when used alone in the text means "H. J. Leak & Co. Ltd."
Electronic "tubes" are referred to as "valves" in the text.
British spelling is adopted for words such as: colour, harmonise, aluminium, etc.

Front Cover:
A Leak Stereo 60 Amplifier.

Rear Cover: The Varislope Stereo pre-amp as styled for the American market.

Opposite Below: The model designation from the chassis of the Leak TL/12 amplifier.

Facing Previous Page: The Leak premises at 470 Uxbridge Rd, Shepherds Bush. (© "Hammersmith and Fulham Archives and Local History Collection". Photo by Christine Bayliss, 1972. Researched by Simon Leak and Elizabeth Perrins.)

for Julie

"POINT ONE"
T.L./12 12W. AMPLIFIER
MADE IN ENGLAND
BY
H. J. LEAK & CO. LTD.
WESTWAY FACTORY ESTATE
LONDON W.3.

Contents

Opposite: The 1948 Leak Moving Coil Pickup and Arm fitted to a Connoisseur three-speed turntable from 1950.

LEAK
THREE SPEED
VARIABLE
Connoisseur
A.R. SUGDEN

LEAK
"POINT ONE"
AMPLIFIERS
KT66
G.E.C.

Introduction

In 1972 my father brought home a Leak "Stereo 20" for use as a rear channel amplifier in an experimental Hafler surround system. Being in my impressionable teens, the under-chassis layout of the Stereo 20 made me certain that the designer of this masterpiece knew what he was doing, as it was the most beautiful piece of electronics assembly I had ever seen. From that day on, I was a Leak addict for life.

My father's sound system was "off limits" to me, but in due course the Hafler experiment was no longer of interest to him, and so the Stereo 20 became part of a second sound system for use by my brothers and I. By 1976 I had learnt of all the other amplifiers that Leak had made and during my holidays from University I built a pair of Leak TL/50 Plus amplifiers using genuine Leak output transformers.

Over the years, more information about Harold Leak and his company came to the surface. One of the biggest mistakes of my life was passing up an opportunity to interview him on the phone in early in 1989. However at the time he was not well and I was too young to know the right questions to ask.

Then in 1998 I was shown a letter from Ted Ashley that he wrote describing some of the early work on FM tuner design at Leak. Ted had been Chief Engineer with Leak from 1952 to 1969. I could scarcely believe this find! Here was a person who had been with the Leak company during the Golden Age of the British hi-fi industry, and who could fill in the background detail of the operations, products and staff at Leak.

After contacting Ted by phone, he kindly agreed to help me compile this book. In a slightly cheeky move, I sent him a list of around 160 questions by mail, which must have looked very daunting. But Ted was able to comment on nearly all of them, with the discussion being recorded by telephone from Australia.

For more than a year following these telephone discussions, Ted generously assisted with reviewing the draft book and refining the information in it. When I first contacted Ted I never envisaged requiring him to be called upon for such an extended period. Thank you Ted, for your patience, friendship, the time you have spent helping me with researching this history, and perhaps most of all for your excellent memory!

Opposite: The KT66 beam-tetrodes used in the TL/12 amplifier, with the Leak logo from a company publication. (Logo graphic reproduced by permission Leak Electronics Ltd.)

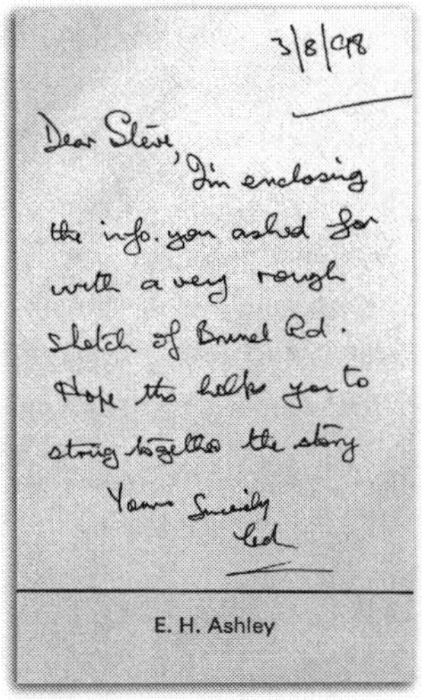

3/8/98

Dear Steve,
I'm enclosing the info. you asked for with a very rough sketch of Brunel Rd. Hope this helps you to string together the story
Yours Sincerely
Ted

E. H. Ashley

This is a note received during the course of research for this book from Leak's Chief Engineer, Ted Ashley, who was instrumental in making this book possible.

Acknowledgements

This book has been made possible by the generous enthusiasm of many people. As mentioned previously, Ted Ashley was pivotal in bringing this book to life. Other ex-staff who were able add their own perspective to the history were: Phil Biggs, John Bolingbroke, Ray Whitcombe, Steve Wilkins, Angus Gilchrist, Norman Rowe, Keshoe Gulhane, Ian Pullen and Jackie Cadby. A number of these people contributed sections of text to the book, making the story of Leak much more tangible. George Wise also provided many details on the projects he undertook for Leak, while additional information on the development of hi-fi after the Second World War came from Peter Stinson, Syd McClory and Joe Middleton in Australia. I have been thrilled to talk to all these people and thank them all very much for their assistance in this project.

Fortunately, the research work for this book led me to Harold Leak's sons, John and Simon. This introduction came about with the help of Ted Ashley. Simon added his own enthusiasm to the project by locating photos and contacting others who were close to his father. John, who worked for the company in the 1960's, contributed many details to the book as well as several extracts of text. I would like to thank the Leak family, and the family of Frank Hawkins for their assistance. The photos of the Brunel Road factory only surfaced after Simon Leak and Elizabeth Perrins approached the developers of the original Westway Factory Estate. Special thanks are due to Douglas Gardner of Brixton Estate PLC for his crucial assistance in locating these photos. The photos of earlier Leak premises were only made possible through the records of the "Hammersmith and Fulham Archives and Local History Collection".

As the compilation of the book progressed it was revealing to learn how many people were able to add to the book directly, or draw on assistance from others. In this regard I thank Sylvia Barlow, Gordon Arbib, Tony Coleman, Ken Barker, Steve Dove and Gordon Provan.

The photographs of equipment, particularly the rarer Leak models, were made possible by Angus and Keith Gilchrist, Jonathan Hill, Steve Mann, Stephen Robinson, George Elliston, Paul Schereck, Stephen Langley and Phillip Spicer. George Elliston and Peter Stinson reviewed the draft of this book as it neared completion – their enthusiasm for the project was very motivating.

Steve Harris and Stuart Perry (from Hi-Fi News magazine, UK), Ben Duncan and John Howse were all influential in the early stages, or in progressing the book towards publication. Without their advice and assistance the project could have taken one of many wrong turns. Richard White (of Hi-Fi World magazine, UK) was also most supportive, being able to provide historical photos, information and

Along with many others, Simon Leak (left, with a Troughline FM tuner) and Ray Whitcombe (right, with a Stereo 70 amplifier) contributed significantly to the information in this book.

advice on publication. Maurice Austin provided technical assistance regarding printing industry software and file formats – as well as having early copies of the Hi-Fi Year books to refer to.

The Leak name has been inherited by International Audio Group Ltd (IAG) of Huntingdon, United Kingdom. Importantly, the Chairman of IAG, Stan Curtis, was most supportive of this book, and as a result it has been possible to include many historical photos and advertisements in this book.

The information that everyone has supplied fills a large gap in a part of the British audio history. There are many other companies that deserve equal treatment, but in my case, H. J. Leak & Co. Ltd. has always been a inspiration for me and it has resulted in this historical perspective.

The final dedication and acknowledgment goes to all the staff that worked at H. J. Leak & Co. Ltd., without whom there would not be any heritage to write about.

At the 1968 Hi Fi Show Janet Leak (the wife of Harold's son John) wore an aluminised Melinex dress as a an eye-catching promotion for the Leak range. Note the Thorens TD150 turntable with Leak pickup arm. (Reproduced by permission Leak Electronics Ltd. Source: HFN, July 1968)

Foreword

Conceived in the 1920's, and born in the 1950's – "hi-fi" was initially an obsession for technically aware music-lovers, but later spread to consumers world-wide. Enthusiasts would spend every creative spare moment building or modifying equipment to gain an edge in sound quality.

This passion for perfection saw many companies in the United Kingdom developing innovative products that incorporated new ideas, and that set performance standards for other manufacturers to aspire to. In many cases the ultimate audio system of the day would comprise equipment from Leak, Wharfedale and Garrard. But it could have also been Quad, Goodmans and Connoisseur; or Rogers, Tannoy and Ferrograph. The list of British hi-fi manufacturers is extensive in number, and impressive for the technical excellence achieved.

Forty years on, collection of vintage audio equipment has developed into a hobby of its own. Chronicled here is just one leading manufacturer of the day – H. J. Leak & Co. Ltd. This firm brought hi-fi products to an international market, and continues to be respected as a company at the apex of British audio.

By the early 1950's, H. J. Leak & Co. Ltd. was recognised by many for making the best audio amplifiers in the world. It was a well deserved reputation, underpinned by the "Point One" trademark that emphasised the low distortion designs pioneered by the firm.

Leak produced affordable high-quality equipment, and after the Second World War turned their expertise to pickups, radio tuners, loudspeakers and turntables. An astonishing range of products was produced from 1948 to 1969 - around 45 distinct designs emerged from the company during this period. The development of these products and the people who contributed to the international success of Leak are presented here for "vintage audio enthusiasts" thirsty for knowledge about how it happened.

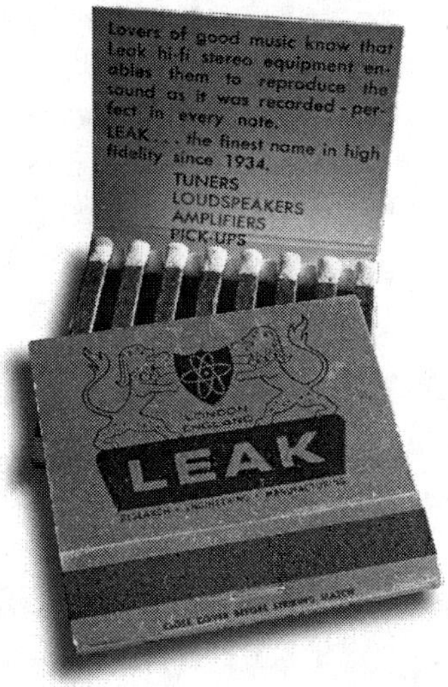

"I inherited an interest in Leak equipment from my Dad, who left me his not un-extensive collection, reaching back to a 'Point One' and thence an example of nearly everything subsequent. He took me up to Brunel Road a couple of times - on one of the occasions, around 1968, we actually met Mr Leak himself - not unusual I'm sure, since it didn't seem a big place, but a very impressive occasion nonetheless.

One curiosity for a speaker manufacturer was that in their auditioning room they were using a pair of Quad electrostatics. Harold Leak was very cordial and gracious; it was noticeable in the quite extensive time he spent with us that he and my Dad's northern accents fed off each other and got progressively broader... As to the Quad electrostatics, he just shrugged and said 'Perspective'."

Steve Dove,
Audio Enthusiast,
PA, USA, 1999

Opposite: Muriel and Harold Leak at the Savoy Hotel, London, 1964. (Photo courtesy of the Leak family.)

1. Personalities

Harold Leak

Harold Leak photographed during the 1940's. His father was a Clerk in local government. (Photo courtesy of the Leak family.)

Harold Joseph Leak lived from January 8, 1907 to August 27, 1989 - a span of eighty-two years. For thirty-five of these years he owned and operated a company that became one of the most prominent names in the British audio industry.

He was born in West London to parents from Stoke-on-Trent. His interest in "wireless" or "radio" became evident while he was still at school in Southall. In 1921 he built a crystal set receiver – a battery-less device for receiving AM radio transmissions that could be heard faintly through headphones. This technical ability came to the attention of his physics master at school, who assisted Harold to set up a "Wireless Society" with the aim of receiving various radio stations, and to learn more about wireless technology – a technology that had undergone considerable advancement during the First World War.

After leaving school Harold was employed full time with a West-end London wine merchant from 1924 to 1929. He started a side-line "Wireless Doctor" repair business in 1926 which was a financial success and allowed him to leave the wine trade in 1929. He spent five years employed by Gaumont British Cinemas gaining experience with modern sound equipment. In 1934 he established his own company – H. J. Leak & Co. Initially a one-man enterprise, it grew steadily through the 1930's and was well positioned after the Second World War to take advantage of a new market for domestic high fidelity sound equipment. This business was sold to the Rank Organisation in 1969 for just over one million pounds.

In 1942 Harold married Muriel Short and subsequently moved his home to Ealing. He had known Muriel for some years – at the age of 15 she had fallen through ice into a pond and Harold had rescued her. As the company became more profitable, Harold was able to move his family home to "Holly Tree Farm" located in Gerrards Cross, Buckinghamshire. (Holly Tree Farm is a Tudor house based on 500 year-old cottages established by squatters rights on common land.)

Ted Ashley, who worked with Harold Leak from 1944 to 1969, describes Harold as: entrepreneurial, fair minded, very intelligent, conscientious in many ways, and on occasions dogmatic about things he believed in. He worked very hard when required, but under normal circumstances he would usually arrive at work just after the morning rush-hour. If he was keen on something he would work through a weekend or even come in during Christmas, if he had a "bee in his bonnet" that needed sorting out. Ray Whitcombe recalls that: "As an entrepreneur and astute business man he was up-front – he always had an eye for a new idea and was in today's parlance a head-hunter. Any idea of interest that appeared in technical circles would invite an enquiry and request to have a chat."

Harold Leak was the initiator of elegant technical solutions for the company's products. Many of the successful products that the Leak company launched could not have been realised without constant surveillance of industry journals for new developments, through which he also located suitable collaborators and potential staff for recruitment (and subsequent product development). Harold Leak identified these opportunities through publications such as Wireless World, the Journal of the Audio Engineering Society, and Audio magazine (USA). He also maintained an extensive library of reference books such as those by Terman, F. V. Hunt and, of course, Langford-Smith's classic "Radiotron Designers Handbook". These publications were also available to all staff at the company, and were often referenced in publications issued by H. J. Leak & Co. Ltd.

Harold Leak, around 1966. He was often, and still is, referred to as "HJL". (Reproduced by permission Leak Electronics Ltd. Source: H. J. Leak (Aust.) Pty. Ltd.)

Harold Leak preferred not to demonstrate his equipment to other people from home. However if there was a requirement (from a marketing or company policy viewpoint) for people to be brought to his home and also listen to a sound system, then he would have an appropriate system at home just for that purpose. This approach to product marketing was not usually preferred by Harold.

Harold's hobbies were bird watching and photography, and later on he very much enjoyed fishing, particularly in New Zealand and Scotland. (On one occasion his involvement with a local fishing club required him to provide bait for a club event. Later, he realised there was a National shortage of the bait, so all staff in the purchasing department at Leak were set the task of locating the necessary bait!)

This boat was built for Harold Leak, launched at Hamble in the UK, and was christened "Point One". It was rigged for shark fishing and capable of 26 knots. (Photo courtesy of the Leak family.)

Harold Leak liked driving very much. In 1944 he owned a Rover 14. Later on, around 1950, he bought a Bentley which he loved driving and owning. Frank Hawkins, a fellow director of the Leak company, felt that really Harold had only gone half-measures and should purchase a Rolls-Royce. After all, for just another 150 Pounds (the retail cost of around six TL/12 amplifiers) the car would have a proper stainless steel grille and the Rolls-Royce name – which Frank Hawkins considered would be beneficial to the company. It took Frank Hawkins quite some time to persuade Harold, but eventually he was convinced to change over to a Rolls-Royce. Harold continued with this marque into his retirement – despite his occasional criticism of Rolls-Royce engineering, and vain efforts by his sons to entice him to buy something a bit sportier like a Bristol of Facel Vega.

Harold Leak enjoyed salmon fishing at Tomintoul, Scotland. (Photo courtesy of the Leak family.)

Harold had no real interest in music, although occasionally he listened to light classical music. While at Ealing, the domestic sound system tended to be a lash-up of equipment, which on one occasion resulted in Harold's eldest son John receiving an electric shock from an unprotected part of the system. John Leak recalls: "we only got the system at Holly Tree Farm when Mother and we children agitated for it". This system comprised a Stereo 30, Trough-Line tuner and Sandwich loudspeakers, all of which are still in possession by the Leak family thirty years on.

Harold Leak didn't see hi-fi reproduction as a way to enhance or improve the emotional experience of listening to music. Rather, he tended to look upon it from a scientific point of view – making the equipment perform to exacting technical standards. But he was always most delighted that people experienced increased pleasure in their music listening when using Leak equipment. In 1958 he wrote:

"Examining the social usefulness of the engineers engaged in hi-fi design and production, I believe that the individual engineer should be able to derive pleasure and satisfaction from the knowledge that has come to him from his endeavours. But he is not being socially useful unless the knowledge is translated into physical apparatus, which can be made available to as many people as possible. To do this the price of the apparatus must be within their spending capacity."

The work that Harold Leak did to introduce very low distortion amplifiers resulted in him being granted membership of the British Institution of Radio Engineers in the late 1940's. Many years later, during his retirement, he received two awards from industry participants. In 1987 he was presented with the "Hi-Fi News Award for Achievements in Audio". The following year the Federation of British Audio presented him with the "Special Award for Lifetime Achievement".

Harold and Muriel had two sons and a daughter. John Leak lives in France and Simon Leak lives in London. His daughter Hannah is living in Jersey.

Harold retired in January 1969, when he sold the Leak company to the Rank organisation. He and his wife moved to the Isle of Jersey off the coast of France. He died on August 27 1989. Muriel remained on Jersey until her death in 1997.

"One of the strongest recollections I have of 'HJL' was his love of fly-fishing – not a common interest I hasten to add. On many an overseas trip he took all his fishing tackle, or so it seemed. We, the staff, did not escape. At least on two occasions he had myself, Phil Biggs and colleagues pay out the full length of his high-tech float lines around the factory perimeter in order to check for defects. This was a total priority over any other work commitments."

Ray Whitcombe, 1999

Harold J. Leak.

Muriel and Harold Leak in 1986. Harold's interest in fine wine and good food remained with him for all his life. In later years he bought his wine from Dolamore's of Baker Street, London, where his contact was Johnny Knight, an ex-fighter pilot from the First World War. (Photo courtesy of the Leak family.)

Frank Hawkins

J. Frank Hawkins was a director of H. J. Leak & Co. Ltd. During the 1930's Frank gained experience in the insurance industry, but joined Leak on a full time basis in 1941, although some part-time involvement prior to this is almost a certainty. Frank was Harold Leak's brother-in-law, being married to Harold's sister Marjorie. Like Harold Leak, Frank was very keen on salmon fishing as a hobby.

Frank Hawkins with Harold Leak at the 1959 London Audio Fair, surrounded by a Leak Point One Stereo pre-amp, Trough-Line tuner, dynamic pickup, and stereo power amplifiers. (Reproduced by permission Leak Electronics Ltd. Source: HFN May 1959)

Harold Leak and Frank Hawkins were the driving force behind the success of the Leak company with skills that naturally complemented each other in the business environment. Harold Leak was the initiator of technical ideas and solutions but it was Frank Hawkins who often pulled the ideas together, to come up with commercially successful products. From this point of view, Harold Leak relied on Frank Hawkins' opinion and would often seek his careful consideration on many issues and decisions.

The contribution of Frank Hawkins to the successful operation of the business, particularly the day-to-day issues which bored Harold Leak, was very significant.

Frank Hawkins also had an ability to foresee commercial changes in the market place. In the late 1940's, when high-fidelity audio products were becoming established as a separate market sector to radio, Frank Hawkins' assessment of this change would have contributed to the decision by Leak to move into this market (and eventually away from sub-contracting work undertaken previously).

Frank Hawkins is remembered by Sydney McClory (who built Leak products under license in Australia) as one of the most genuine people that he has ever met. Frank always provided the commercial support necessary for Sydney's operations in Australia and was usually waiting at the airport to pick up Sydney McClory on his various business visits to Leak. John Leak recalls that Frank's questioning skills were outstanding and would have impressed any barrister.

Frank remained with Leak until 1971 – just two years after the Rank takeover of Leak. He had agreed to stay with Rank-Leak while the factory remained in London. When Frank retired, he and his wife Marjorie subsequently lived on the Isle of Man. Frank died in the mid 1970's whilst on a fishing holiday in Scotland with Harold Leak.

J. Frank Hawkins. (© E. H. Ashley 1951. Reproduced by permission.)

Ted Ashley

Frank and Marjorie Hawkins (left) with Alma and Ted Ashley (right) on holidays in Shanklin, Isle of Wight, August 1951. They had met for a dance that evening. (© E. H. Ashley 1951. Reproduced by permission.)

This book would not have been possible without Ted Ashley, who worked as Chief Engineer for Leak and was able to supply much of the background information that can't be gleaned from normal company publications.

Ted Ashley attended grammar school, qualifying with Matriculation. After he left school, various jobs and opportunities became available, most of which didn't suit Ted's interest in radio and electronics. But in 1944, soon after leaving grammar school, a job became available in the production section at H. J. Leak & Co. Ltd., and Ted started immediately.

In his first six years with Leak, Ted undertook part time study which was funded by the company. This required about 10 hours per week during evenings. However in March 1946 Ted left the United Kingdom to serve for the Army in Singapore, returning in August 1948. During this time he was with the Royal Signals Corps and was able to remain in touch with communications technology. As a result of this time away from Leak, he completed his part-time study later than anticipated (in 1952), gaining an Associate Membership of the British Institution of Radio Engineers. Eventually Ted went on to become a Fellow of the IEE.

"His honest approach to life and his sense of humour made Ted Ashley easy to get along with. His hard work contributed significantly to the success of the firm in the 1950's and 1960's."

George Wise, 1999

When Ted started with Leak he was living at Westville Road, Shepherds Bush, which was about a mile or so from the Leak premises at 470 Uxbridge Road, Shepherds Bush. In these days, travelling to work would usually mean an early-morning bicycle ride, although after Ted married, in 1950, he travelled by car.

From 1944 to 1946 Ted Ashley was primarily involved in production aspects at Leak. When he completed his service with the Royal Signal Corps in 1948, he returned as an assistant to Harold Leak. At this time his task was to investigate stability problems on feedback amplifiers being supplied to STC for cutting records.

Ted has always enjoyed listening to music and over the years has kept his sound system at home up-to-date. Around 1960 one would have found an amplifier based on pre-production models of the Stereo 20, a customised Leak pickup and Sandwich Speakers. Before that he had a Goodmans Axiom 80 (with a free-edge cone). Ted recalls that the Axiom 80 was a very good loudspeaker. It had a low free-air resonance of around 25Hz but it was only rated at about 6 Watts and was expensive. By mounting this loudspeaker drive unit in the "serving hatch" between kitchen and dining areas, the best low frequency response was obtained. Of course the sound from the back of the speaker went into the kitchen but it gave the desirable effect of a very large speaker enclosure.

The pickup Ted was using in 1960 was based on the original Leak moving coil pickup from 1948. Ted and a friend modified this pickup for home use so that the head could be detached from the arm. A second head with a 1 mil (0.001") stylus was made for playing micro-groove records, while the original head was for playing 78's. Ted never used a stereo Leak pickup (or prototype) at home, even though the company did a lot of development work on these during the 1960's.

Above: Ted Ashley photographed around 1965. (Courtesy E. H. Ashley.)

Ted Ashley has been a keen listener to music over the years – particularly jazz recordings – and he recognises that his work with Leak has been of benefit to the quality of reproduced sound that he and other people listen too. His older jazz recordings certainly do not rate as high-fidelity recordings but it is still of benefit to reproduce them as well as possible. Ted's favoured loudspeakers are the 3090 Sandwich units; these are much under-recognised high performance speakers from Rank-Leak.

Ted had intimate involvement with the design and production of the famous Trough-Line tuner series by Leak, and went on to become Chief Engineer at Leak. During the 1960's another major project was the Stereo 30 transistor amplifier and following that, the Stereo 70 and Stereo 30 Plus. After Rank bought Leak in 1969, Ted continued on in a General Management role and was responsible for assembling an engineering team to design and produce the 2000 series of Rank-Leak equipment in the mid 1970's.

In 1976 he was appointed Manufacturing Manager of Leak and Wharfedale and he continued until 1981, when Rank closed the operation. Subsequently, Ted undertook consultancy work for several years until 1984, when he retired. Ted had kept all his technical reports and lab books at the Rank Service Department in Yorkshire, but unfortunately during the late 1970's the Marketing Department discarded them, much to his dismay. This documentation would indeed make fascinating reading if it had been archived.

Ted and his wife Alma retired to South Devon.

Above: Leak's Works Manager, Phil Biggs, with his wife who undertook component assembly for Leak. (Photo circa 1998 and courtesy P. Biggs.)

Phil Biggs, Fred Blackman and John Leak

At the age of 17, Phil Biggs joined the company on October 14 1940. He had met Fred Reynolds (a friend of Harold Leak) in the Local Defence Volunteers, and Fred advised Phil of the vacancy at Leak. Phil remained with Leak until the Brunel Road factory was closed in 1971, except for a period when he served for the Royal Signals Corps in the 1940's. Phil became the Works Manager for Leak, but like other employees, started out wiring amplifiers. One of his first tasks was to clean up the workshop, for which he received a "rollicking" when Harold couldn't find anything. Other staff recall that Phil was a dedicated employee and applied high standards to his work – a real asset to the firm.

Fred Blackman (left) and Phil Biggs (right) photographed in 1952. Phil Biggs recalls: "Fred put his soul into any work he was given. He reached his forte when he was appointed Sales and Purchasing Manager – he certainly had the personality and drive for the job." (Photos © E. H. Ashley. Reproduced by permission.)

Another employee who remained with Leak for many years was Fred Blackman. He commenced in 1941, served with the Royal Air Force as a Flight Engineer on Lancaster bombers during the War, then rejoined Leak and progressed into Sales, Marketing and Purchasing. Phil Biggs recalls that "Fred was a great buddy for the whole of my time at Leak". He was loyal staff member who contributed very significantly to the success of the firm. Fred left the company in 1969.

Harold's son John recalls that Phil Biggs and Fred Blackman were key contributors to the team-effort at Leak. John Leak was with the company from 1962 to 1969 and considers the team-culture at Leak crucial to the success of the firm. John had managerial responsibility for the Leak loudspeaker factory at Downham Market, and in his other role as Marketing Manager, brought a new focus to product image and promotion during the later part of the 1960's.

Phil Biggs

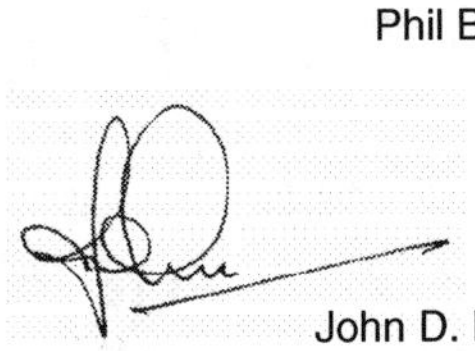

John D. Leak

2. Inception

The Beginnings 1924-34

When Harold left school in 1924 he initially steered away from a career in radio, opting instead to work as a trainee for Hankey Bannister wine merchants in the West-end of London. Harold was in charge of the blind testings and spurned many a proffered half-crown for inside knowledge. He spent long enough there to learn a lot about Burgundy and, as he used to say "if you only know one kind of wine, what else?".

But in his spare time Harold was being drawn towards his true interests in radio and audio technology. In what could be called a "joint-venture" with a friend (who was possibly his school-friend Arthur Gay, later employed at Leak) he started a small side-line operation in 1926 called the "Wireless Doctor". With an initial capital of £22, a hand cart and a bicycle, the small business aimed to provide repair and service work for people with troublesome "new fangled wireless sets". Harold's policy was "No Cure, No Pay". There were many other people providing similar radio repair services in the late 1920's, as can be seen from a survey of advertisements from Wireless World, so Harold would not have been without competition.

The Wireless Doctor business turned out to be quite popular, filling a need at the time. It's success occasionally led to opportunities for designing complete radio receivers. In one instance, a "portable" radio receiver was constructed that weighed

These advertisements show how Harold Leak might have promoted his radio repair business in a competitive environment of other "Radio Doctors" during the late 1920's. (Reproduced with permission of Electronics World magazine, UK. Source: WW 1926 to 1929. The Harold Leak insert is simulated.)

in at 20 lbs, most of which was from the various batteries required to run the circuit. (As Harold's radio set required batteries for portable operation, power consumption would have to be minimised to reduce the size of the batteries, but battery powered household sets were not uncommon in the late 1920's.)

Another project undertaken by this early enterprise was the construction of a moving coil loudspeaker in 1926 – very soon after the invention of this device by Chester Rice and Ed Kellog. The following year Harold's interest in audio was beginning to show even more, when a two-stage amplifier and a pickup for playing records were designed and constructed.

By 1929 the Wireless Doctor was a financial success and Harold decided to capitalise on this. He left the wine trade, in which he had been employed for five years, to work for Gaumont British Cinemas, who required people skilled in sound reproduction technology for the relatively new "talking pictures". During his time there, Harold gained much experience with Western Electric sound systems and also became acquainted with a leading variety performer, Jack Buchanan. Jack offered Harold Leak some space in his dressing room for him to set up a lab-bench and small facility for repairing and making specialist audio equipment. In 1931, in conjunction with his friend D. W. Turpin, Harold experimented with remote reception of stereo sound using a stereo microphone set-up. In a London theatre he placed microphones near the performers on stage, and arranged to listen to stereo sound though speakers elsewhere in the theatre.

Harold Leak with a "robot" he built for an exhibition in 1934. Inside there was a "memory-man", a loudspeaker and one of Harold's amplifiers. People attending the exhibition could ask the robot three questions - if any of the answers were incorrect they could win a prize. (Photo courtesy of the Leak family.)

1934 to 1945

In 1934 Harold left employment as sound-film engineer and founded his own company. This was H. J. Leak & Co., founded on September 14, 1934 and operated from premises in West London. The company grew steadily through the 1930's and during this time Harold lived nearby in Hanwell, West London. In 1955, Leak company literature reviewed the beginnings of the company as follows:

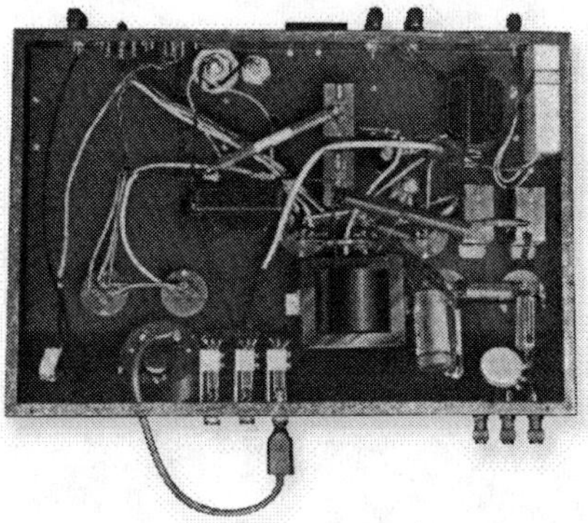

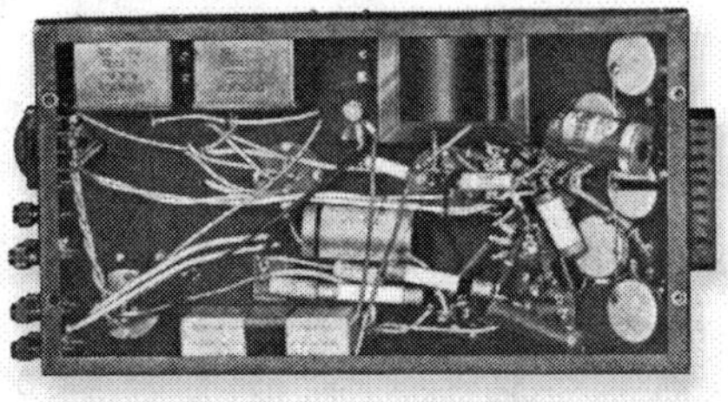

Under chassis view of the Partridge Economy 28 (upper), and the Partridge 45 Watt amplifier (lower). Point to point wiring was used, possibly because these amplifiers were low quantity "specials" built for Partridge, or because component-board construction had not yet been adopted by Leak.

Phil Biggs recalls that when he commenced with Leak in 1940, component board construction was being used by Leak. (Photos by courtesy of Transformers and Rectifiers Ltd, Guildford, United Kingdom. Source: Partridge Amplifier Circuits, circa 1938.)

> "The beginning was modest. The aim of the founder was to earn a living in the way which appeared to hold the best promise of a happy life; by making sound reproducing equipment better, if possible, than anyone else."

The modest beginnings meant that Harold could not afford to carry stock, so this necessitated collecting supplies when they were actually needed. He would often pickup up an amplifier chassis from his supplier on the way home – and sometimes the paint would still be tacky.

Before the Second World War the Leak company built amplifiers for many different purposes such as: telecommunications, public address systems and sound re-inforcement. These amplifiers were usually of Leak design and were sold to large corporations, for re-sale under their own name, or to government departments. Communication companies such as Telephone Rentals were buying amplifiers from Leak and using them in the telephone transmission system.

The Partridge "Economy 28" amplifier as built for Partridge by Leak. (By courtesy of Transformers and Rectifiers Ltd, Guildford, United Kingdom. Source: Partridge Amplifier Circuits, circa 1938.)

The "Partridge 45 Watt" amplifier built by Leak used very large DA30 power triodes. Anode currents for the output valves, and the total output stage current could be measured by plugging the meter cable into one of three sockets. (By courtesy of Transformers and Rectifiers Ltd, Guildford, United Kingdom. Source: Partridge Amplifier Circuits, circa 1938.)

Another company which ordered amplifiers from Leak in the 1930's was "Partridge Transformers". Before the War this company provided an extensive service for enthusiasts building public address amplifiers. Partridge published many amplifier circuits and also the "Partridge P.A. Manual". In a later edition of this manual, two amplifiers are described that were built by Leak for Partridge – the "Economy 28" and the "Partridge 45 Watt Amplifier". (One of the earliest amplifiers built by Leak followed a design in the Partridge handbook and Harold sold one to the Dorchester Hotel, Park Lane, London.)

The "Economy 28" used pentode connected Pen428 valves (from Mullard) in push-pull Class AB_1 with cathode bias and 375 Volts high tension. It featured a novel screen regulation circuit, called the "Partridge Control Valve", which was later used by Leak in the TL/25A power amplifier. (The screen regulation circuit was effectively a DC cathode follower with a reference voltage on the grid and the

H. J. LEAK & CO.

H. J. LEAK

MANUFACTURERS OF SPECIALISED HIGH-FIDELITY SOUND REPRODUCTION EQUIPMENT FOR THE ELECTRICAL AND ENTERTAINMENT INDUSTRIES

124A DALLING ROAD, HAMMERSMITH, W.6

TELEPHONE: RIVERSIDE 5981

A facsimile of the company letter-head in 1940. The description of the company's main business differs from that shown in letter-heads after 1944 – see also Chapter 3. (Original letter-head courtesy Phil Biggs.)

SOUND EQUIPMENT by

LEAK.

The standard 13-watt chassis introduced at Olympia, 1936-37.

13 GNS. (ex works).

Complete with Osram valves, less meter.

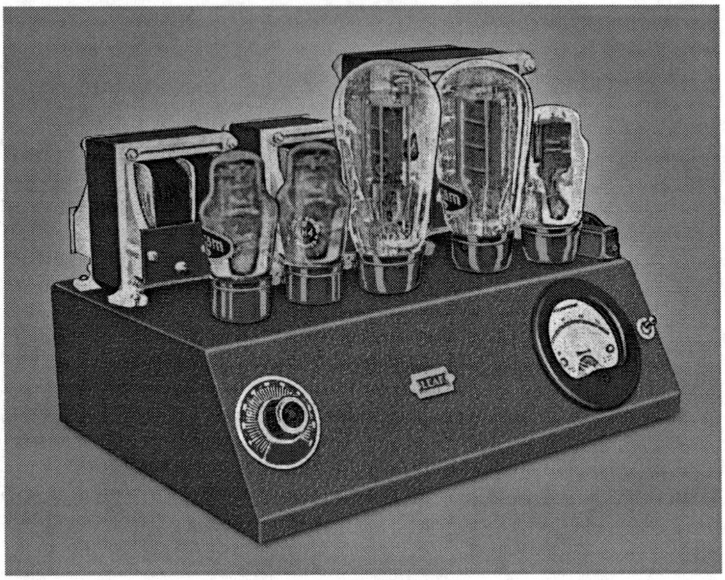

Behind the design of this amplifier lie six years specialised experience of Public Address work. Its capabilities in practice may be heard during the run of Jack Buchanan's show "This'll Make You Whistle", at the Palace Theatre, Shaftsbury Avenue. Used in conjunction with our M.2* moving coil microphone and D.3* speakers the amplifier easily fills this 2,000-seater. It is, of course, equally suitable for radio or gramophone reproduction.

*The M.2 Microphone lists at £10, and its shielded transformer, 30/-. * The D.3 Speaker, D.C. 200/250 volt, £4/12/6. AC/DC 200/250 volt, £5/15/0.

SPECIFICATION. For A.C. operation, 200/250 volts, 40/100 cycles. Consumption from mains, 100 watts. Circuit details: MH4 valve r.-c. coupled to another MH4, shunt-fed transformer coupled to two PX25 valves in Class A push-pull, U14 rectifier. Speech output, 13 watts for 5% total harmonic distortion under working conditions, the design allowing for peak outputs up to 19 watts with imperceptible distortion. Frequency range, 40/14,000 cycles within plus or minus 1 db. Input volume control of 250,000 ohms, with graduated dial. Output impedance standardised at 8 ohms, or wound to individual requirements at no extra charge. To ensure utmost reliability no electrolytic high voltage condensers are used, and all components (excepting valves) carry one year's free replacement guarantee against breakdown. Mains switch. Fully protected fuse holder. Engraved insulated terminals. The layout leaves ample room for the addition of a tone control (5/- extra), and Ferranti output milliammeter (30/- extra), if these are desired. Overall dimensions, 13" wide, 9-1/2" deep, 10" high. Weight 23 lbs. Finish, black crystalline, or as desired.

As so many alleged P.A. Amplifiers appear to have been designed by engineers with obviously little or no working experience of P.A. requirements, we venture to give you four cardinal "DONT'S". DON'T use for P.A. work an amplifier:-

(1) With pentodes in the output stage. Under working conditions the varying load imposed by the speaker with change in frequency will at least halve the output stage wattage measured under laboratory conditions.

(2) With a "high-efficiency" output stage, for the same reason.

(3) With condenser and grid-leak feed to the output valves. The peak wattage will be only slightly above the laboratory figure for 5% distortion.

(4) With electrolytic smoothing or de-coupling condensers. Paper types approach more nearly the 100% reliable mark.

H. J. LEAK, Sound Engineer-Technician,

124A DALLING ROAD, HAMMERSMITH, W.6.

Telephone:RIVerside 5981

cathode supplying stable DC to the screens of the output stage.) The Economy 28 also had tone controls and mixing facilities for general public address work.

The "Partridge 45 Watt Amplifier" built by Leak used Osram DA30 power triodes in push-pull Class AB_1 and was based on a GEC circuit. With 500 Volts on the anodes and a huge grid bias of -145 Volts, the power output was 45 Watts. The large grid bias would have required swings of nearly 300 Volts peak-to-peak on the DA60 grids, so the driver stage would have been transformer-coupled to the output stage.

Standard 13 Watt Chassis

The decision by Leak only to supply amplifiers to large corporate customers meant that the company didn't require a high profile with the retail market and as a result there is little evidence of Leak advertising left from this era. It would not be until the release of the TL/12 in 1948 that Leak's marketing approach would shift towards domestic consumers.

There is one exception, however, and that was the "Standard 13 Watt Chassis" public address amplifier launched by Leak in 1936 and demonstrated at the 1936 and 1937 Radio Olympia Exhibitions in London. This amplifier did carry the Leak name and was advertised for sale to retail customers, rather than to large organisations. The promotional leaflet produced to promote the amplifier shows Harold advertising himself as "Sound Engineer-Technician". Throughout the brochure the emphasis on high quality is evident in many places, with references to: a push-pull triode output stage; distortion specification of 5% under working conditions; high reliability paper condensers in preference over electrolytic types; and suitability for reproduction of "radio or gramophone". Along with the amplifier, Leak was also selling a dynamic microphone (Model M.2) and an electro-dynamic loudspeaker (Model D.3). The advertisement also discusses the ability of the amplifier to deliver up to 19 Watts, probably by a transition from class AB_1 working to class AB_2 for powers beyond 13 Watts. This would have been possible as the output stage grids were transformer-coupled to the driver stage, allowing grid current to be drawn without disrupting the bias setting.

The "Standard 13 Watt Chassis" amplifier does not however feature any use of negative feedback for distortion reduction. This technique was invented in 1934 but was slow to become established as a popular method for improving amplifier performance, although by the late 1930's magazines such as Wireless World were featuring amplifier designs that included some negative feedback.

During the Second World War, Leak continued to make amplifiers. Whilst the War did not affect the type of product the business was producing, it did result in the premises at Dalling Road being bombed by the German air-force. The company moved a short distance to Hide's Yard, Dalling Road where they would remain

"Before I joined, Harold Leak had another assistant (or 'boy') like myself. Two days before I started this boy was killed from a bomb raid. . . When I joined at Dalling Road we were doing a few theatre jobs. We would go out and survey the theatre, then come back and build the amplifier. It was just Harold, myself, an AVO meter, and an oscillator that we had dug up from somewhere. That was about it. . .

My first roles were many – apart from making the tea, I learnt how to solder, how to read a circuit, how to answer the telephone, and to drive Harold's car around London collecting stores to make amplifiers. Occasionally, with thorough instruction from Harold I went out and did the odd servicing job. As they say, you name it I did it – every day was a new experience that went on for 31 years."

Phil Biggs, 2000

Opposite: A facsimile of a brochure from the late 1930's describing the Leak 13 Watt public address amplifier. (Reproduced by permission Leak Electronics Ltd.)

until 1941. At the time of the move to Hide's Yard, Phil Biggs (later to become the Works Manager) joined Harold Leak.

By 1944 the business had grown from a one-man operation to one employing 14 people. At this time a bright-eyed new recruit was employed to assist with the production operations of the company – this was Ted Ashley, who had just left his schooling and would go on to become the company's Chief Engineer in the 1950's.

"I was in the production side until I went into the Services. When I came out of the Services one of the first things I was concerned with was getting stability using STC cutter heads."

Ted Ashley, 1998.

Standard Telephones and Cables

Standard Telephones and Cables (STC) was a regular customer for Leak-produced amplifiers in the mid 1940's. STC were providing cinema sound systems for its customers but, more importantly for Leak, STC were also interested in development of cutter heads for 78rpm records. Through a contact at STC (M. L. Gayford, later Chief Engineer of STC Acoustics) Leak started to supply amplifiers for driving STC cutter heads.

The work undertaken for STC on record cutter heads became very significant for the Leak company, as it brought about the application of significant amounts of feedback in Leak's designs. One of the problems in high quality record cutting is controlling the frequency response of a cutter head, which can be impaired by the high frequency resonance of the cutting assembly. To solve such problems Leak employed negative feedback to control the cutter head. (Typically, a cutter head would be a low-impedance transducer, requiring about 15 Watts of power – much like a loudspeaker – except that some types of cutters would have a tertiary winding on the head allowing a feedback signal to be generated, giving the possibility of reducing distortion in the cutter head as well as in the amplifier driving it.)

However the practical application of feedback needs a sound understanding of issues affecting stability. This was a new area of expertise for the company, so in the early 1940's Frank Hawkins encouraged Harold Leak to review the theoretical basis of feedback by establishing a small library of books and publications on the topic.

By 1944 the knowledge of negative feedback theory, combined with experience from project work with STC, enabled Leak to apply a feedback magnitude of about 26dB over four stages, provided that care was taken with the design of the amplifier and cutter-head system. Whilst STC had never specified to Leak that distortion should be as low as 0.1%, it became apparent from the Leak development work that such figures were attainable. As a result of the projects undertaken for STC, Leak was able to launch the first "Point One" amplifier (the Type 15) on which the future success of the company would be built.

In 1944 when this negative feedback development was being undertaken, Harold Leak undertook all the design work whilst Jack Turner was his Laboratory Assistant, assembling prototype amplifiers.

A Partridge advertisement showing a transformer similar in style to that used on the Leak "Type 15" Point One amplifier. (By courtesy of Transformers and Rectifiers Ltd, Guildford, United Kingdom. Source: WW June 1945)

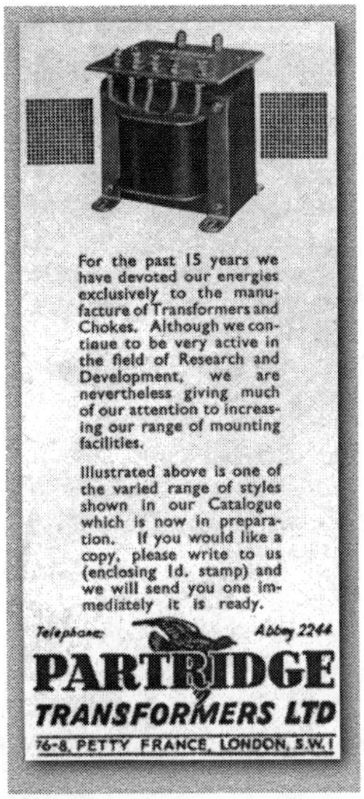

The Type 15 had a pentode first stage, three Osram L63 single triodes and triode-connected KT66 output valves for 0.1% distortion. A paper capacitor block was used for power supply filtering. The chassis had spare capacity for additional valves that increased the sensitivity to 5mVrms if required. (© Sunrise Press, Devon. Photography by Jonathan Hill. Reproduced by permission. Source: Audio! Audio!)

GEC and the Type 15 Point One

Another large company buying amplifiers from Leak was GEC, who were using them for cinema sound systems and telecommunications applications. As development work proceeded on negative feedback amplifiers, GEC asked Leak if they would like to tender for supply of amplifiers. To assist GEC assess the Leak tender proposal, they asked if Leak could supply mock-ups of the amplifiers – these would perform perfectly well, although the physical appearance would leave much to be desired. If the mock-up amplifier was acceptable to GEC, they would then nominate the number of amplifiers to be supplied by Leak. Harold Leak was open to build the amplifiers to his own design provided the overall performance was acceptable to GEC.

Ted Ashley recalls that Harold Leak took a four-stage amplifier to the GEC-Osram Laboratories, located in Brook Green near Hammersmith. Graham Woodville, who was widely recognised an industry expert on valve applications, was one of Harold

Leak's contacts at GEC, and may have received Harold on this occasion. The amplifier that Leak demonstrated to GEC had a significant amount of negative feedback applied around the four stages and the output transformer. The harmonic distortion at rated output was just 0.1%. This was made possible by a design that took stability factors into account and by the use of high quality output transformers that were supplied by Partridge Transformers. As a result, an order was placed by GEC for these Leak amplifiers.

A derivative of this amplifier was eventually advertised and marketed by Leak as the Type 15 amplifier – being advertised in September 1945 in the proceedings of the British Institution of Radio Engineers. To launch these amplifiers the company used the phrase "Point One" in all its advertising, emphasising the low harmonic distortion achieved. The phrase subsequently became a famous trade mark which both set a yardstick for others to aspire to, and for the Leak company to adhere to when transistor amplifiers were introduced in the 1960's.

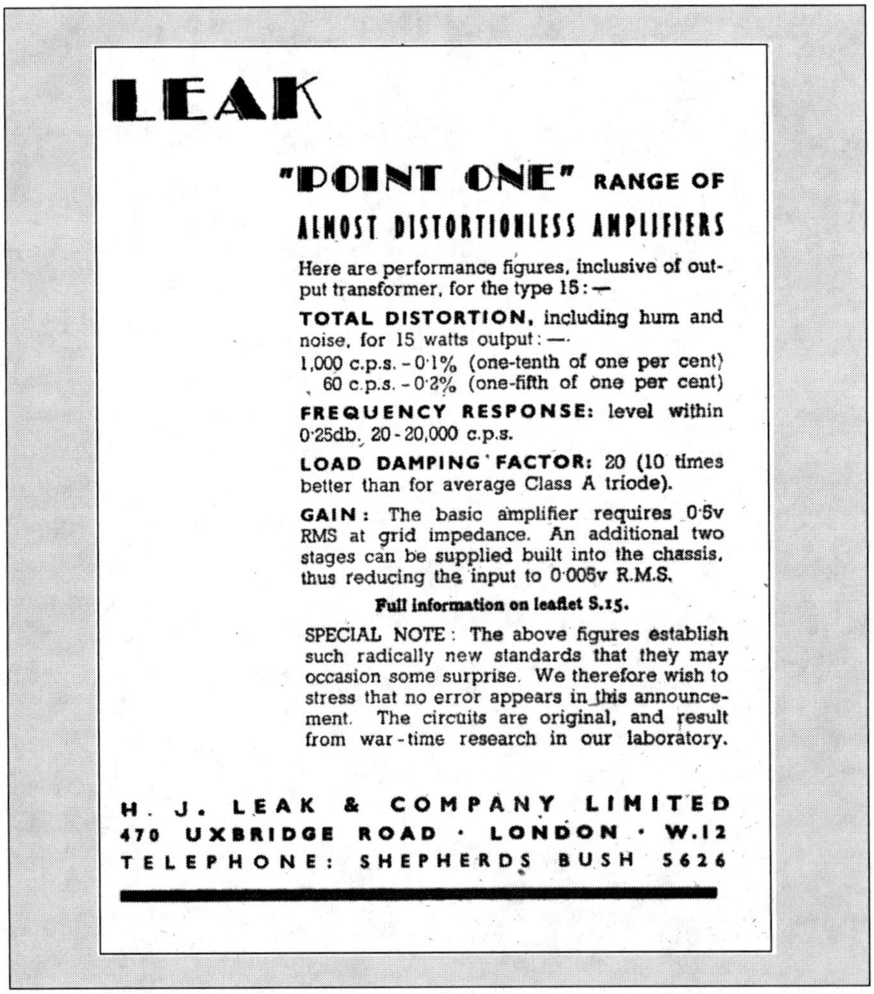

This advertisement from Wireless World, December 1945, describes the Type 15, and emphasises the low distortion of 0.1%. At the time, the accepted standard for laboratory amplifiers was 2%, so the specification published by Leak was met with some incredulity by Engineers in the communications industry. (Reproduced by permission Leak Electronics Ltd. Source: WW Dec 1945)

This amplifier is a 30 Watt version of the "Point One" Type 15 amplifier. It uses KT66 valves operating in tetrode mode for higher power, with the screen supply being derived from an S130 gas-regulator valve. This particular unit is also fitted with additional stages for microphone and music inputs, with mixing and tone control facilities.

The circuit for this amplifier shows a secondary with four windings. In addition, the feedback is applied from a tertiary winding comprised of two sections in parallel, indicative of the effort required to obtain stability with the output transformer included in the feedback loop. (Photo and amplifier courtesy Keith and Angus Gilchrist, 1999.)

The Type 15 amplifier had a tertiary feedback winding that applied approximately 26 dB of feedback over the four stages and the output transformer. Tertiary feedback windings make the amplifier less affected by the variations in load impedances and so improve the stability of a design. This would have been a contributing factor in the design that allowed the high level of feedback required without stability problems. (Even so George Wise, who was closely associated with the company for many years and owned a pre-production Type 15, recalls that each production amplifier was fine-tuned on test to ensure stability.)

The under chassis layout of the "Point One" 30 Watt amplifier shows components and construction techniques similar to the later TL/12 "Point One" amplifiers. (Photo and amplifier courtesy of Keith and Angus Gilchrist, 1999.)

The power amplifier section of the Type 15 used the following valves:

- EF37 pentode input stage capacitively coupled to the second stage
- L63 triode concertina phase splitter
- 2 x L63 triodes acting as a push-pull driver stage
- 2 x KT66's for a push-pull triode connected output stage, and individually self-biased.

The overall topology of the Type 15 is somewhat similar to the famous Williamson circuit published a few years later in 1947. However, Williamson elected to use a triode first stage that was direct coupled to the second stage, included an AC-balance control on the driver stage, and used a common cathode resistor for the self-biased output stage that included a DC-balance trimmer.

If necessary, two additional stages could be built into the Type 15 amplifier to increase its sensitivity from 500mV to 5mV for full power output. All stages were provided with facilities for reading valve currents via an octal socket, or meters could be mounted on a panel if required. There was also a 30 Watt version of this amplifier, the only external difference being a gas-filled voltage stabiliser valve to operate the tetrode output stage. Other production variations included:

- up to three inputs with or without mixing facilities
- bass and treble controls
- low impedance output, or 50 Volt and 100 Volt line outputs
- front panels for free-standing, or for rack-mounting

As with the later TL/12's, power supply filtering was with oil-filled paper dielectric capacitors, which were regarded as the last word in reliability.

The Type 15 was expensive to produce and is reputed to have sold for the massive sum of £70. The total production of Type 15 amplifiers is not known accurately but it is estimated that several hundred would have been made.

"These amplifiers were supplied to such companies as STC and GEC, who had their own logos put on them, and to many others under our own name. A number went to the Royal Festival Hall, while Firestone Tyres had a bank of eight amplifiers just to communicate from one end of a machine to the other. . . The noise level was very high and there was a lot of vibration – the amplifiers were on 24 hours a day and they just kept going for several years without attention. In short they were supplied for all sorts of situations such as industrial, commercial and laboratory work."

Angus Gilchrist, 1999.

Negative Feedback and Stability

Negative feedback principles had been known before the War, being invented in 1934 by Harold Black of Bell Telephone Labs in the USA*. However it took many years for the practical application of feedback to become commonplace, despite designs for feedback amplifiers appearing in industry journals such as "Wireless World" during the late 1930's.

Amplifiers with small amounts of feedback are easier to design than those with large amounts, however the latter is attractive to the designer as it will give an amplifier a better specification. The effects of feedback are well known and will only be summarised here. For a typical triode power amplifier, application of 20dB negative feedback (ie: a loop-gain of 10, where loop gain is the product of forward path gain and feedback network gain) will result in:

- a reduction of harmonic distortion by ten times (20dB)
- a reduction of the output impedance by 10 times, giving flatter frequency response on practical loads such as a loudspeaker
- a reduction of hum and noise by 20dB
- an improvement in the frequency response
- an alteration in the overload characteristic such that the onset of clipping

* Negative feedback had been applied previously in other engineering fields. The use of a governor on a steam engine to regulate the rotational speed regardless of load relied on the same underlying principles. In this case rotational speed and torque are analogous to voltage and current from an amplifier.

is hard, rather than soft

• the possibility of oscillation (of a continuous or transient nature) if the feedback becomes positive at certain frequencies

The last characteristic listed above is the primary design difficulty faced by negative feedback amplifier designers. To overcome it the designer must ensure that the feedback is negative for all frequencies where the amplifier has a net gain around its feedback loop. Over a restricted frequency range this task is relatively easy but audio amplifiers are required to operate over a very large range of frequencies – from 20Hz to 20kHz. For stability purposes however, the working range is even greater. The designer needs to consider frequencies from a few Hertz to around 200kHz. At the frequency extremes phase shifts are introduced as a result of the roll-off in frequency response (for every 6dB per octave in the rate of roll-off, another 90 degrees of phase shift accumulates). When the phase shift reaches 180 degrees the output signal is no longer useful in a negative feedback amplifier – if it was returned to the input, it would no longer subtract as it should; it would instead be additive, giving positive feedback and result in the amplifier oscillating. When the phase shift approaches 180 degrees, many feedback designs are marginally stable and can be affected by the load connected to them or by the power output demanded.

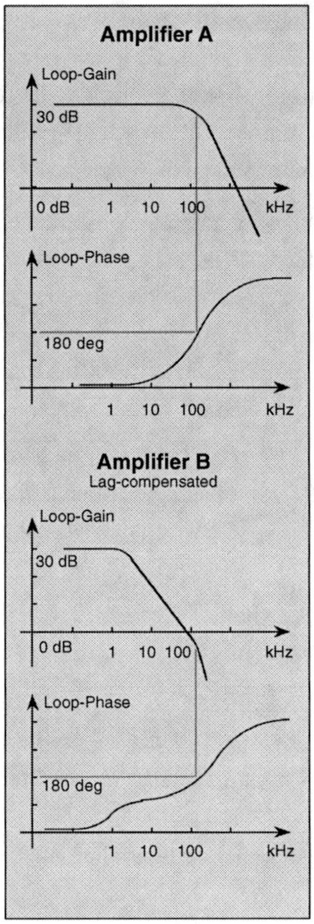

Amplifier A will oscillate because the gain around the feedback loop is greater than one (0 dB) when the phase shift around the loop reaches 180 degrees. Amplifier B is stable because a compensation network reduces the loop-gain slowly (at 6dB per octave) thus limiting the phase shift to 90 degrees for frequencies where the loop gain is greater than unity. (Note: Loop-gain is often referred to as the amount of feedback.)

The fundamental limitations that need to be considered when applying feedback were described by H. W. Bode in his book "Network Analysis and Feedback Amplifier Design". This publication, along with others by F. E. Terman and H. S. Black were used by Leak to establish the correct operating conditions for his feedback amplifier designs.

In design work undertaken by Harold Leak, the stability of mock-up designs was assessed using tools such as Nyquist diagrams and double-beam oscilloscopes. These enable a designer to plot the gain and phase characteristic versus frequency and thus determine the gain margin and phase margin (the gain margin is that additional amount of feedback required to cause oscillation and the phase margin is the additional phase shift required to cause oscillation – larger gain and phase margins represent a more stable design). Leak was one of the few manufacturers to publish gain and phase margins for their amplifiers.

The Brunel Road factory also had facilities for testing amplifiers under various load conditions that could be encountered in practice. Extremely long loudspeaker cable runs were installed that enabled stability to be assessed, even under conditions where the possibility of spurious oscillations from complex amplifier loads could occur.

The Williamson Amplifier

From a paper, "The Williamson Amplifier of 1947", by P. R. Stinson, Australia.

The impact of the Williamson amplifier on the growth of the hi-fi market was enormous and as it was designed around the same time as the Leak Type 15 a short description is presented here.

Theo Williamson joined the Applications Laboratory of Marconi-Osram in 1944. Outside his professional role with the Lab, he set about development of a high quality amplifier design. By 1945 the performance Williamson was obtaining with his design impressed many of his colleagues, so Herbert Smith of Marconi-Osram arranged for a demonstration of the design to the company Board. As the project was not officially sanctioned by Marconi-Osram, an internal report on the amplifier was written in late 1945 to support the demonstration to the Board, following which they decided the project was not of commercial interest to Marconi-Osram or the parent company GEC (despite the fact that GEC were purchasing Leak "Point One" amplifiers of similar performance around this time).

In 1946, Williamson left Marconi-Osram for Ferranti in Edinburgh. His high quality amplifier design lay fallow for a period, until Williamson's internal report was forwarded to Wireless World magazine by F. E. Henderson – a senior sales executive who felt that publication might increase sales in KT66 valves.

In May 1947 the design was published and it attracted an unusual amount of interest. Within three months it was published in Australia and a version was developed by Fritz Langford-Smith using 807 output valves. The design became popular in America through Radio-Electronics and Audio Engineering magazine articles. Many commercial manufacturers around the world provided components meeting Williamson's tight specification, or adopted the design for sale as a kit or fully built.

Amplifiers meeting Williamson's design criteria must have the following topology:

- The first stage is a common cathode triode gain stage
- The second stage is a triode concertina phase splitter
- The third stage is a push-pull triode driver operating in common cathode mode
- A push-pull output stage with power triodes, cathode-biased for Class A operation via a common cathode resistor
- High quality output transformer capable of inclusion in a feedback loop
- Approximately 20 dB feedback to give 0.2% distortion at full output

The design provided very high performance, but at low cost through the availability of parts or kits. This brought hi-fi performance to a mass market of world-wide audio enthusiasts, thus spurring the development of the hi-fi industry.

In the USA, British Radio Electronics Ltd. (operated by Colonel Harris, a friend of Williamson's) released an "authorised" version of the Williamson amplifier called the Hallmark Williamson in 1953. It was manufactured by Acoustical Manufacturing Co. in the UK and featured a separate power supply chassis, but did not sell in large numbers as it was expensive, at US$169, compared to American-built Williamsons at around US$120.

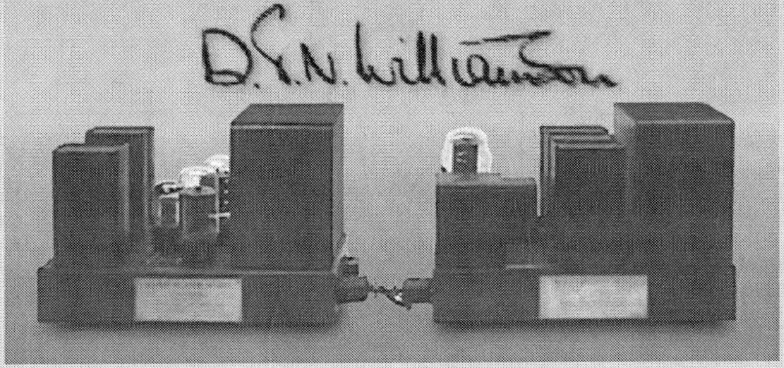

3. Operations

Company Structure and Staff

Initially the firm was named "H. J. Leak & Co." with Harold Leak being listed as the proprietor on company correspondence. In the early 1940's, around the time Harold married Muriel Short, the firm expanded its structure, began to employ more people, and became a limited liability company with the official name "H. J. Leak & Co. Ltd." This firm had three directors:

- Harold Leak (Chairman)
- Muriel Leak (Director).
- J. Frank Hawkins (Managing Director).

Frank Hawkins was Harold's brother-in-law, being married to Harold's sister Marjorie. His primary role with the company was with the commercial and financial aspects (he had previous experience in the insurance industry). Frank started full time with Leak in about 1941, although he may have had some part-time involvement prior to this. Harold's wife Muriel was in the main a silent director although she worked for the company in the early days, assisting with clerical work. Another director of the company in the early 1940's was John Short (brother to Muriel Leak) but company letterheads show that he had left this role by 1945.

By 1960 the company structure was such that three staff were reporting directly to Harold Leak. These were: Don Barlow (Loudspeaker Development), Ted Ashley (Chief Engineer), and Frank Hawkins (Managing Director). Harold Leak maintained hands-on decision-making control in regards to:

Various logos were used by Leak over the years. The initial design for the logo with the lions came from the American agents for Leak, and was not much liked by Harold Leak or Frank Hawkins, although it was used for sales in the USA. Later, John Leak had this logo redesigned with properly designed heraldic lions as shown here. By the late 1960's the Union Jack was being used to fend off competition from the Japanese industry. (Logos reproduced by permission Leak Electronics Ltd.)

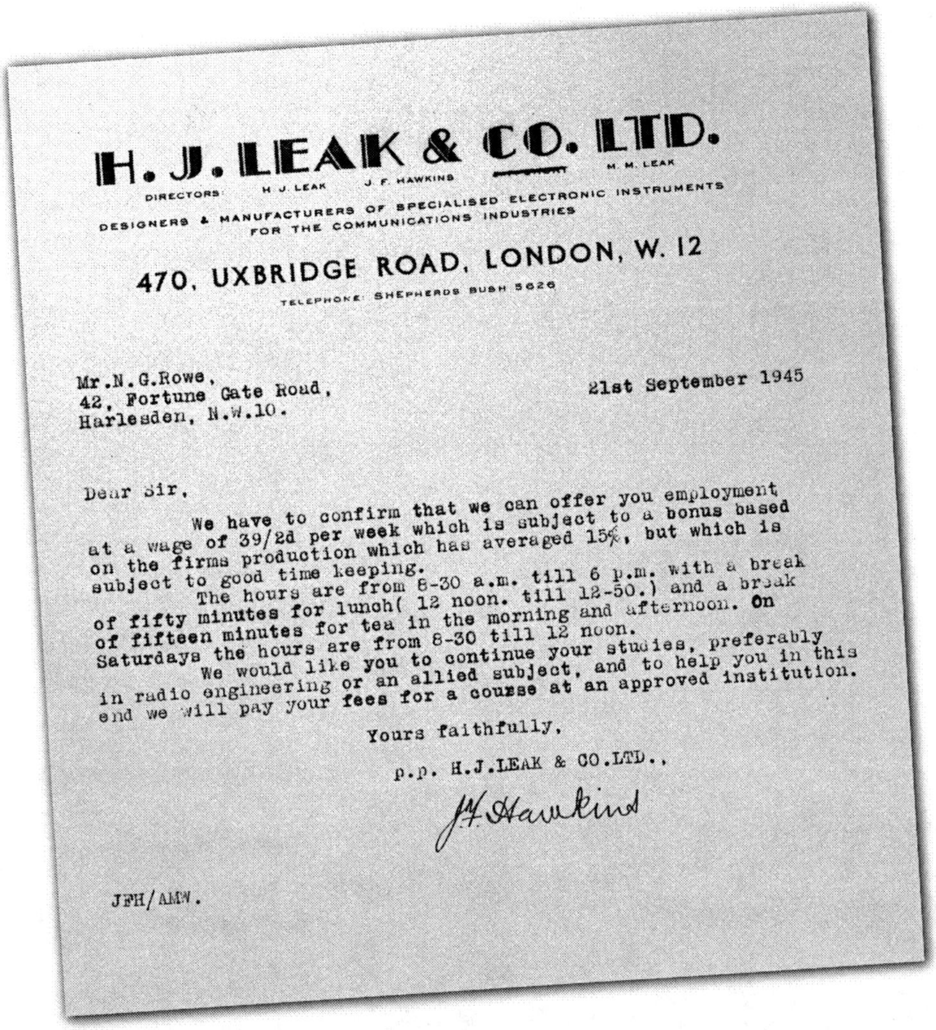

H. J. LEAK & CO. LTD.

DIRECTORS: H. J. LEAK J. F. HAWKINS M. M. LEAK

DESIGNERS & MANUFACTURERS OF SPECIALISED ELECTRONIC INSTRUMENTS FOR THE COMMUNICATIONS INDUSTRIES

470, UXBRIDGE ROAD, LONDON, W. 12

TELEPHONE: SHEPHERDS BUSH 5626

Mr.N.G.Rowe,
42, Fortune Gate Road,
Harlesden, N.W.10.

21st September 1945

Dear Sir,

We have to confirm that we can offer you employment at a wage of 39/2d per week which is subject to a bonus based on the firms production which has averaged 15%, but which is subject to good time keeping.

The hours are from 8-30 a.m. till 6 p.m. with a break of fifty minutes for lunch(12 noon. till 12-50.) and a break of fifteen minutes for tea in the morning and afternoon. On Saturdays the hours are from 8-30 till 12 noon.

We would like you to continue your studies, preferably in radio engineering or an allied subject, and to help you in this and we will pay your fees for a course at an approved institution.

Yours faithfully,

p.p. H.J.LEAK & CO.LTD.,

J.F. Hawkins

JFH/AMW.

This offer of employment for Norman Rowe was issued by Frank Hawkins on the company letter-head paper. With Saturday mornings included, 44 hours and 5 minutes were worked each week. Wages were just under £2 per week – three years later the TL/12 would cost just over £25, making it worth about three-months wages for a young technician. In later years, the hours were: 8:15am to 5:30pm for the factory, and 9:00am to 5:30pm for office staff, with a one hour lunch-break.

- Company policy
- All product designs, especially during final stages
- Company publicity material – Harold would always ensure that company publications, service manuals and advertisements were of a quality consistent with the high standards set by the products themselves. Both Harold and Frank Hawkins considered this aspect to be very important.

Staff Recruitment

As a director of the company, Frank Hawkins was involved in all recruitment, while Harold Leak would attend second interviews or senior appointments – for those people who would be reporting directly to Ted Ashley or Frank Hawkins.

Whilst Harold Leak didn't conduct all job interviews, he always ensured he was briefed on potential employees. In the case of Don Barlow (developer of the Sandwich loudspeaker), Harold visited him personally and made him an offer on the spot.

Staff Numbers

Staff numbers grew slowly in the early years but there was a period of expansion around 1942, and again after the release of the TL/12 amplifier in 1949:

- 1934 to 1940: 1 to 2 staff (Harold Leak, plus an assistant)
- 1944: 14 staff (4 design staff, 8 assembly, 2 sundries)
- 1949: 20 staff
- 1950's: approximately 40 staff
- 1960: 40 to 45 staff
- Late 1960's: approximately 50 staff at Brunel Road factory, and 15 staff at Downham Market factory. Product design and development was undertaken by about 5 staff.

The Leak Athletic Football Club for the 1951/52 season. Staff not named are from other companies. (Photo courtesy E. H. Ashley.)

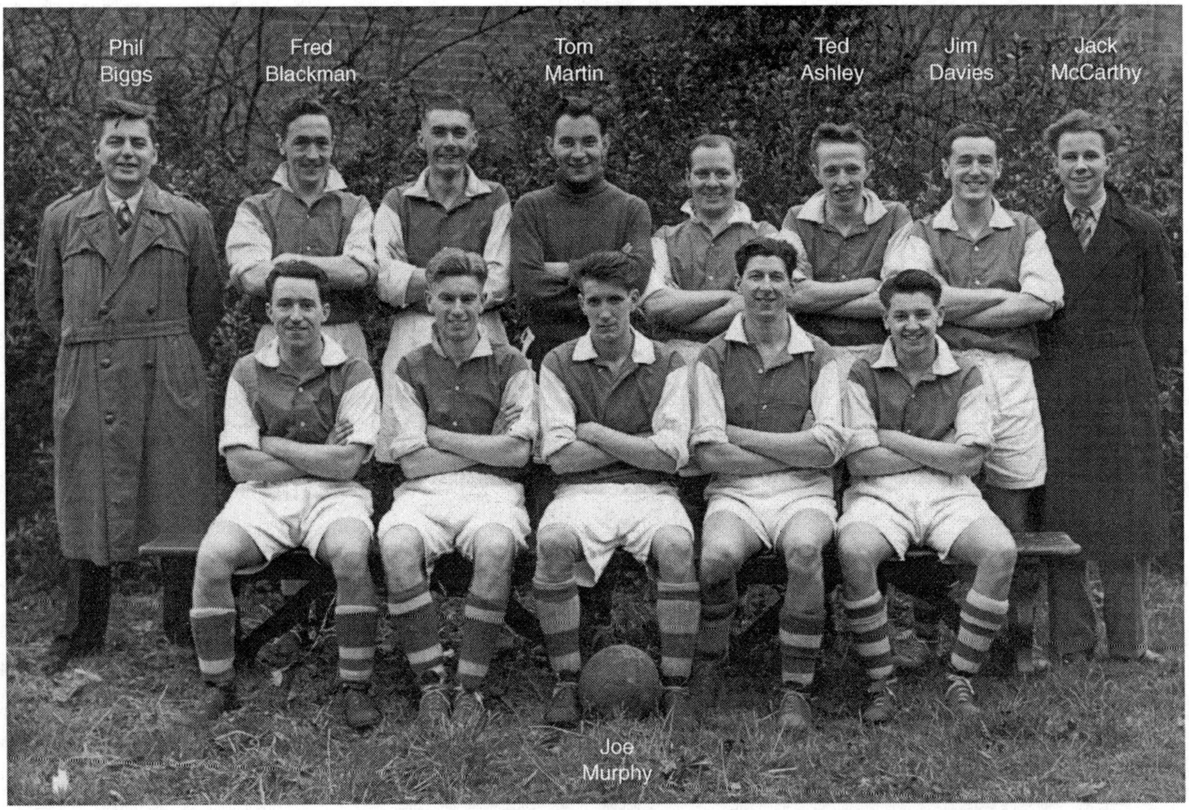

The Leak Athletic Football Club for the 1949/50 season. Staff not named are from other companies. Trevor Holman worked at Leak in Inspection, and on Mark 1 pickup assembly. He used to cycle a long way to work, used a newspaper inside his jumper to keep out the cold, and insisted that oil lamps on his bicycle were far more efficient than battery lights. (Photo courtesy E. H. Ashley.)

Sporting Clubs

Some staff participated in the Leak Athletic Football Club, in which about half the players were Leak employees. Frank Hawkins was Chairman and Ted Ashley was a "half-back" player and Secretary. Phil Biggs and Fred Blackman were also players, with Phil taking on the additional role of Treasurer.

The team was part of the Brentford and District League, and competed against other teams from local companies on Saturday afternoons (e.g. Bush Radio Football Club). Frank Hawkins would always attend the games, no matter what the weather conditions were. In the late 1940's the team often played at Wormwood Scrubs (even carrying the goal posts there for the game) but later on many games were played at Gunnersbury Park about 3km (two miles) south-east of the Leak factory. The team met with some success – one year playing in the Premier Division and going on tour to Belgium and Swanage (Dorset, UK) in 1952, '53 and '54.

Another sporting activity for staff was table tennis. Norman Rowe and Ted Ashley both played for Leak in the North Acton Table Tennis League, along with other staff. Cricket was also played by staff on vacant land to the west of the Brunel Road factory.

A promotional poster for the Leak Athletic Football Club's 1953 tour – this match was lost 1-3. (© E. H. Ashley 1953. Reproduced by permission.)

Leaving for Henley: 1953

On a Friday afternoon, staff pose before the bus departs Brunel Road for the 1953 annual trip to Henley. This view faces south-east, so the Leak factory is out-of shot to the right. The Electro Alloys factory is on the right, and Hampton's Furniture behind the bus. Left to right: Bill Kemp (maintenance/general work), Fred Westoll, Angus Gilchrist, – –, Tom Martin with Jim Davies on his back, Brian – , Gerry Willis, Bob Buckledee, Albert Street (at rear), – –, Phil Biggs, Cyril Bezant (at rear), – – (carpenter), Fred Blackman, and Frank Hawkins. Albert Street featured in some of Harold Leak's demonstrations of live versus reproduced sound as he had quite a good singing voice. The trip to Henley usually required a toilet seat for use while camping out – Phil Biggs was responsible for fabricating it! (© E. H. Ashley 1953. Reproduced by permission.)

The Annual Henley Weekend

From around 1947 to 1953, the staff at Leak organised a weekend away during the summer – usually July or August. Leaving on a Friday afternoon they would travel to Henley, stay near Marsh Lock on the River Thames, then return to London on Sunday afternoon. Ted Ashley recalls: "We had an annual week-end on the river for the males in the early days which involved sleeping in punts over two nights and canoeing, swimming, playing cricket and the like during the days. We visited the local pub or dance hall according to choice during the evenings."

By 1953 the group travelling to Henley had grown to the extent that a bus was required. On occasions other guests joined the Leak staff for the weekend trip.

Top: Weekend trip to Henley, July 1950. Left to right: Tom Martin, Frank Hawkins, Bob Buckledee, Fred Reynolds (Winston Electronics), Jack Turner (of Winston Electronics, previously Leak), Fred Blackman and Phil Biggs.

Middle: At Henley, July 22 1951. Left to right: A guest, George Garside, Fred Blackman, Frank Hawkins, Norman Rowe, Phil Biggs. Kneeling: Jim Davies (left) and a guest.

Bottom: Fred Blackman (left) and Phil Biggs (right) en-route to Henley 1952. George Wise recalls that "Fred Blackman was a very loyal hard-working man who did a lot for the success of the firm" and "Phil Biggs was honest, fair, hard-working, and kept the standard up - a real asset to the firm". (All photos © E. H. Ashley 1950-52 and reproduced by permission.)

At 11:00pm on September 23 1940, bombs from the German Luftwaffe exploded in Dalling Road near where Leak was located. The Leak premises were in the terraced house at the far left of this photo which is taken looking north-east. A large bomb crater is to the left of this photo. (© "Hammersmith and Fulham Archives and Local History Collection". Researched by Simon Leak and Elizabeth Perrins.)

Company Premises

From 1934 to 1940 it seems likely that the company may have operated from a number of premises in the West London area, but these locations are now unknown.

In 1940 Leak was located at 124A Dalling Road, Hammersmith – Phil Biggs was interviewed in an attic here: "I can picture it now, as I walked up the stairs I had to climb over boxes, old speakers and the like. It was a dust-pit." The building was located at the corner of Atwood Road and Dalling Road, at the back of a terraced house shared with a dental practice run by Fredrick Sage at 124 Dalling Road.

On Monday September 23, bombs from the German Luftwaffe fell over Hammersmith at 11:00 pm. Alongside 124A Dalling Road, the terraced houses a suffered a direct hit from a high explosive bomb, but the Leak premises remained relatively undamaged. Twenty two people were injured and seven were killed, with gas, electricity and transport services being disrupted, as reported in "The West London Observer" of September 27, 1940. Simon Leak remarks how fortunate it was that his father was not working that evening, as he often worked very late.

After the bombing the company moved a short distance north to Hide's Yard, Dalling Road (approximately opposite Furber Street) where there was a mews with other small businesses. Phil Bigg's recollects: "The premises were more or less like a large garage. It had a galvanised iron corrugated roof that regularly leaked."

The bomb-crater in Dalling Road, September 23 1940. (© "Hammersmith and Fulham Archives and Local History Collection". Researched by Simon Leak and Elizabeth Perrins.)

Photographed in October 1972, 470 Uxbridge Road was used by Leak from 1940 to 1947. This is the site where the Type 15 Point One amplifier was developed and put into production. Harold Leak's office and lab was located upstairs. The building is now demolished. (© "Hammersmith and Fulham Archives and Local History Collection". Photo by Christine Bayliss. Researched by Simon Leak and Elizabeth Perrins.)

In 1941, the fledging two-man company moved to another mews at Percy Place, Percy Road, Shepherds Bush, where they had two rooms upstairs. Phil Biggs continues: "It was here that we really started to expand. Frank Hawkins joined, then Fred Blackman, a boy called Bert Jones, followed by Bunny Phipps (my cousin), then Margaret Trainer, Joe Murphy and John Short."

Uxbridge Road

Then in 1944 Leak moved again to 470 Uxbridge Road, Shepherds Bush. The building was actually a converted double-fronted house, but it functioned well for the company's purposes – all the internal doors were removed to allow interconnection of various rooms, which assisted efficient production. In 1944 there were about fourteen staff at the Uxbridge Road premises. About four of these were on design, eight were on assembly, and two in other areas such as stores-collection.

Harold Leak used to describe his company as a fabricator rather than a manufacturer. It was clear in his mind that processes such as sheet metal work, cabinet making

Inside (the original Brunel Road factory) approximately one-third of the north-end was divided into relatively small rooms for offices etc., these were constructed of cement-rendered breeze-blocks with asbestos sheets used as ceiling boards. At this time the factory-end of the building was an extremely cold place to work in the winter. Later, breeze-block partitions were built to separate the working area from the rest of the building, and a floor-area was built onto the bottom of the roof-trusses to provide a stores area.

Angus Gilchrist, 1999

and transformer winding were best left to specialist suppliers. Supplies would flow into the factory on a "just in time" basis, long before the expression became popular.

Out Workers

The Leak company used "out-workers" for assembly of certain sub-components. A good example here is the component boards used in amplifiers produced by Leak. These boards were made of an insulating material and were designed to hold all the resistors and capacitors as well as interconnect them according to the circuit. The boards would be drilled out at the Uxbridge Road factory (not on a multiple drill, but using a drill press arranged to drill out three or four boards at a time, one hole at time). After drilling the boards, they would have tags riveted onto them before being sent to "out-workers" who would work at home to wire the connections on the back of the component board.

Leak used cable looms extensively for interconnection of major components on an amplifier chassis. As the skills of various out-workers developed, some of them were employed to make up cable looms, using a form to accurately place wires within the loom. Other sub-assemblies such as tone and filter switches were also made by out-workers

Ted Ashley notes how well this process worked. It was beneficial to both the company (by keeping overheads down) and to the out-workers who could work their own hours from home. As well as training full-time production staff, Phil Biggs was responsible for training the out-workers. The use of out-workers continued until around the mid 1960's, with the maximum number employed at one time reaching 48.

A staff photo taken in the driveway at 470 Uxbridge Road. Standing, left to right: Margaret Trainer, Frank Hawkins, Joe Murphy, Harold Leak, Roy Edwards, Cyril Penn, Don Loades, Rodney Cook (behind) and Fred Westoll. Kneeling, left to right: Jim Davies, Albert Street, Billy –, Norman Rowe and (inset) Ted Ashley. For a short time Rodney Cook was Harold Leak's Lab assistant after Jack Turner left. (© N. Rowe 1947. Reproduced by permission. Compiled from three photographs.)

The original portion of the Brunel Road factory can be seen here behind the office block that was added later. (Photo courtesy of Douglas Gardner, Brixton Estate PLC. Researched by Simon Leak and Elizabeth Perrins.)

Some Staff Names: 1940 to 1971

40-69	Harold Leak	Chairman of Board of Directors
(34-69)	(Muriel Leak)	(Silent Director)
40-71	J. Frank Hawkins	Managing Director Finance and Commercial Director Chairman of "Leak Athletic" Football Club
44-71	Ted Ashley	Lab Assistant / Chief Engineer / General Manager. Also Secretary of "Leak Athletic" Football Club
41-69	Fred Blackman	(ex R.A.F) Sales, Marketing, Purchasing
40-71	Phil Biggs	Works Manager
58-68	Don Barlow	Acoustic Engineering (Sandwich Loudspeaker, Stereo Pickup) Responsible to HJL
45-71	Norman Rowe	Test & Service Dept. / Service Manager
51-71	Angus Gilchrist	Pickup Department / Sales Office Manager
50's-73	Ernie Hord	Speaker Production / Manager of Downham Market Norfolk Factory Responsible to PHB / EHA
62-69	John D. Leak	Marketing Manager / Overall Manager of Downham Market
44-71	Jim Davies	Dispatch / Manager of Dispatch Responsible to FB
60-71	Brian Collins	Storeman / in charge of Purchasing & Progressing. Responsible to FB
50's-71	Les Kirkpatrick	Progressor / Deputy Manager of Downham Market.
50's-70	Elsie McGuiness	Accountant, Wages etc. Responsible to JFH
53-70	Jackie Cadby	Secretary to all, and General Office Manager
60-65	Steve Wilkins	Moving Coil Pickup Production
45-50 & 54-55	John Bolingbroke (ex R.A.F.)	Development Engineer on Loudspeakers, and Testing of Electronics
46-47	John Short	Loudspeaker cabinets
67-71	Ian Pullen	Electronics Development Engineer Designed Stereofetic FM Tuner
47-50's	Arthur Gay	Project Management, General work & testing, school friend of HJL
60-71	Ray Whitcombe	Senior Electronics Development Engineer
67-71	Mike Deadman	Graduate / Development on Loudspeakers, Pickups and Turntables
65-66	Simon C. Leak	Test Department (for 14 months)
55-56	Ananta Sarkar	Loudspeaker development Engineer

** Some staff undertook National Service at various times, resulting in a period of absence from H. J. Leak & Co. Ltd.*

Norman Rowe recalls that all staff referred to each other with Christian names, apart from Mr Leak. Even Frank Hawkins was referred to as Frank.

The listing presented here is not exhaustive.

Other Staff
Jack McCarthy
Bill Kemp (maintenance)
Miss C. West
Mrs. B. Fitzgerald
Mr K. Farkas
Cyril Penn

Development:
Jack Turner
Roger Paton
Bob Skegg

Test:
Trevor Holman
Ted Snoad
Richard Karpoinski
Jim Neill
Len Grimshaw
Dillip Sinha
Keshoe Gulhane
Mr Rehal
Mr Sawney
George Hill
Sid Mason

Production:
Bob Buckledee
Cyril Bezant
Joe Murphy
Fred Westoll
Ron Innes (also Draughtsman)
Douglas Lord
Gerry Willis
Albert Street
Roy Edwardes
Don Loades
Rodney Cook
Archie Miller
Julian Walcuch
Jan Prosniefskwi
Gordon Bartlett
George Garside
'Tubby' Beer
Bunny Phipps

Telephonists:
Doris Gunter
Avril Hord (wife of E. Hord)
Pam Gilkes
Maureen Collins
(sister of Brian Collins).

Office:
Dana Jelen
Jennifer Clarke
Jean Attwell
Vi Greenway
Tom Martin (Draughtsman)
Margaret Simpson
Margaret Trainer
Miss T. Risby (Accounts)
Margaret Hawkins (briefly)
Mrs. Hill (wlfe of G. Hill)

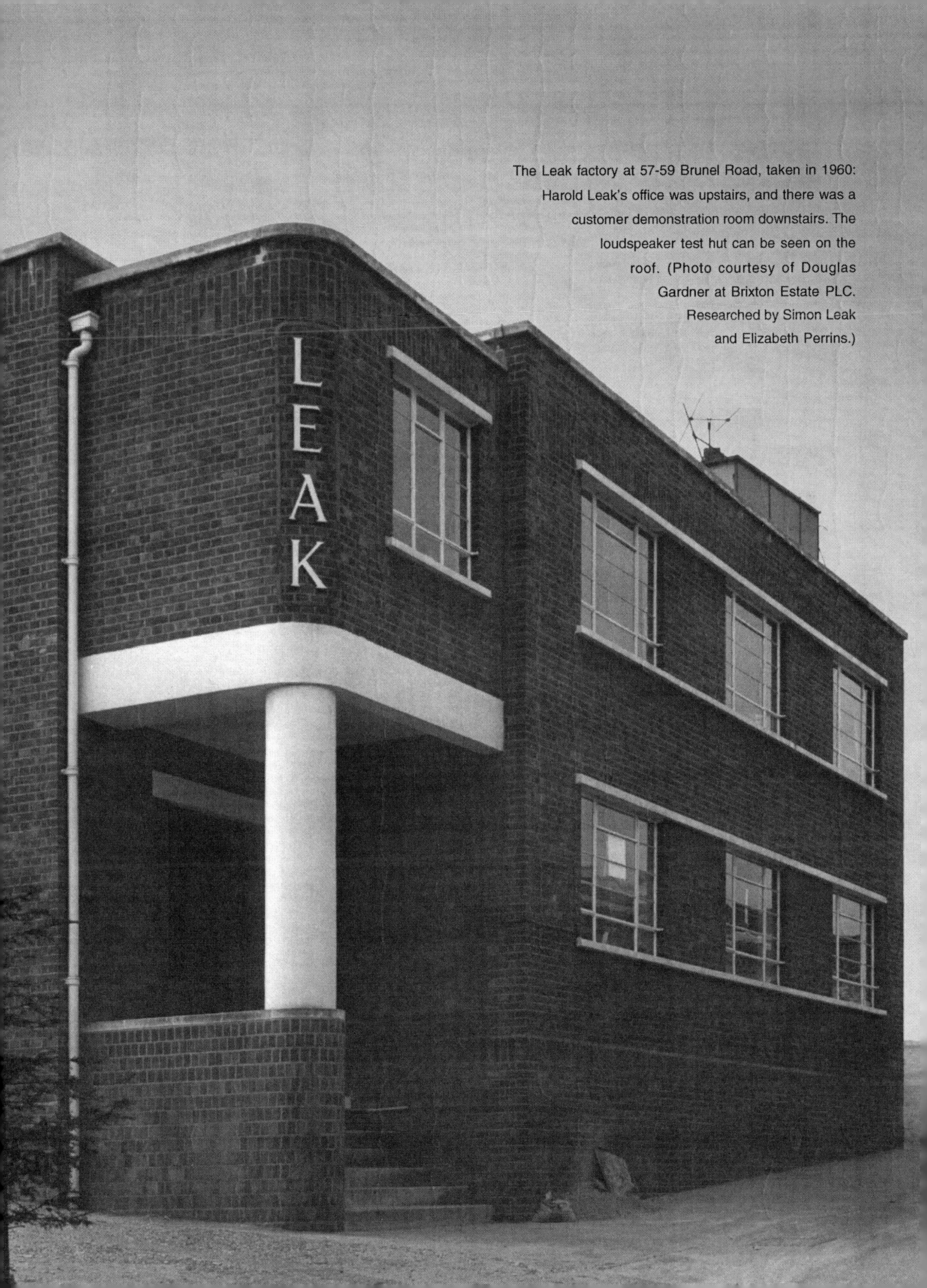

The Leak factory at 57-59 Brunel Road, taken in 1960: Harold Leak's office was upstairs, and there was a customer demonstration room downstairs. The loudspeaker test hut can be seen on the roof. (Photo courtesy of Douglas Gardner at Brixton Estate PLC. Researched by Simon Leak and Elizabeth Perrins.)

Norman Rowe recalls that he and other staff from Leak undertook out-work to supplement their wages, even though he was studying at night and working 44 hours per week at the factory.

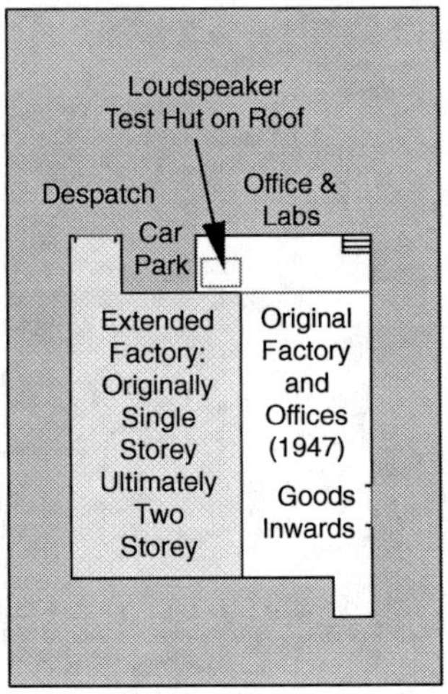

The factory layout at Leak: an office block was added in the 1950's, and factory extensions added in the 1960's. (Source: E. H. Ashley)

Brunel Road

By 1947, there had been an increase in production and staff numbers to the extent that the accommodation at 470 Uxbridge Road was becoming cramped and inefficient. As a result, the Leak company decided to move to new premises at numbers 57 & 59, Brunel Road, Westway Factory Estate, Acton. Phil Biggs recalls that the company bought the vacant blocks of land, and that subsequently the new factory was built (in-part from recycled building materials). Angus Gilchrist recalls that originally the rear section of the factory was an extremely cold place to work in during winter. The choice of site for the new factory allowed for future expansion, as the original factory building only occupied about one-third of the site.

At the end of the gentle rise travelling east up Brunel Road (name after Isambard Brunel, a famous Engineer of the nineteenth century), there were then three factories, with a heavy engineering works in the middle and Leak in the left-hand corner. To the left of Leak, just before the end of Brunel Road, were Electro Alloys Ltd and Hampton's Furniture (later occupied by Williams Cook printers, who were engaged by Leak for many years to print company publications). At the rear of the heavy engineering works a foot bridge over the railway line provided a link to the Western

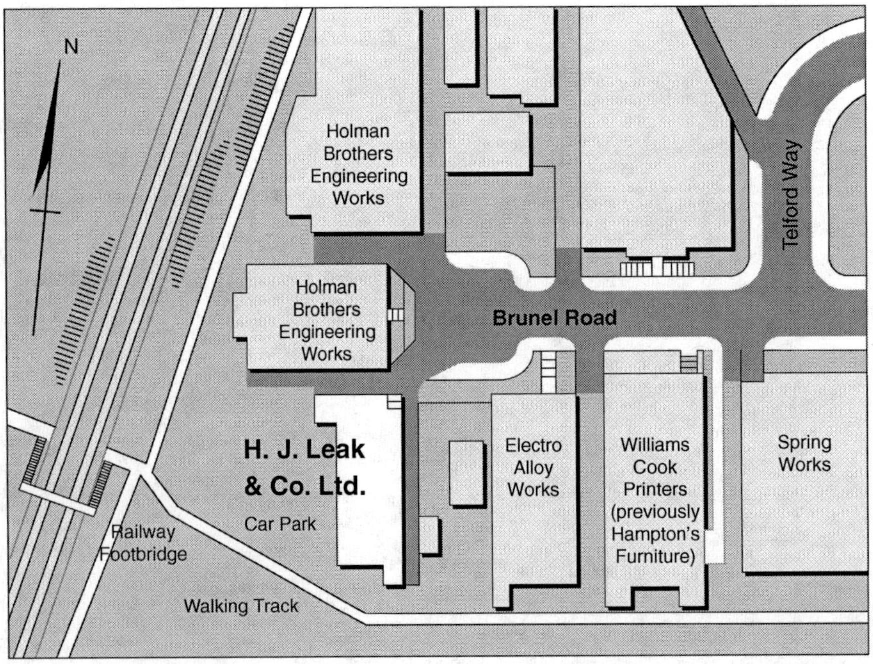

Leak was situated in a "cul-de-sac" at the end of Brunel Road – at numbers 57 and 59. Originally the factory only occupied about one third of the site, but was extended over the years. This plan from 1960 was prior to extensions for Sandwich speaker production. The depth of the factory from front to back was about 33 meters. (Layout and plan courtesy of Douglas Gardner and the archives of Brixton Estate PLC, with supplementary information from E. H. Ashley and A. Gilchrist.)

Avenue (the A40), and from there it was just a short distance to the Telegraph Condenser Company (TCC) located in Wales Farm Road. (Extensive use was made of TCC capacitors in most Leak amplifiers.)

In the 1950's extensions were undertaken at Brunel Road to increase the space for production. More importantly though, a new double-storey office block was added at the front of the building – originally the office accommodation was provided by simple partitions adjacent to the production area (taking up about one-third of the factory). Subsequent extensions converted the newer part of the production space to double-storey and added a loading bay at the front (north side).

Parking at Brunel Road was provided by a small car-park at the front, and on land rented from British Rail that was located between the railway line and the side of the factory. Harold and a few others had allocated places, although they were not marked with names. If a visitor happened to arrive and unknowingly park in these places, usually Harold would get Phil Biggs to go down and sort it all out. Phil recalls one particular incident: "HJL parked his brand new Bentley outside the inwards-goods bay. The storeman complained to me that he was expecting a van load of chassis to arrive, and could the Bentley be moved. I consulted HJL on his error of parking and was given the keys to move it. I was unaccustomed to the power of the Bentley and its automatic transmission – the result being that I drove it straight into the rear of Arthur Gay's Ford Console, shunting it three feet up the drive! I must say HJL took it very well – I felt a little silly".

The Brunel Road factory was closed in 1971 and operations were moved to Rank's Yorkshire facility. It was demolished in 1996 along with others in Brunel Road.

Leak didn't have a cafeteria – just a tea bar. Shown here is Jackie Cadby (left), with Jennifer Clarke. Staff often used the "British Restaurant" in East Acton for lunch, or the canteen of Hampton's Furniture, nearby to Leak in Brunel Road. (© N. Rowe 1960. Reproduced by permission.)

Jennifer Clarke at Brunel Road, 1960.
(© N. Rowe 1960. Reproduced by permission.)

Secretary to Harold Leak, Jackie Cadby, models for a photo of the Southdown cabinet. (Reproduced by permission Leak Electronics Ltd.)

Downham Market

A Leak TL/10 during final assembly at the Brunel Road plant. Leak standardised on the used of Ersin Multicore "Savbit" solder in 1956. (Reproduced with the permission of Multicore Solders, UK. Source: WW Jul 1956)

In 1964, a new factory at Downham Market, Norfolk was established – primarily because of insufficient production space at Brunel Road. The production work between the two sites was efficiently organised by splitting the type of work undertaken between each factory. From this time on, Sandwich loudspeaker assembly was undertaken at Downham Market, whilst Brunel Road was primarily used for manufacture of electronic products and for some assembly of Sandwich drive units.

Harold's son, John Leak, who joined the company in 1962 was responsible for the selection and purchase of the factory at Downham Market. John Leak was keen on motor racing and used to go up to Snetterton on the A11 motorway. During these trips he learnt that Downham Market would make a good site for a loudspeaker production facility. One of the attractions of Downham Market was a good source of labour, making it possible to select suitable staff, who could be well trained. Subsequently, the necessary land was purchased from an old friend and associate Fred Winston Reynolds (see also "VS-Tuner", Chapter 6). The new Leak factory was established by his company, Pineview Developments.

Ray Whitcombe recalls: "There were very serious plans to move the whole of Leak's operation to Downham Market during 1967. John Leak was very keen on this move and arranged for all the senior personnel to be given a tour of the area in order to get an idea of how many would re-locate. In spite of all this, the idea went down like a lead Zeppelin, and the only division to move was the Loudspeaker production group. I remember Downham Market as being very remote and isolated – all those flat dark fields and water everywhere!"

Staff at the Brunel Road factory in the mid 1960's. (© K. Gulhane, and reproduced by permission).

With the establishment of the loudspeaker production facility at Downham Market, Ernie Hord was appointed as the Works Manager. He frequently kept in touch with Ted Ashley and Phil Biggs at Brunel Road regarding progress of work and projects. John Leak had managerial responsibility for the Downham Market operation, and also wrote some papers on marketing which were extensively discussed at Board meetings.

The guarantee card issued with Leak Sandwich loudspeakers.

Factory Output

Amplifier chassis had a label attached to track progress of each unit through the factory, and for the calculation of team bonus payments. An adhesive sticker placed inside the chassis recorded the production details as well. A three-digit code was used that recorded the clock-card numbers of: the chassis assembler; the charge-hand of the production team that wired the chassis; and the number of the quality control inspector. (After 1964 amplifier chassis were indelibly marked with a sequence serial number, commencing in "Z".)

By the 1960's there were at least six production lines managed by Phil Biggs, excluding those established for speaker production. Ted Ashley estimates that around 1960 the Brunel Road factory had a capacity of about 50 units per day total output – this being a mix of pre-amplifiers, power amplifiers and Trough-Line radio tuners. On a yearly basis this equates to about 10,000 units.

Quality Control

Everything produced at Leak went through a strict test procedure. Rigid quality control was part of the reason for Leak's success, and it began at check-in. A cabin in the stores area housed two or three people who performed 100% inspection on almost all the visible elements of the equipment, metalwork, front panels, knobs etc. The process continued through production with assembly workers watching out for scratched transformer shrouds and the like – if anything had to be changed later it could affect the assemblers' team bonus. This attention to aesthetics, as well as the engineering side, was by no means universal at the time.

Production amplifiers were tested against demanding standards before leaving the factory and, in addition, samples from the production output were taken every few weeks for extensive and rigorous testing using more sophisticated laboratory equipment. This ensured that the quality control and testing of production amplifiers was being carried out correctly. After testing, other staff members did the final inspection.

A photo of extension work to the Test Lab happens to show tested and tagged pre-amps and tuners ready for packing and dispatch. (© N. Rowe 1960. Reproduced by permission.)

Power amplifiers were first tested without any negative feedback connected, so that the amount of feedback could be determined after it was applied. (Sometimes, out of tolerance variations in the 100Ω cathode resistor for the first stage could cause unacceptable variations in the feedback level and overall gain, so the component would be replaced). In the case of the TL/12 amplifier, the factory "pass" figures for TL/12's on final test before leaving the factory were:

Harmonic Distortion:	0.1% maximum at 10 Watts, 1000Hz
	0.19% maximum at 10 Watts, 60Hz
Hum and Noise:	-80dB referred to 10 Watts.
Sensitivity:	160mV or better
Load Damping Factor:	20 minimum
Frequency response:	20Hz to 20kHz +/- 0.1dB

Individual inspection tags were printed for each model. Several tear-off sections at the bottom of each label were used to track the progress of a unit through the factory, and for the calculation of team bonus payments. A testing officer checked the measured performance prior to a final inspection by another officer. (Upper: © G. Shore 1998. Lower: © T. J. Dale 1999. Reproduced by permission.)

Norman Rowe

Following a suggestion by the Headmasters Association at his school, Norman Rowe commenced with Leak at the age of 16 in 1945. He had no particular interest in audio prior to this, but during his first few years at Leak he undertook further study (paid for by the company) in radio, telecommunications and maths at a City Guilds night school.

"When I joined the company it was located at 470 Uxbridge Rd. At that time there were about 15 people employed. All the production staff were of a similar age to myself. I went through the various stages (of training): tagging and wiring circuit boards; fitting components to them; assembling amplifiers; and finally wiring them."

Frank Hawkins was already with the company and he managed the financial affairs of the business:

"He got the most from us lads by running a bonus scheme. We had a target number of units to produce and if we exceeded the target we would get a bonus payment. The quality of our work was high – anyone making a wiring mistake or producing a dry joint received a bit of stick from the rest us."

In September 1947 Norman left Leak to do National Service with the R.A.F. On his return in 1949 he was posted to the Test Department where all final testing of production amplifiers was performed, along with servicing. In this area he went on to become the Service Manager at Leak.

"Service and Test were always run as an integrated department. Staff (numbers) increased as the factory expanded and product range increased. I would think the maximum number of staff (in this Department) would be around 15. Angus Gilchrist ran the Pickup Department as a separate unit for both production and service."

In the late 1960's rumours surfaced amongst staff that the company was going to be sold. Norman was upset at the thought of this, as he was concerned about the future of the company: "I always enjoyed my time with Leak from 1945 to 1971" he recalled, but the sale to Rank made the future uncertain.

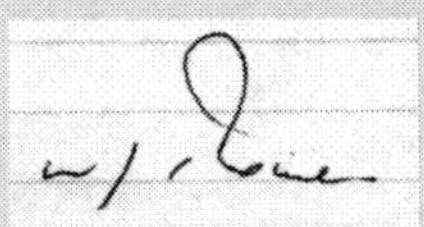

Every Leak valve amplifier produced since the TL/12 passed through this facility – the Test Department at Brunel Road. Some visible Leak products are: a Stereo 30 being held at the left, a Trough-Line FM tuner (including the US gold-fascia model) and a Point One pre-amp. (© N. Rowe 1964. Reproduced by permission.)

When stereo pre-amplifiers were introduced, Harold Leak was concerned about variations between channels being introduced from the ganged potentiometers used for tone and gain controls (due to normal manufacturing tolerances). To reduce the impact of this, Leak designed a test instrument that limited the variation in production parts to 3%. Initially potentiometers were tested at Leak but later on, suppliers such as Morganite and A. B. Metals provided matched potentiometers to Leak at a premium price.

One approach that was used to test production pre-amplifiers was the use of an automated system that could test performance, diagnose faults, and identify probable faulty components (it required many test points in the pre-amplifier to be connected to the system). Norman Rowe recalls how unsuccessful this approach was as the test system itself was found to be very unreliable – nearly all the faults it diagnosed were not in the pre-amplifiers being tested, but within its own unreliable circuits!

Another area that required a careful approach to quality control was solid state pre-amplifiers. Variations in performance parameters were common in early transistors, as semi-conductor manufacturing processes were not as refined as they would be in later years. In the test area for transistor amplifiers, AC107 transistors were tested for noise so that the lowest noise ones could be used in the first stage of the pre-amplifier. (Similarly, EF86 valves were tested – if samples from a batch proved excessively noisy or microphonic, then the whole batch would be tested.)

Bob Buckledee with a Stereo 30 amplifier and a Sandwich loudspeaker at the rear. (© K. Gulhane, and reproduced by permission).

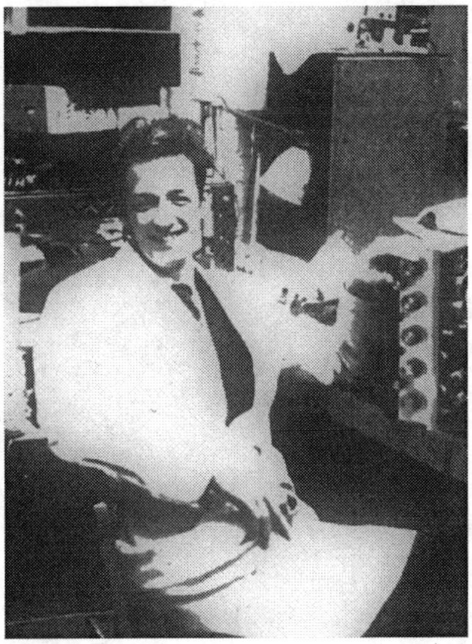

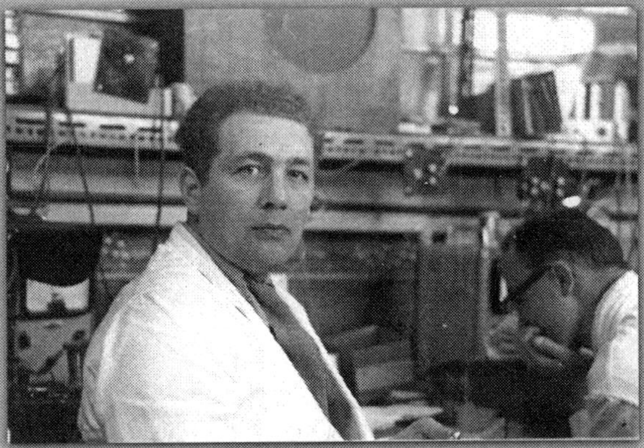

Test and Service Department:

Above Right: Len Grimshaw in the Test Room, 1958.
A Point One pre-amp and Trough-Line tuner are just visible, being built into the wall behind the test gear.

Above: Stephen –, one of the Polish workers at Leak, 1958.

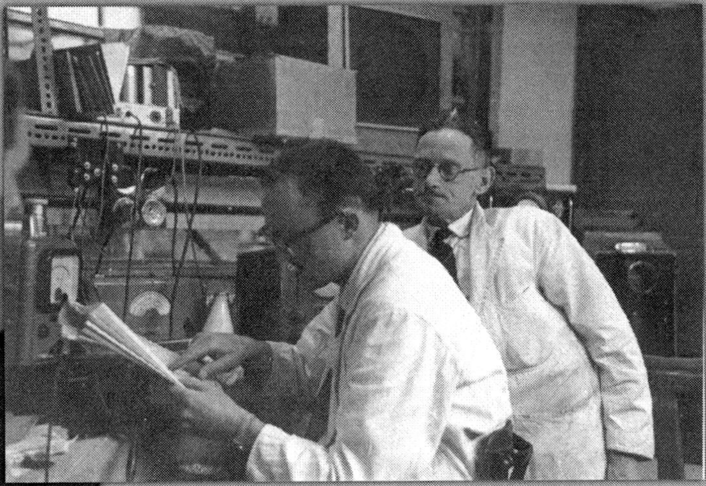

Richard Karpoinsky (left) with George Hill, 1958.

(All photos © N. Rowe 1958-1960. Reproduced by permission.)

Len Grimshaw in the Test Room during a tea-break, 1958.

Left to right: Richard Karpoinsky, George Hill and Sid Mason in 1958.

Test Lab Extensions, 1960.

Whilst quality was a fundamental aim for the company, Harold Leak considered that his products were closer in nature to Jaguar cars, rather than Rolls-Royce cars – emphasising the requirement to produce large volumes as well as quality.

"I worked for H. J. Leak & Co. from 1953 to 1970 as Secretary for Mr Leak and others. They were good years – we all got on so well with one another."

Jackie Cadby

Jackie Cadby, 1999.

Company Publications

Harold Leak wrote several technical papers that were published in journals such as Wireless World, the Journal of the British Institution of Radio Engineers, and The Gramophone. These papers discussed various studies that the Leak company had made into suitable transducers for a high quality loudspeaker (see Chapter 10 for details). The publication of technical papers by authors working for companies in the audio industry was not uncommon in the 1950's. Such publications were potentially beneficial to the public image of these companies and Wireless World was a common place to publish them. In this way Wireless World provided a forum for presentation and discussion of new ideas, that clearly enhanced the development of the audio industry in Britain.

In 1949 the Leak company published a booklet titled "Leak Point One Amplifiers" that described the TL/12 amplifier and RC/PA pre-amp. Whilst this was intended more as a marketing strategy, it was a detailed description of the design and performance of these amplifiers, with an explanation of the effects of negative feedback. The booklet was fully referenced to the standard technical publications of the day and it was updated in 1951 after the RC/PA/U was released.

Overleaf: Letters addressed to the company from around the globe, with the 1945 company letterhead (top-right), and a late 1960's letterhead (middle-right). At the bottom-right is the 1950's letterhead, with the text from a letter Harold Leak wrote to Wireless World included to show how it might have appeared. (Envelopes and letterheads courtesy: N. Rowe, R. Whitcombe and the Leak family.)

A 28 page booklet by Harold Leak describing the "Point One" amplifiers was available on request from the company.

AIR MAIL
PAR AVION
空郵
MALTA
Messrs H.J. Leak & Co Ltd.,
Brunel Road,
Westway Factory Estate,
London W.3.
England
PAR AVION
BY AIR MAIL
PER LUGPOS
JOHANNESBURG
13.30
14 -X
1961
For attention : MR. F. HAWKINS.
Messrs. H.J. Leak & Co.,
Brunel Road,
Westway Factory Estate,
London, W.3.
ENGLAND.
Mr. J.F. Hawkins,
H.J. Leak & Co., Ltd.,
Brunel Road,
Westway Factory Estate,
LONDON, W.3,
England.
AIR MAIL
MELBOURNE
30MAR
1962
VIC-AUST
ADDRESS MAIL TO
PRIVATE BOX
IT EXPEDITES
DELIVERY
AUSTRALIA
Mr. J.F. Hawkins,
H.J. Leak & Co. Limited,
Brunel Road,
Westway Factory Estate,
LONDON. W.3. England.
Simon Gray
PTY. LTD.
28 ELIZABETH ST., MELBOURNE, C.1. VIC.
PALACE
TELEGRAMADRESS
PALACE
TELEFON
VÄXEL.....17 42 40
SVERIGE
H.J. Leak & Co Ltd.
Brunel Rd
Westway Factory Estate
London W.3
England.
Rangoon RADIO Service
RANGOON
PER AIR MAIL
OTAN
Mers. H.J. LEAK
Brunel Road. Westway Factory Estate
LONDON W.3
INGLATERRA
H. J. Leak, Esq.,
c/o H. J. Leak & Co
Holly Tree Farm,
Gerard's Cross,
Buckinghamshire, Eng
Messrs.
Brunel
Westway
LONDON,
Capt Ann J Previto
Room 151, I G Farben Bldg
Frankfurt, Germany
DEUTSCHE BUNDESPOST
VIA AIR MAIL
Durch Eilboten
Exprès
Mr. J. F. Hawkins
H. J. Leak & Co. Ltd
Brunel Road, Westway
London, W. 3.
ESPAÑA
PALMA
MALLORCA
Mrs. H. J. LEAK & CO. LTD.
Brunel Road
Westway Factory Estate
L O N D O N W 3
BERMUDA
POSTAGE
HAILE ATHA
15-X
1960
H. J. Leak & Co
Brunel Rd,
Westway Fa
London

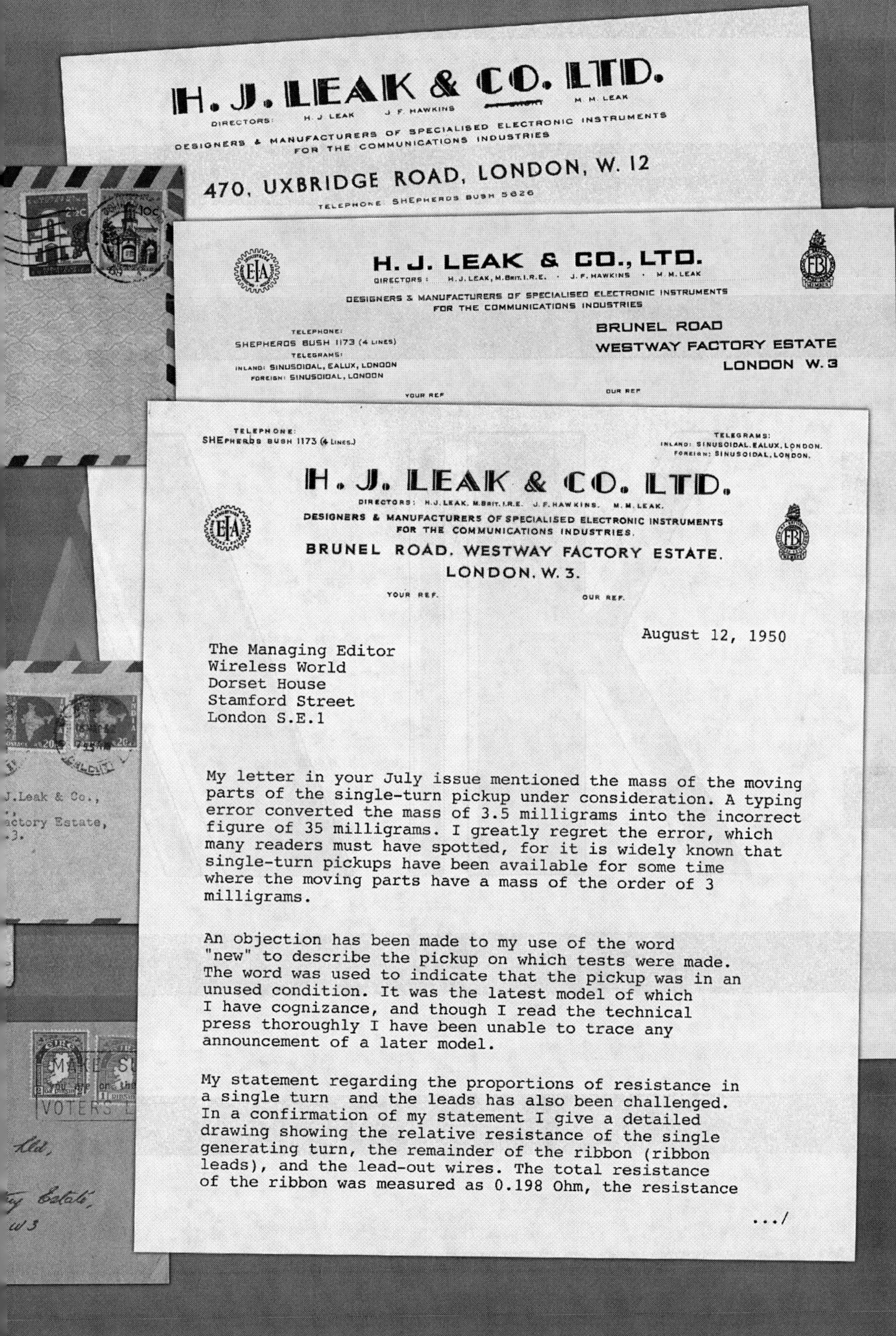

H. J. LEAK & CO. LTD.
DIRECTORS: H. J. LEAK J. F. HAWKINS M. M. LEAK
DESIGNERS & MANUFACTURERS OF SPECIALISED ELECTRONIC INSTRUMENTS FOR THE COMMUNICATIONS INDUSTRIES
470, UXBRIDGE ROAD, LONDON, W. 12
TELEPHONE: SHEPHERDS BUSH 5626

H. J. LEAK & CO., LTD.
DIRECTORS: H. J. LEAK, M.BRIT.I.R.E. · J. F. HAWKINS · M. M. LEAK
DESIGNERS & MANUFACTURERS OF SPECIALISED ELECTRONIC INSTRUMENTS FOR THE COMMUNICATIONS INDUSTRIES
TELEPHONE: SHEPHERDS BUSH 1173 (4 LINES)
TELEGRAMS: INLAND: SINUSOIDAL, EALUX, LONDON FOREIGN: SINUSOIDAL, LONDON
BRUNEL ROAD
WESTWAY FACTORY ESTATE
LONDON W. 3
YOUR REF OUR REF

TELEPHONE: SHEPHERDS BUSH 1173 (4 LINES.)

TELEGRAMS: INLAND: SINUSOIDAL.EALUX, LONDON. FOREIGN: SINUSOIDAL, LONDON.

H. J. LEAK & CO. LTD.
DIRECTORS: H.J.LEAK, M.BRIT.I.R.E. J.F.HAWKINS. M.M.LEAK.
DESIGNERS & MANUFACTURERS OF SPECIALISED ELECTRONIC INSTRUMENTS FOR THE COMMUNICATIONS INDUSTRIES.
BRUNEL ROAD, WESTWAY FACTORY ESTATE.
LONDON, W. 3.

YOUR REF. OUR REF.

August 12, 1950

The Managing Editor
Wireless World
Dorset House
Stamford Street
London S.E.1

My letter in your July issue mentioned the mass of the moving parts of the single-turn pickup under consideration. A typing error converted the mass of 3.5 milligrams into the incorrect figure of 35 milligrams. I greatly regret the error, which many readers must have spotted, for it is widely known that single-turn pickups have been available for some time where the moving parts have a mass of the order of 3 milligrams.

An objection has been made to my use of the word "new" to describe the pickup on which tests were made. The word was used to indicate that the pickup was in an unused condition. It was the latest model of which I have cognizance, and though I read the technical press thoroughly I have been unable to trace any announcement of a later model.

My statement regarding the proportions of resistance in a single turn and the leads has also been challenged. In a confirmation of my statement I give a detailed drawing showing the relative resistance of the single generating turn, the remainder of the ribbon (ribbon leads), and the lead-out wires. The total resistance of the ribbon was measured as 0.198 Ohm, the resistance

.../

KT66
"POINT ONE"
T.L./12 12W. AMPLIFIER
MADE IN ENGLAND
BY
H. J. LEAK & Co.
WESTWAY FACTORY ESTAT
LONDON. W.3.
LOUDSPEAKER
A.C.MAINS

4. The TL/12

The Birth of Hi-Fi

The term "high fidelity" was coined by H. A. Hartley (of Hartley-Turner Radio Ltd.) in 1927. During the 1930's the phrase was frequently used by other manufacturers to describe their products but at this time the term was mainly used in the context of radio reception – the notion of "high fidelity audio" (abbreviated to "hi-fi") as a hobby in its own right, or as a separate market segment had not matured.

"In the middle 1930's there were three great names: Paul Voigt, P. K. Turner, and Frank Murphy. They were the only men in the radio industry who had the vision that high quality sound could be made available to the general public"

John Bolingbroke, 1999. Formerly of H. J. Leak & Co. Ltd.

A survey of articles from Wireless World before the Second World War shows that the seeds of "high fidelity audio" were beginning to grow though. In 1936 a negative feedback amplifier design was published, and in 1939 Fritz Langford-Smith wrote a two-part treatise on triode, beam tetrode and pentode output stages, loudspeakers and negative feedback. M. G. Scroggie published a design for an "Electric Gramophone" in 1939, and took care to explain why the gramophone amplifier had not been combined with the radio, as was the normal practice.

The War doused consumer interest in hi-fidelity gramophone reproduction but manufacturers such as Leak were able to undertake research and gain experience with new tools like negative feedback during this time.

For the first few years after the War, advertisements for sound equipment were mainly for public address applications. In just a few issues of Wireless World from 1947 many names that would become dominant in hi-fi during the 1950's were advertising public address equipment:

Wharfedale:	public address W12 loudspeaker
Tannoy:	omni-directional ceiling speaker
Acoustical Manufacturing:	a booklet titled "Addressing the Public", and the L15 labyrinth speaker for music societies and quality PA work
Celestion:	18 inch, 40 Watt public address speaker
Goodmans:	Type AL/5 Public Address loudspeaker

Gilbert Briggs of Wharfedale was at the vanguard hi-fi developments in loudspeakers. (Reproduced by permission IAG Ltd, UK.)

But there is also evidence of high-fidelity audio starting to become of wider interest. At the 1947 Radio Olympia exhibition there were "amplifiers for the quality enthusiast, designed to reduce distortion below the usually accepted commercial limits" being shown by Lowther (Donald Chave) and Acoustical Manufacturing (Peter Walker). The QA12/P from the latter had harmonic distortion of around

Opposite: Leak entered the Hi-Fi market with the TL/12.

0.5%. Also on show was the "Concentric Dual" 15-inch speaker from Tannoy that would, in later versions, become a standard in many recording studios. Also in 1947, on GEC's initiative, Wireless World published D. T. N. Williamson's 1945 paper "Design for a High Quality Amplifier". This would prove to be a landmark paper on audio amplifier design that would bring very low distortion performance within financial reach of audio hobbyists all over the world.

Norman Partridge photographed in the 1930's. Partridge Transformers are recognised for high performance that sets them apart from others. (By courtesy of Transformers and Rectifiers Ltd, Guildford, United Kingdom. Source: Partridge PA Manual.)

So by 1947, the War was in the past and people were starting to become interested in getting the best sound possible from records and also in sound recording. This was a new field of interest for people, that differed from radio and which before the War would have been considered a poorer cousin to radio. Ted Ashley remembers that people would come to the Leak factory and spend many, many hours listening to demonstrations – at the end of which they would usually buy an amplifier – but it wasn't a very efficient way to market audio products.

In the early days after the War, the hi-fi industry was of course quite small, so this resulted in a friendly atmosphere where many people were keen to share and discuss new ideas. There were a number of organisations where people met and talked about radio and audio, examples being: the Association of Public Address Engineers (APAE), Radio Society of Great Britain (RSGB), and the British Sound Recording Association (BSRA). From the activities of these organisations a new specialised interest in audio began to grow in the wider public. The Leak company was a member of the BSRA and Harold Leak would give lectures at the meetings to other members of the BSRA. The BSRA had yearly meetings at which various industry people demonstrated equipment. In time these meetings got larger, until eventually the idea of holding a public Audio Fair was proposed. And so, in April 1956, the first Audio Fair in Britain was held at the Washington Hotel with 37 manufacturers exhibiting (the first New York Audio Fair was in 1949).

Donald Chave of Lowther, who had been manufacturing quality sound equipment before the War. (Reproduced with permission from Hi-Fi News & Record Review, UK. Source: HFN Aug 1956)

The Move Away from Sub-Contracting

Until the TL/12 was introduced, Leak's main business was still in manufacturing amplifiers for large companies, as it had done before and during the War. GEC and others were purchasing amplifiers from Leak that were based on the "Point One" technology of 1945. An example here is the GEC Type 2430A amplifier which was identical to the Leak 30 Watt "Point One" amplifier using KT66 valves in tetrode operation.

Another example of post-War work done under sub-contract was an amplifier that had a power output of 1,000 Watts. Several were supplied to the Vauxhall Motors factory located in Dunstable. Norman Rowe recalls how difficult it was to test the amplifier owing to the higher than usual voltages used.

Harold Leak with other industry figures in the early 1950's, including Leonard Carduner (behind Harold) and Gilbert Briggs (second from the right). (Photo courtesy of the Leak family.)

However, despite these commercial successes with sub-contracted work, Frank Hawkins and Harold Leak were becoming aware of post-War developments in the market place and the way things were changing in the economy. By the late 1940's it was apparent that domestic hi-fi was a growing market segment – and that Leak was in a strong position to move away from its traditional sales base to this new market.

The growing numbers of people visiting the factory in search of high quality sound equipment would have been another factor that influenced the company to re-assess its commercial strategy, and subsequently alter focus towards the manufacture of sound equipment for the home, marketed through retail outlets. Fortunately the demand for hi-fi was just starting to grow at the time the TL/12 amplifier was released, so Leak was in a good position to capitalise on this new market with an amplifier that had stunning specifications and performance.

Leak continued with some sub-contracting work beyond 1948, although this work became less significant to the company's success. Even during the mid-1960's, the Leak company letterheads still carried the signature phrase "Designers and Manufacturers of Specialised Electronic Instruments for the Communications Industry", which was more a reflection of the past than the company's success with hi-fi products.

"One of the weirdest 'special' contracts of H. J. Leak and Co Ltd was the Firestone one. This stemmed from the fact that Leak had installed intercom systems for the production lines of the Firestone Tyre Company at the famous Great West Road factory. ...Leak had the maintenance contract for the intercoms right into the 1960's and I remember Engineers going to Firestones and bringing back very dirty TL/12's and control boxes (for microphone switching etc.). These artefacts were covered with many layers of black carbon dirt and silver paint that had been applied by the Firestone factory maintenance people to keep the factory (looking) clean! I was 'conscripted' to this duty, which was very unpopular and required one to wear old clothes which could be then discarded, so I was told!"

Ray Whitcombe, 1999

Industry Personalities

The early hi-fi industry in Britain was small and friendly – Harold Leak associated with many people in the industry, a few of whom are presented here:

• Norman Partridge was a friend of Harold's, and the Partridge company supplied all transformers to Leak, up until the TL/12 series was launched. Leak also built amplifiers for Partridge before the War.

• Harold didn't know D. T. N. Williamson until he was introduced to him by Graham Woodville at the Osram Laboratories located in Brook Green. In 1945 Harold Leak took the "Point One" Type 15 amplifier around to Osrams, and at this stage D. T. N. Williamson was employed by Osrams.

• Peter Walker was of course operating the Acoustical Manufacturing Company at this stage, and was a fellow-competitor. In the 1940's the Acoustical Manufacturing Company was making amplifiers for public address and commercial use (making at least 10 different model amplifiers between 1941 and 1952 for such use).

• Harold Leak gave many public demonstrations of live - versus - reproduced sound, and lectured on high fidelity reproduction in general. Even before the TL/12 was released, Harold was giving public demonstrations of prototypes at meetings of various societies, and continued with these

Gilbert Briggs of Wharfedale Wireless Works (left) and Harold Leak (right) with Muriel Leak (left) and Gilbert Brigg's wife (right). (Photo courtesy of the Leak family.)

Below Middle: A very impressive sound-system installation of the mid-1950's that incorporated a Leak amplifier and pickup, with Ferrograph tape recorder and Garrard turntable. This system was owned and designed by Richard Arbib of Multicore Solder, who was a close friend of Harold Leak. At the bottom left, the popular Bib tape splicer (invented by Richard Arbib) is seen attached to the tape recorder. Richard Arbib's son Gordon shares a birthday with John Leak and both undertook schooling at Stowe. (Photo courtesy Gordon Arbib.)

At Carnegie Hall on October 3 1956 the equipment used for the Gilbert Briggs presentation comprised: a Garrard 301 turntable, Leak pickup, Point One pre-amp and three Leak TL/25A amplifiers with Wharfedale corner systems. Gilbert Briggs is seated centre-left and facing away, his daughter centre, and Harold Leak next to her at the rear. Danny Daniels is tap-dancing. (Reproduced by permission IAG Ltd, UK. Source: HFN Nov 1956)

into the 1950's – for example, at the following venues:

- The British Kinematograph Society, G-B Theatre, Wadour St, London, in 1946. (Albert Street, a young employee at Leak, provided his singing voice to allow a live-versus-reproduced demonstration of the Type 15 amplifier.)
- The Royal Society of the Arts, and reported on in the BSRA Journal, 1948.
- The Institution of Radio Engineers at venues in Manchester, and Newcastle-on-Tyne during 1950.

• Harold Leak had discussions with Gilbert Briggs regarding loudspeakers and in particular loudspeaker surrounds and suspension systems as early as 1946. Harold Leak also provided amplification, pickups and technical assistance for some of the famous Gilbert Briggs series of public hi-fi demonstrations. One of these was given to an American audience in Carnegie Hall on October 3, 1956. However the majority of the Gilbert Briggs lecture/demonstration series used Quad amplifiers from Acoustical Manufacturing Co.

• In September 1949, following the New York Audio Fair, Leak appointed British Industries Corporation (BIC) as distributors in the USA. An influential person here was Richard Arbib from Ersin Multicore Solders, who already had distribution arrangements with BIC. Gilbert Briggs also arranged for distribution of Wharfedale through BIC.

Above: Harold Leak and Norman Mordaunt with Acos Pickup promotional material in the foreground. Norman was with Decca in the late 1940's, then with BSR and Tannoy before forming Mordaunt-Short. He died in April 1999. (Photo courtesy of the Leak family.)

Harold Leak with Peter Walker at the 1957 London Audio Fair. At the left is Richard Arbib of Ersin Multicore Solders. (Reproduced by permission Leak Electronics Ltd. Source: HFN May 1957)

Above Right Middle: Harold Leak with Arnold Sugden (second from the right) of Connoisseur motors, and Donald Aldous (Journalist) in 1952. A Connoisseur turntable and amplifier is also shown. (Photo courtesy of the Leak family.)

Bottom: Max Krain (left) with Harold Leak and other industry figures. (Photo courtesy of the Leak family.)

"During the war I carried out certain development work on negative feedback which could obviously be applied to high-fidelity home-music amplifiers. By October 1945, I was advertising in the British technical press a four-stage feedback amplifier using push-pull KT66's for which I claimed a distortion of 0.1 per cent at 10 Watts. . . . Four stage circuits had margins of stability which, in my opinion, were not sufficient, and the obvious solution was to improve the phase characteristics by omitting one stage. The problem then was to devise a three-stage high gain circuit, and this was accomplished by replacing the conventional phase-splitter valve by a circuit having gain seven times as great and by replacing the input triode tube with a pentode. The distortion . . .was 0.03 percent at 10 Watts output. Not only was the distortion lower but the stability margin was greater and the cost of manufacture was lower than for the four-stage circuit."

Harold Leak, 1953.
(From "The New High Fidelity Handbook", by I. Greene & J. R. Radcliffe, by courtesy of Crown Publishers Inc., NY, 1955.)

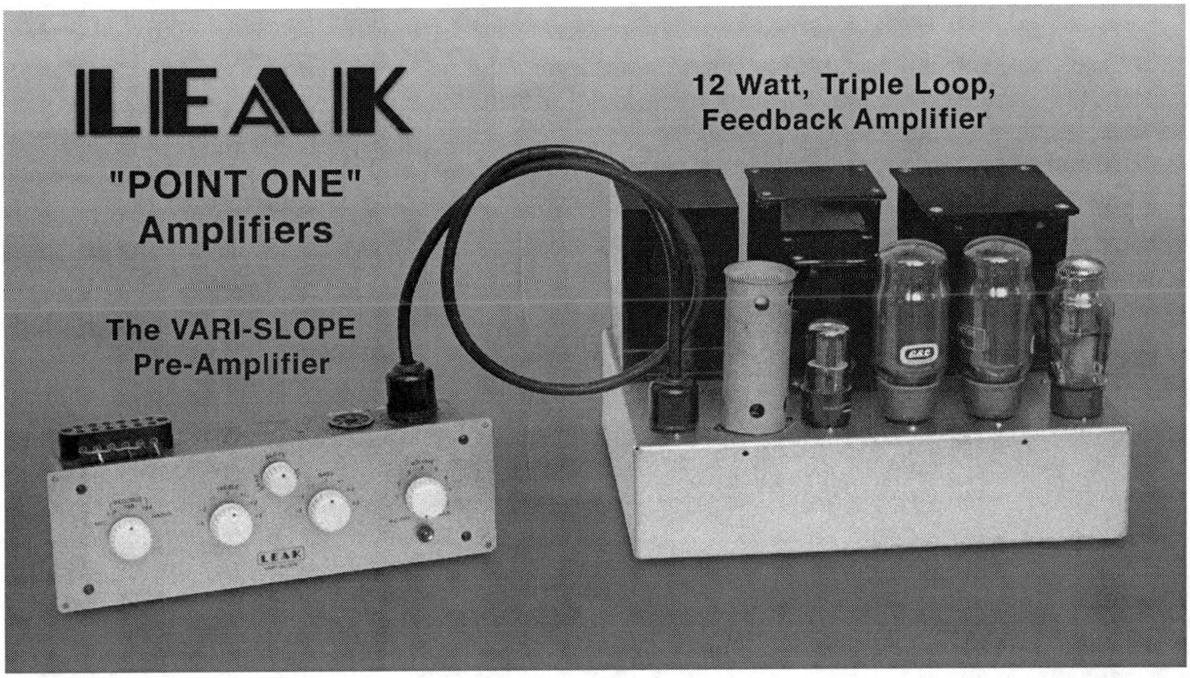

The TL/12

The "Type 15 Point One" amplifier, being a four stage circuit was expensive to produce – the purchase price was very high at around £70*. In addition, each stage in an amplifier has the potential to introduce phase shifts that lessen the stability margin when feedback is applied. By 1946/47, Harold had decided to revise the design of the Type 15 to counter these problems and to target a new market – that of the music lover or hi-fi enthusiast. The Leak company published the following perspective in 1951 about the TL/12 design, which owes its heritage to project work on amplifiers for driving feedback record-cutting heads:

> "In 1948 the problem of costly production was surmounted when we abandoned our original 4-stage circuit of 1945 for the simpler, more sensitive and completely stable 3-stage circuit of the TL/12.
>
> "The basic TL/12 circuit was evolved in 1947 by H. J. Leak for use with feedback cutter-heads, and the circuit was so stable that feedback magnitudes greater than 40dB could be obtained over the loop containing the three stages, the output transformer and the electro-mechanical circuit of the cutter-head. We realised that with this new circuit we had the opportunity of offering a superlative amplifier, but with our high-grade labour it was essential to use the most advanced techniques of manufacture and to ensure quantity production if the price was to be kept low. We knew that the very large

Many TL/12 advertisements followed this style. When first released the associated pre-amp was the RC/PA, but shown here is the later Vari-Slope. "TL" was a reference to the "triple loop" feedback design, although the secondary and tertiary loops are not significant from a conventional circuit analysis viewpoint.

Opposite: Harold Leak in New York in the early 1950's. (Photo courtesy of the Leak family.)

* Un-confirmed price

orders necessary for quantity production would come only if the amplifier gained world-wide acceptance from three main classes of users:

1. Professional audio engineers representing the communications industries (broadcasting and recording companies, telephone, telegraph and cable companies, Public Address and diffusion operators) and also those engineers in the distributing branches of the radio industry whose main interest is high-fidelity reproduction.

2. The non-technical music-lover or gramophile.

3. The knowledgeable amateur enthusiast (the "audio hobbyist" of the U.S.) who had previously built his own amplifiers for reasons of economy and/or personal satisfaction.

Below: This TL/12 is finished in gold, has enclosed transformer casings, and dates from some time after 1953. On the left side the holes for mounting the pre-amplifier to the TL/12 chassis can be seen. Below Opposite: A TL/12 finished in bronze, and fitted with open-frame transformers, dating from some time prior to 1952.

"With these requirements before us further development work was carried out on the basic circuit and on the physical details, and the TL/12 in its present form was launched in December 1948, together with the RC/PA pre-amplifier.

"Today, late 1951, it is well known that our projects of three years ago have proved successful, and that the RC/PA and TL/12 amplifiers have a prestige and market that is world-wide. During 1951 orders to date totalling over 2,000 have been received from the U.S.A. for RC/PA/U and TL/12 amplifiers for high quality radio/gramophone reproducers for the home." *

"When Harold Leak sent the TL/12 to the National Physical Laboratory for tests, he told me that he was worried about the frequency response because he did not have an accurate meter to measure it."

George Wise, 1999.

* Extract reproduced by permission Leak Electronics Ltd.

The launch of the TL/12 was accompanied by a high quality and expensive advertising campaign featuring full-page advertisements in Wireless World. The campaign was most effective with 500 units being sold within three months, and

advertising costs being about £1 per unit at a time when the retail cost of the amplifier was £25/15/0. In 1949 the performance of the amplifier was verified by the National Physical Laboratory, and their report was published in advertisements for the TL/12 accompanied by the phrase "Britain's Best Audio Amplifier".

The design of the TL/12 was undertaken by about four staff, and took around 18 months to two years to complete. At the time there were fewer than 20 staff working for the company and Harold Leak did the electronics design work and component layouts himself, with Jack Turner being his assistant.

The amplifier uses a pair of KT66 valves in the output stage that are connected in a triode configuration with individual cathode bias. This enables the amplifier to use un-matched valves in the output stage and makes maintenance of the amplifier easy for non-technical owners, as the bias will adjust itself as valves age. A neat side-effect is that the R/C cathode network also acts as a feedback compensation network at low frequencies to control the stability of the amplifier (the gain of the output stage reduces at low frequencies, without accumulation of phase shift from the "step" in frequency response).

The TL/12 was the first three-stage circuit by Leak to use the long-tailed pair as the phase splitter. For a perfectly balanced output signal a long-tailed pair actually requires an infinite resistance to signal variations in the tail resistor (the resistor from the junction of the two cathodes to signal ground). In practice, a resistor 20 to 50 times larger than $1/g_m$ will give adequate balance (where g_m is the transconductance of stage). Any residual un-balance can be negated by using slightly different values in the anode resistors of the stage and this is what is done in the TL/12.

The first stage is a pentode voltage gain stage, which is capacitively coupled to the second stage. As it happens, the stage could have been direct coupled although some minor design alterations would have been needed. But the requirement for the amplifier to operate correctly with or without a pre-amplifier meant that the high-tension supply to the first stage varied too much between these two conditions (as the pre-amp needed a well filtered supply, which was taken from the first stage high-tension supply in the TL/12).

Telegraph Condenser Company were nearby to Leak in North Acton and supplied many capacitors to Leak, including the oil-filled paper capacitor block for the TL/12. (Source: WW June 1945)

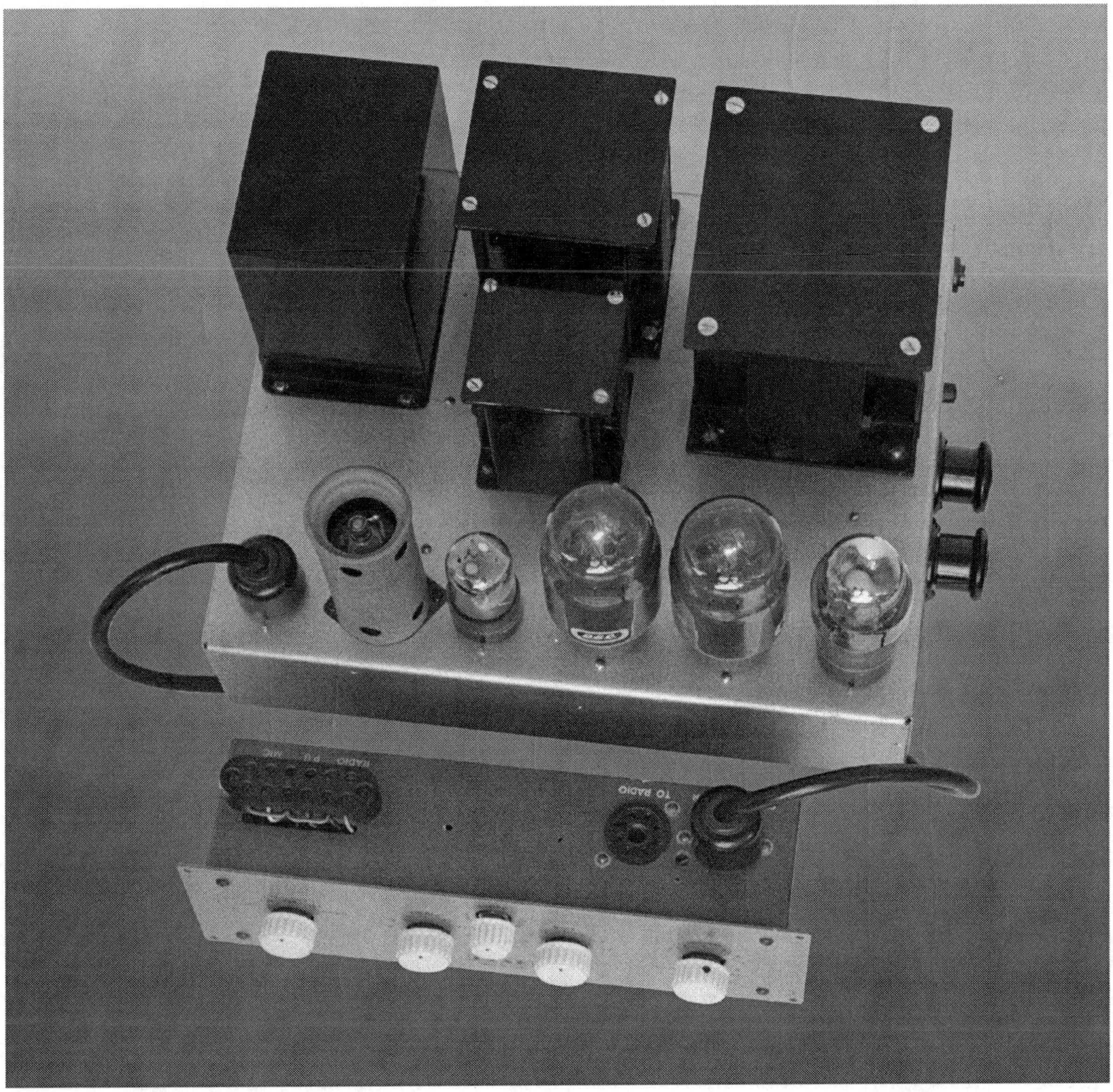

A common combination – the TL/12 power amplifier paired with a Vari-Slope pre-amplifier.

To achieve the distortion figure of 0.1%, the main feedback loop encompasses all three stages and the output transformer. The loop gain is 20 (ie: 26dB) so the distortion present in the amplifier without feedback would be 2% – a figure typical of push-pull triodes. In addition Leak published stability margins for the TL/12, emphasising their claim that the application of heavy amounts of feedback cannot be done without due consideration to the stability of the circuit. The TL/12 had a published gain margin of 10dB +/- 6dB, and a phase margin of 20 degrees +/- 10 degrees.

The amplifier circuit was a landmark for the company, as it was used as the basis for all Leak designs over the following 20 years. Its commercial success set the stage for Leak to expand into other complementary products that would be developed during the 1950's. It is of interest that several amplifiers from other manufacturers released after the TL/12 have used a circuit topology very similar to this – one of which is the Mullard 5-20.

Leak promoted the TL/12 at many exhibitions, including the 1951 Association of Public Address Engineers at Horse Shoe Hotel, Tottenham Court Rd, London. In the centre is Frank Hawkins, and at the right is Arthur Gay. The TL/12 on the left appears to be fitted with a heat shield, a modification for housing the amplifier in cabinets with restricted ventilation (such as in the base of the BBC monitor loudspeaker). An up-turned TL/12 is shown in the middle, and a portable system with cover is on the right, with an open pre-amp in front. (Reproduced with permission from Electronics World magazine, UK. Source: WW July 1951)

Very soon after the TL/12 was released, staff numbers increased very quickly as world-wide demand for the amplifier grew. All TL/12's and TL/25's were made at Brunel Road, with the TL/12 being released in December 1948, and discontinued in 1957.

The TL/12 was an eye-catching amplifier, as the chassis was finished in bronze enamel (later gold) with black transformers sporting shiny bakelite plates on top. Leak were fastidious about the chassis finish, the process being:

- 16 gauge silver-finished steel, gas-welded, and passivated zinc-plated
- one coat of stoved primer
- first finish coat of stoved enamel
- second coat of stoved enamel
- silk-screen lettering
- final stoved transparent coat

Most TL/12's have retained a good appearance due to the quality and durability of the chassis finish.

At the time the TL/12 was designed, Leak were not using industrial design consultants for determining the appearance of products (this would come later). The appearance of the TL/12 was primarily determined by Harold Leak; however Frank Hawkins was always involved with such decisions, as he had constructive contributions to make in this regard.

One visible feature of the TL/12 is the large capacitor block for power supply filtering. The Telegraph Condenser Company (TCC) produced this 4 x 4 microFarad oil-filled paper capacitor to custom order for Leak. These capacitors have virtually an indefinite life, although they cost three times that of an equivalent electrolytic capacitor. TCC were located just half a mile away on the Western Avenue (A40), North Acton. Wego were also used later for supply of capacitor blocks.

During the production run for the TL/12, Hafler and Keroes from the USA published a paper on the "ultra-linear" (or distributed load) configuration for output stages using beam-tetrode or pentode valves (Audio Engineering, Nov 1951). This type of circuit has a similar power output to the tetrode connection, with the benefits of low distortion and low output resistance characterised by triodes. As a result, the possibility of converting the TL/12 to an ultra-linear design that would have given 20 to 25 Watts output was considered by Leak. Accordingly, a revised output transformer with screen-taps was fitted to all TL/12's from this point on, however Leak never released the TL/12 in an ultra-linear configuration (it was superseded by the ultra-linear TL/12 Plus later on).

Today, 50 years on, the TL/12 amplifier is much sought-after and a loved addition to the best quality hi-fi systems, just as it was in 1949.

Jack Turner photographed at Henley, on one of the annual weekend trips by Leak staff. Jack was Lab Assistant to Harold Leak at the time the TL/12 was developed. (© E. H. Ashley, 1951. Reproduced by permission.)

Production Variations

During the production run of the TL/12, which lasted 8 to 9 years, several variations were made to the physical aspects and components of the amplifier, however the circuit was always the same. Some of these variations are:

The cable loom and tag-board construction of the TL/12 was of the highest standards.

• Early units featured a bronze finish enamel, and later ones gold finish. The change-over date was around 1952/53.

• Early units had open-frame transformers, wound on fully-cheeked bobbins (not un-like those used today), and the wiring from the transformer coil to the terminal plate can be seen above the chassis. Around 1952/53, transformers with pressed-steel shrouds were introduced, and were potted for tropical operation. Some TL/12 amplifiers around this time had a mix of open and potted transformers on the one chassis.

In this 1949 TL/12 advertisement, the power transformer core is parallel to the output transformer core. (Reproduced by permission Leak Electronics Ltd. Source: WW Oct 1949)

• Later TL/12's had an output transformer which included taps for ultra-linear operation, but these were never used on production units.

• A very early advertisement shows the power transformer core parallel to the output transformer core. It is possible that the unit concerned was a mock-up for promotional purposes only, or that changes were made in early production so that the two cores were at right angles. (This change would have required an accompanying alteration in the wiring loom, as the power transformer terminations would have been in a different position.)

• The paper high tension capacitor was originally 4 x 4 microFarads in the one block. Variations include: 2 x 4 μF, plus 1 x 8 μF with various earthing arrangements that commoned-up some of the earths inside the can (but there was always a separate earth for the single 4 microFarad section acting as filter input capacitor, thus keeping heavy ripple currents out of the audio circuits). Some capacitor blocks were of a more conventional appearance, with a ridged-lip around the top, but these are uncommon.

Opposite: This close up shows the chassis layout and high quality of finish attained with the TL/12.

• Special purpose units were produced for communications organisations. These featured transformer-coupled inputs and an attenuator to operate with professional

The BBC incorporated the TL/12 amplifier into the base of its studio monitor loudspeakers. A special input transformer gave the amplifier a balanced 600Ω input, and an attenuator was provided. A heat shield between the valves and power transformer has been removed to clearly show the amplifier. (Reproduced by permission Leak Electronics Ltd. Source: WW May 1951)

O.S.C.

audio circuits using balanced 600 Ohm interfaces. One such example was the BBC monitor loudspeaker which consisted of a Lockwood cabinet, a loudspeaker drive unit (typically a Tannoy 15-inch dual concentric), and a Leak TL/12 modified to BBC design needs.

• The standard output transformer (designated TL/12/T2/2) had four separate secondary windings that could be strapped for nominal loads of: 2Ω, 8Ω, 17Ω and 32Ω. An optional transformer (designated TL/12/T2/1) also had four secondaries, but these could be strapped for nominal loads of: 1Ω, 4Ω, 8.6Ω and 16Ω. Another less common transformer is the TL/12/T2/140 which has a high output impedance and is unsuitable for driving a loudspeaker directly.

RC/PA

With the release of the TL/12 an associated pre-amplifier was also launched – being designated the "RC/PA" (an abbreviation of "Remote Control Pre-Amplifier"). It could be used with a microphone, radio tuner or pickup and provided the user with bass and treble controls. The RC/PA did not provide any equalisation for discs, and the gain on the pickup input was quite low (sensitivity was only 50mV when used with the TL/12, but most pickups of the day were of high output). The list price of the RC/PA was £6/15/0 (the TL/12 alone was £25/15/0). Leak described the pre-amplifier as follows:

> "The pre-amplifier contains a single miniature pentode valve (an EF40) arranged to give a voltage gain of 100 times, or 40dB, before feedback is applied. As the valve has to provide an output voltage of only 160mV to load the TL/12 power amplifier the distortion is extremely small. When feedback is applied, the gain falls from 100 times to 3 times, and distortion is reduced to an immeasurable figure.
>
> "The feedback reduces the input impedance from grid to cathode, and the output impedance from anode to cathode, to values of a few thousand ohms. These low output impedances bring two major advantages: an exceptional freedom from hum pickup within the valve, and an extremely flat frequency response. The low input impedance does not appear across the signal sources, these being padded out by a resistor across which condensers are switched to provide 3 steps of treble rise.
>
> "Bass rise is achieved by releasing some of the total feedback in such a manner that the rise is asymptotic to 6dB per octave over the 3 octaves between 400Hz (cps) and 50Hz (cps) when the bass control is set at +3. This setting is, of course, only for use on record reproduction. By deliberate engineering the rise is retarded below 35Hz and the response is then made to fall, thus obviating a major disadvantage of the usual bass rise circuits

The earliest RC/PA's produced were shown advertised with dark knobs, black screen-printing, and a light chassis colour. The colour-scheme was subsequently altered and production shifted to the darker bronze colour used on the TL/12, with cream knobs and cream screen-printing. Note the lack of a pilot lamp.

which often rise continuously to 20Hz or below. This disadvantage, together with the fact that many pickups have a resonance of considerable magnitude between 35Hz and 20Hz, results in excessive 'motor rumble'.

"Treble loss is provided by increasing the feedback at high frequencies, and by additional capacity in the input circuits when switched to the -2 and -3 positions. These methods decrease harmonic distortion, whereas the more usual method of shunting a valve by a capacitor tends to increase distortion. Bass loss is provided by a resistance-capacitance network in the output circuit, feedback being fully maintained." *

Leak TL/12: List Prices and Optional Facilities

TL/12 Power Amp	£25/15/0
1250 Watt Mains Switch	5/6
100kΩ attenuator and Input Jack	£1/7/6
Flush Baseplate	10/0
Overlapping Baseplate	10/0
Top Cover, for Power Amplifier Only	£1/10/0
Top Cover, for Pre Amp and Power Amp Portable Assembly	£2/2/0
Input Transformer Microphone or Line Level	£2/15/0
Steep Slope Filter in Mu-Metal Case, for 6, 8 and 10 kHz	£5/10/0
Portable Conversion Kit for RC/PA/U including: non-overlapping front panel, short inter-connecting cable, and Bulgin 3-pin input connector plate.	£1/18/6

Note:
The TL/12 and TL/25A amplifier are packed in a strong returnable case for which a charge of 30/- is made. This charge is fully refunded on return of the case.

* Extract reproduced by permission Leak Electronics Ltd.

5. The Leak Idiom

Construction Practice

Leak amplifiers have a very strong reputation for excellent build-quality. Anyone who turns an unmodified Leak amplifier upside down never fails to be impressed with the construction practice. Whilst the component-board and cable loom approach was used by other sectors of the professional electronics industry, it was most uncommon to see this approach taken in domestic electronic equipment. Harold Leak used this construction system in his amplifiers more or less from the beginning. Whilst it gives the impression of being costly, if a production system is well thought out for this technique, it actually ends up saving the company money – the assembly of all the components on a chassis becomes quicker, easier and much more reliable, giving fewer assembly failures. In summary, advantages of this approach are:

The under-chassis view of the TL/12 amplifier was a big selling feature, giving customers tangible assurance of attention to detail throughout the whole design of the amplifier.

Haddon Transformers (who were dissolved in 1983) supplied transformers for the Leak TL/12. (Source: WW Sept 1956.) Hinchley Engineering were suppliers of transformers to Leak from the TL/10 onwards. (Reproduced with the permission of Tamura-Hinchley Ltd, Devizes, UK. Source: WW April 1956.)

- a reduction of induced interference and hum between circuit components
- loops in the wiring are avoided (e.g. earth loops)
- ease of assembly, ensuring repeat-ability and high quality
- reduced assembly time
- easier servicing and access to components.

All the internal wiring on Leak amplifiers is done using single-core solid conductor wire, which has several advantages: it is easier to strip, easier to solder (making assembly time quicker), and much easier to keep straight, giving the finished product a professional well-finished appearance.

When assembling the component board, Leak made sure that the orientation of components was consistent: all capacitors were soldered into position with the markings showing the value and rating visible and all resistors were oriented with the tolerance bands at the same end. This again demonstrates the attention to detail that Leak pursued.

"The people wiring up the main amplifiers during the early 1950's took great pride in their work. One person would wire the whole set, including cutting and straightening the cable-form wires – the component board having been previously assembled. On the rare occasions that a set was sent back from the test/inspection department for untidy or incorrect wiring, the other 'wirers' would gather round to see if they could spot what was wrong and then laugh and make jokes at the miscreants expense!"

Angus Gilchrist, 1999.

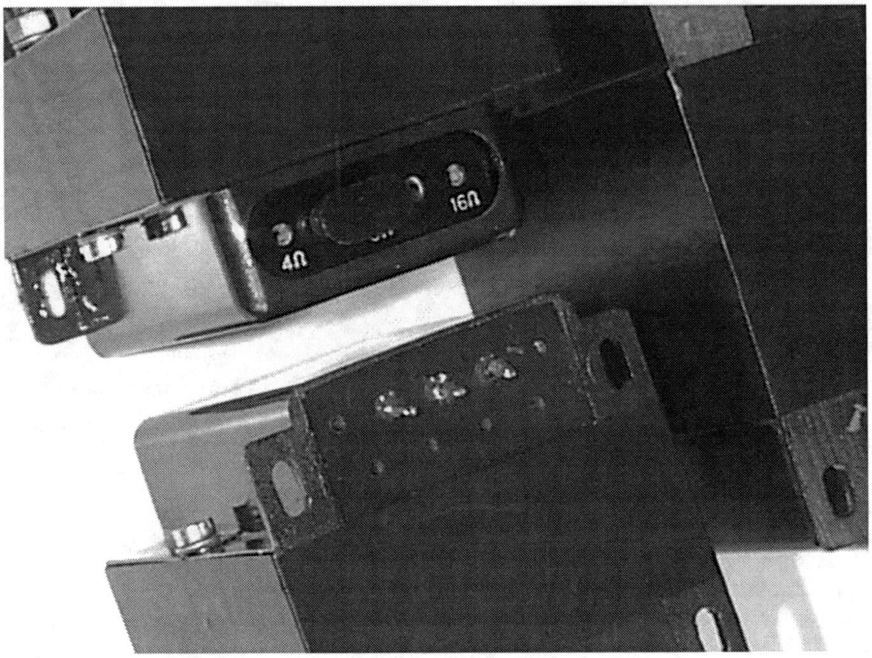

The excellent cast-zinc end-covers on the Hinchley transformers contributed to an attractive appearance, and allowed for a terminal strip to be included in the base of the transformer.

Transformers

Leak never had an in-house production facility for winding transformers so transformers were purchased from external suppliers such as Partridge, Haddon, Hinchley and Sowter.

Input transformers or microphone transformers were usually supplied by Sowter Transformers (still operating in 1999 from Ipswich). Some early GEC designs had inter-stage transformers although as feedback became an integral part of an amplifier design it would have been necessary to dispense with these and move to resistor-capacitor coupling.

Up until the TL/12, Partridge had supplied all the transformers for Leak (e.g. Type 15, STC and GEC amplifiers). Whilst Partridge transformers are beyond reproach from a quality standpoint, it became apparent that transformers with similar performance could be supplied from Haddon Transformers. As component cost was becoming an issue in keeping the TL/12 design competitive it was decided to use Haddon Transformers for the TL/12, TL/25, and TL/25A.

Haddon Transformers Ltd. was located in Hertfordshire, and operated by Eric Haddon. Some issues of Wireless World during the 1950's feature advertisements for Haddon Transformers, although in later years the name disappeared. Haddon Transformers supplied all the transformers for the TL/12 amplifiers – both the open-frame and enclosed-case potted types (which had the ultra-linear screen taps).

An advertisement for transformer frames that are similar to those on the Leak TL/12 transformers. (Source: WW 1945)

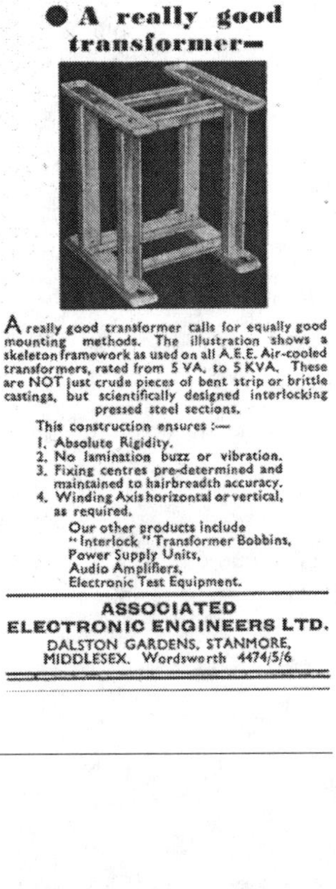

Eventually Leak changed suppliers, this time to Hinchley Engineering Co. Ltd. of Devizes, Wiltshire, who could supply transformers at less cost again than Haddon Transformers. At the time, the Managing Director of Hinchley Engineering was S. Reg Poole and the Technical Manager (who was the primary contact for Leak) was Ken W. Barker. Also employed with Hinchley was a young apprentice, D. A. (Tony) Coleman, who was involved in the design of the Leak transformers and remained with the firm for over 40 years.

Hinchley Engineering Co. Ltd. supplied all Leak transformers from the TL/10 and TL/12 Plus onwards – including transformers for solid state amplifiers. The Hinchley transformers were wound using the paper-interleaving technique which was less labour-intensive than making bobbin-wound coils as used by Haddon Transformers. The technique involves winding the coils for a number of transformers (say 10) on to a wide roll comprised of a former, paper insulation and copper windings. The individual coils are cut from the roll on completion (in contrast bobbin winding requires individual strips of insulation to be cut for each coil former).

To obtain satisfactory high frequency performance the output transformers were wound in five sections (either 2 or 3 primary sections plus 3 or 2 secondary sections respectively, depending on the model). Further details regarding winding layouts are given in Chapter 12.

The Hinchley transformers were very well finished. The end-covers (shells or shrouds) were made from cast zinc, being supplied to Hinchley Engineering by Burden and Miles. To moisture-proof the windings the transformers were impregnated with wax (which normally migrates a little under normal operating conditions in a power transformer). In the first series of Hinchley transformers, the laminations were finished in "Brushing Belco" cellulose paint, however it was difficult to obtain a good finish and the technique required labour-intensive cleaning of the laminations to remove the wax before painting. Ken Barker suggested the use of black vinyl tape glued over the laminations. Despite reservations from Reg Poole and Tony Coleman that heat would destroy the adhesion over time, the new technique worked well and it became a feature on Leak amplifiers from the TL/12 Plus onwards.

Hinchley Engineering Co. Ltd. (now Tamura-Hinchley) still operates from premises in Devizes, Wiltshire. The paper-interleaving technique for transformer manufacturing has now been phased out in favour of bobbins (which can more easily satisfy modern safety requirements).

At Hinchley Engineering the Leak transformers were wound using the paper-interleaving technique. A number of coils are wound simultaneously, and paper insulation is interleaved as a wide roll (not being cut until completion of the winding operation) giving lower labour cost. In the bottom photo six coils are being wound, and a connecting strip terminating the windings can be seen along the top. (Photos courtesy of Sue Alexander and Tony Coleman, Tamura-Hinchley Ltd, UK.)

Laboratory Facilities

The first distortion factor meter that Leak had was one built by Harold Leak, mainly because commercial manufacturers had not seen the need to make instruments capable of measuring down to 0.1% distortion. One of the commercial meters available at the time had more residual hum than distortion products originating from the amplifiers under test at Leak.

In 1947 Leak approached the National Physical Laboratory (NPL) so that the company's distortion measurements could be verified, but NPL did not have this capability at that time. By 1949 though, NPL were able to undertake these tests and Leak subsequently published the report in company advertisements for the TL/12 amplifier.

By the 1960's typical Test Department and Laboratory equipment included:

- Furzehill oscilloscopes which were popular before Tektronix oscilloscopes appeared
- Marconi wave-analysers, which could analyse the distortion spectrum
- Radford distortion factor meters (by Arthur Radford)
- Radford low distortion oscillators
- Telequipment S51 oscilloscopes
- Advance oscillators
- Bruel and Kjaer sound measurement equipment.
- Rank-Cintel wow and flutter meter

In production testing, the Telequipment oscilloscopes were coupled to the Advance oscillators and valve volt meters to make it quicker and easier to do measurements. For example: if the range-change switch was adjusted by 10dB on the oscilloscope, the oscillator output would be simultaneously changed by 10dB, as would the voltmeter sensitivity, giving a stable display on the oscilloscope and meter.

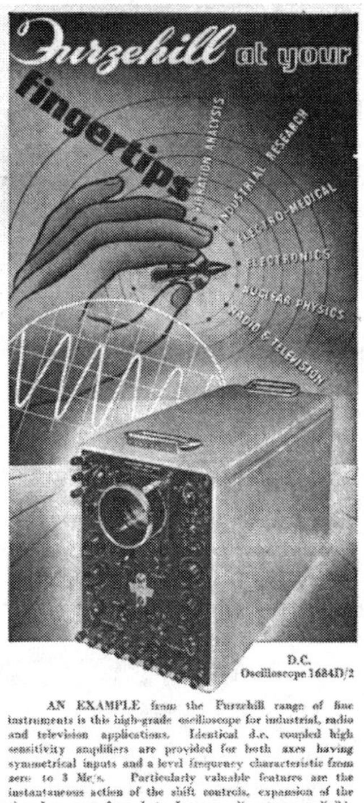

Leak used Furzehill oscilloscopes in their laboratory. This advertisement gives an idea of the type of equipment Furzehill made at the time. The model shown features a bandwidth of 3MHz. Furzehill Labs were dissolved in 1984. (Source: WW 1949)

To do acoustic measurements in a reflection-free environment, Leak installed a loudspeaker test hut on the top of the Brunel Rd office block, which contained the necessary equipment. The side of the hut that was flush with the office block wall had an opening in it for positioning a loudspeaker drive unit. A calibrated microphone was suspended on axis with the speaker under test which was driven

Part of the facilities in the Test Lab at Brunel Road where the quality of Leak products was ensured. A Trough-Line tuner with final inspection tag is on the bench at the left and a Varislope III pre-amp on the right. A Cossor oscilloscope, an AVO multi-meter and Advance oscillator are part of the test equipment. (© N. Rowe 1958. Reproduced by permission.)

with a test signal. Quite good measurements can be done this way, provided the weather is fine. Another practical problem at Brunel Road was noise from the close proximity of the train line behind the factory.

Leak never had an anechoic chamber, except when the loudspeaker operation joined with Wharfedale under the Rank organisation. However from the early 1950's onwards the factory had a listening facility that simulated a domestic lounge room.

Industrial Design

Up until 1955, Harold Leak and Frank Hawkins, in conjunction with other staff, designed the physical appearance of products. For example, pre-amplifier control layouts and finishes were determined in-house at Leak.

Around 1955 it was recognised that the appearance of pre-amplifiers might be enhanced by using an external industrial design house. Frank Hawkins was instrumental in the move to use an external design house and Leak used several over the years from 1955 to 1969 including: Rozental, Richard Lonsdale-Hands and Noel Haring Associates.

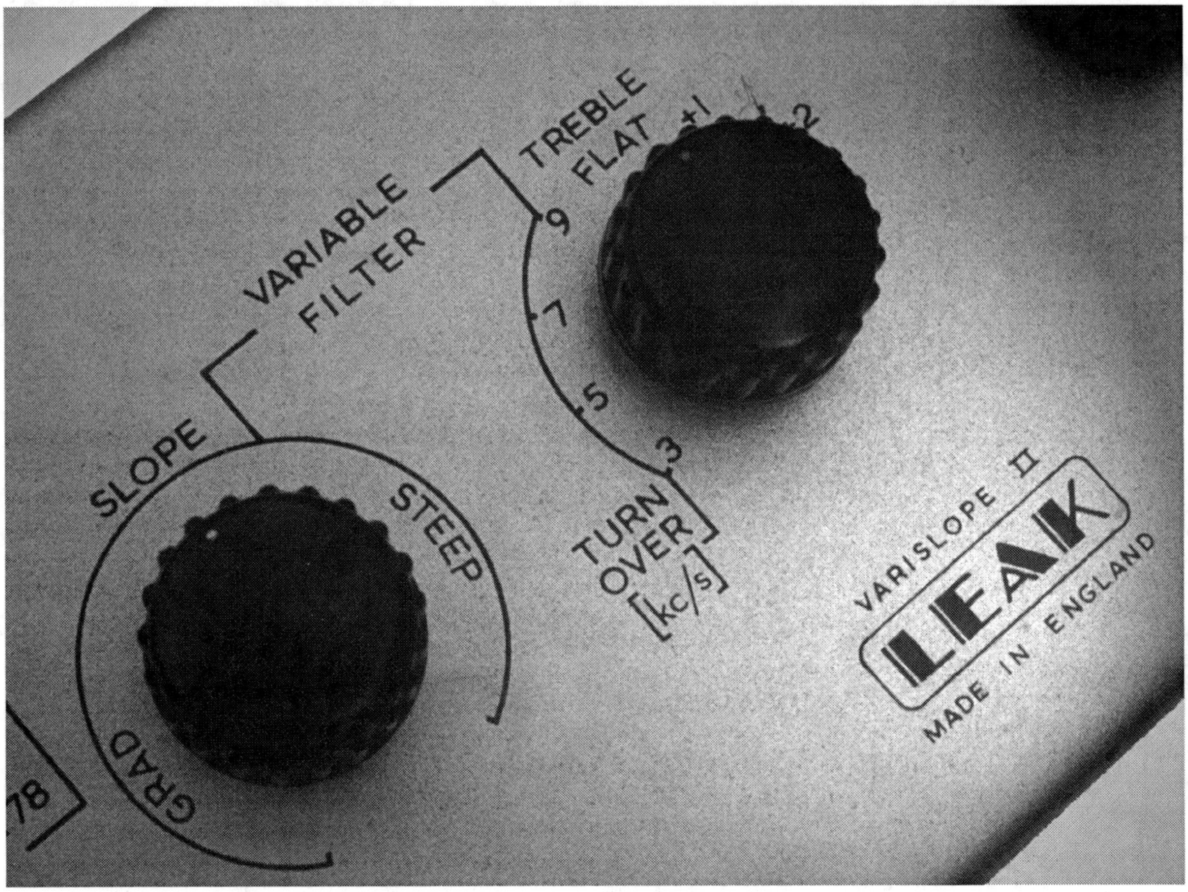

Up until 1955 Leak designed product appearance in-house, with primary input from Frank Hawkins and Harold Leak, as would have been the case with the Varislope II pre-amplifier.

Richard Lonsdale-Hands designed the plastic front panel that first appeared on the Varislope III pre-amp, and continued with the Point One Plus, Varislope Stereo and Point One Stereo pre-amps. The fundamental idea behind the design was to achieve a three-dimensional effect in the lettering and graphic design, by moulding the rear of the panel appropriately: the recesses were painted cream, and the remainder in brown. The moulding used a hard plastic known as diakon in the trade, but the design was later revised to a flat panel of softer perspex that was screen printed on the pack with gold and maroon (this was probably one of Frank Hawkins' many successful improvisations). However the softer perspex and the screen printing proved more susceptible to damage during production.

The knobs used with the Lonsdale-Hands front panel were initially black, but later they were changed to cream (both with gold inserts). Batches of these sometimes proved troublesome, particularly on selector switches, where they cracked from the increased forces required.

The Noel Haring Front Panel Design
by John D. Leak

The black and silver front panels caused more problems than all the previous panel designs. As originally conceived they were not in fact black and sliver at all, the first mock-up having a fawn-brown strip at the bottom, and a gold-anodised aluminium insert panel. The final appearance was hit on by Frank, influenced by Harold Leak's Leica M3. We actually experimented with a black leather strip along the bottom but could never get the edges straight.

As usual the industrial designers overlooked many of the practicalities of making these to the required standard in large numbers. The design involved first getting chromium plating to adhere to the substrate of the die-casting, and then getting the paint to stick to the plating. Initially the black section at the bottom was smooth, but this high-lighted imperfections and shrinkage in the casting. So a stippled finish was introduced, partly to give a leather effect, but more to hide imperfections in the casting and paint.

If the cast panel was difficult, the aluminium insert was a disaster. We tried double anodising, both matt and polished, mechanical brushing, and nothing was right. Some of the panels fitted to production units were matt anodised or mechanically brushed all over, and there were even a few painted metallic grey which were very unattractive. Lunds of Wandsworth, who had supplied all the metalwork since about 1955, and normally did a superb job, admitted defeat. Eventually a small silk-screening business in South London was able to resolve the problems – on my insistence they were given the chance after they had heard about our problems through industry contacts. This supplier was able to meet the required quality, although the occasional batch suffered from screen printing that could be easily dislodged.

In 1962 Noel Haring Associates industrial design house was engaged to design a new appearance for the imminent launch of the Stereo 30 transistor amplifier. The design would later be echoed in a face-lifted range of valve pre-amps and tuners, and later again with the Stereo 30 Plus and Stereo 70. Noel Haring had proposed several designs using various colours but eventually it was Frank Hawkins who proposed the black and silver finish, as he had been inspired by the desirable

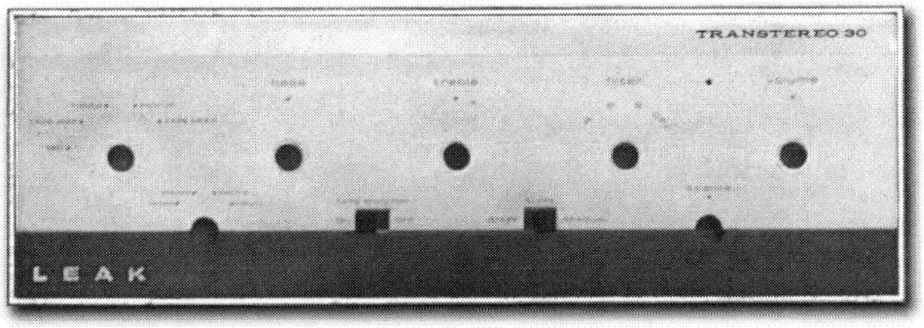

This prototype front panel for the Stereo 30 shows the proposed name "Transtereo 30". The aluminium panel insert is gold-anodised, and the strip at the bottom is fawn-brown, and lacks the stippled finish finally adopted. The outer escutcheon is chrome plated as was the final production version.

appearance of the Leica camera. After having some prototypes made up, it was an easy decision to go with the black and silver idea commercially. Ted Ashley considers these black and silver tuners and amplifiers to be a Leak "classic" from an appearance point of view. He also recalls that it was often the case that staff at Leak would suggest useful alterations to design ideas proposed by industrial design consultants. (The castings for this front panel were supplied by Joseph Fray Ltd. of Birmingham).

The 1949 New York Audio Fair was held at the Hotel New Yorker. (© Hachette Fillipacchi Magazine Inc, USA. Reproduced with permission from Audio magazine, USA. Source: Audio Engineering, Oct 1950.)

In the early 1960's, Ercona (then Leak agents in the USA) asked Leak to produce a special aluminium gold-anodised front panel for the "Varislope Stereo" and "Point One Stereo" pre-amps. A novel feature of this front panel is the provision for fitting a coloured plate behind the input selector knob, with matching knob centres. Different colours were available from which the owner could select (orange, grey, white and red), and thus match the appearance to the environs where it was being used. The panel was also screen-printed so it could be used without the "decor-inserts".

The design was not particularly practical as the raised escutcheon had to be milled from a solid section of aluminium and then fastened to the main panel with tiny screws. The knobs and main panel were also made from aluminium, anodised in gold. There were large variations in the finish and reject rates sometimes ran as high as 50% which is why, apart from the cost, they were not sold in Britain.

Leak found that such requirements for the American market were often an early indication of trends in the industry. Another example of the American distributor affecting product design occurred when Leak designed the Varislope II pre-amp. Six different playback equalisation curves for records were incorporated into the design at the request of British Industries Corporation.

Export Markets

As soon the TL/12 and RC/PA were launched in December 1948 they started to sell well. They became known internationally for their quality even before overseas distribution was arranged, as people travelling to the United Kingdom would take amplifiers home with them. During 1949, which was a very eventful year for Leak, international enquiries made it necessary for Leak to start appointing agencies – and most of these agencies would remain with Leak for a long time.

Leonard Carduner of British Industries Corporation, with Harold Leak. (Photo courtesy of the Leak family.)

American Distribution

After the Second World War, the market for quality sound reproduction equipment for use in American homes was rapidly growing. Without knowing it, the audio industry was moving into the "Golden Age" of the 1950's and early 60's, where quality products were being developed that would influence the industry for decades to come.

Harold Leak had released the "Point One TL/12" amplifier in Britain in December 1948, and so by 1949 was in a position to take advantage of the growing American market for the new "Hi-Fi". In late 1949, one Saturday morning, Harold Leak read about the coming Audio Fair in New York and decided to attend. Because of exchange control regulations in post-War Britain, he had to apply to the company's bank for the necessary funds to travel to the Fair.

The 1949 Audio Fair was held on the sixth floor of the "Hotel New Yorker". The Fair hosted 56 exhibitors, and about 1500 people attended. Harold Leak travelled by himself from Britain and took the following Leak products for promotion :

- Three TL/12 amplifiers with RC/PA pre-amplifiers
- Three Leak dynamic pickups and arms for 78 rpm records
- Three Leak two-way "550" moving-coil loudspeakers

Prior to the Audio Fair, the company placed its first ever American advertisement in the respected industry journal "Audio Engineering" (October 1949). At this stage the company did not have any distribution arrangements in America, but stated in the advertisement that a suitable agency for distribution would be sought whilst at the Audio Fair. Harold Leak was available for commercial discussions at the Hotel New Yorker from October 20 to November 6, 1949.

The amplifier was very successful indeed at the fair, and many Americans were astonished at the build-quality as well as the technical performance. As a result, Leak established distribution through British Industries Corporation (BIC) – an American company focused on the import of British electronic and sound equipment (e.g. Garrard turntables, Ersin Multicore Solders, Marconi/Osram valves under the Genelex label and from 1950, Wharfedale loudspeakers). Harold knew

British Industries Corporation sales literature for the TL/12.

At the 1953 New York Audio Fair – Harold Leak on the left, Gilbert Briggs at the rear, and Leonard Carduner (of BIC) on the right. Seated is possibly Dr. Ed Kellog, who developed the dynamic paper-cone speaker with Chester Rice. Note the Leak promotional material behind. (© Hachette Fillipacchi Magazine Inc. Reproduced with permission from Audio magazine, USA. Source: Audio Engineering, Dec 1953.)

Richard Arbib at the 1954 New York Audio Fair. The Arbib family were close friends of the Leaks, living nearby in Gerrards Cross. Richard was influential in Harold Leak's decision to arrange distribution of Leak products through BIC in the USA. Note the promotional display for BIC's products which emphasised an association with icons of British quality, like Rolls Royce. (Photo courtesy Gordon Arbib)

of BIC prior to leaving Britain for the Audio Fair through Richard Arbib of Ersin Multicore Solders. BIC was operated by Leonard and Bill Carduner, with Eugene Carduner being the sales agent for the New York area.

Opposite: In Oct 1949, Leak placed its first advertisement in the prestigious magazine "Audio Engineering", with the aim of establishing a distribution agency in the USA. (Reproduced by permission Leak Electronics Ltd and Audio magazine, USA. Source: Audio Engineering, Oct 1949)

The News Chronicle in the United Kingdom (November 4, 1949) reported on Harold Leak's visit to the New York Audio Fair in glowing terms. The report focused on the post-war British dollar-drive for export markets and how Harold's small company in Acton was successfully selling into markets in the USA.

By the time Leak introduced the Sandwich speaker in 1961, British Industries Corporation had built up a strong market for Wharfedale products in the USA, so the entry of Leak in the loudspeaker market presented somewhat of a challenge if BIC were to distribute Leak speakers as well as Wharfedale. Harold Leak decided that sales of the Sandwich loudspeaker in the USA would be stronger if another company took on the representation of Leak. As a result, Ercona Corporation was selected for distribution of Leak products in the USA for the remainder of the 1960's.

Leonard Carduner, founder and President of BIC, at the BIC stand, 1954 New York Audio Fair. (Photo courtesy Gordon Arbib.)

BRITAIN'S BEST AUDIO AMPLIFIER
DISTORTION : 0.1%

Investigate its quality of reproduction and workmanship at

ROOM 650, THE AUDIO FAIR

H. J. Leak M. Brit. I. R. E., will be in attendance to discuss technicalities and to settle distribution rights. He will be at The Hotel New Yorker from Oct.20—Nov.6.

LEAK
"POINT ONE"
AMPLIFIERS

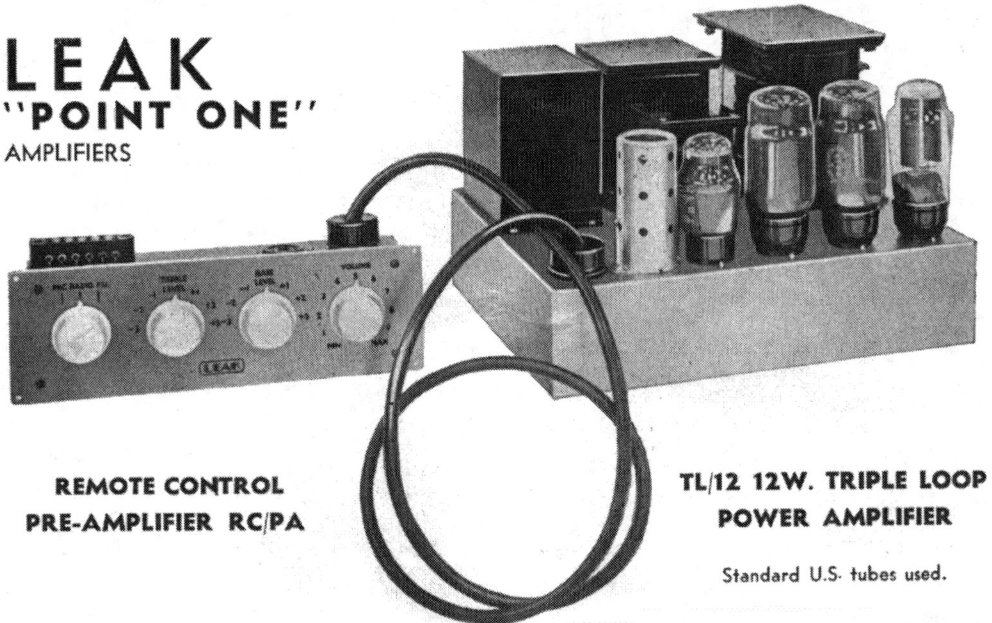

REMOTE CONTROL PRE-AMPLIFIER RC/PA

TL/12 12W. TRIPLE LOOP POWER AMPLIFIER

Standard U.S. tubes used.

An original feedback tone-control circuit.

No resonant circuits employed.

- Distortion: Less than 0.05%.
- Switching for Pick-up, Microphone and Radio, with automatic alteration of tone-control characteristics.
- High sensitivities. Will operate from any moving-coil, moving iron or crystal P.-U.; from any moving coil microphone; from any radio unit.
- Controls: Input Selector; Bass Gain and Loss; Treble Gain and Loss; Volume. Output Impedance: 0-30,000ω at 20 kc.p.s.

The unit will mount on motor-board through a cut-out of 10⅛ in. x 3⅛ in., or it can be bolted to the power amplifier, when, with a top cover, the whole assembly becomes portable.

For use only with LEAK amplifiers.

A Leak triple loop feedback circuit, the main loop giving 26 db feedback over 3 stages and the output transformer.

- Push-pull triode output stage. 400 V. on anodes.
- No H.T. electrolytic smoothing or decoupling condensers.
- Impregnated transformers; tropically finished components.
- H.T. and L.T. supplies for pre-amp. and radio units.
- Distortion: at 1,000 c/s and 10 W. output, 0.1%: at 60 c/s and 10 W. output, 0.19%; at 40 c/s and 10 W. output 0.21%.
- Hum and Noise: —72 to —80 db on 10 W.
- Frequency response: ±0.1 db, 20 c/s-20 kc/s.
- Sensitivity: 160 mV.
- Damping Factor: 20. Input impedance: 1 Meg. Output impedances: 2ω; 7-9ω; 15-20ω; 28-36ω. Phase margin 20° ± 10°: Gain margin 10 db + 6 db.

25 W. model available.

The TL/12 Specification is bettered by check of the National Physical Laboratory (equivalent National Bureau of Standards) and their certificate will be on view.

Write for 16 page brochure "A".

We will be demonstrating with Leak Dynamic Pick Up & "550" Loudspeaker.

H. J. LEAK & CO. LTD. Westway Factory Estate, LONDON, W. 3.

Australian Distribution

From 1949 to 1966, Leak equipment was distributed in Australia solely by Simon Gray Pty Ltd. In 1966, primary distribution was transferred to a new company established to manufacture Leak products in Australia under license – H. J. Leak (Aust) Pty Ltd. Another consideration was that Simon Gray Pty Ltd had put considerable effort into distribution of Wharfedale in the 1950's, so successful distribution of the Leak Sandwich loudspeaker would have required a change in marketing emphasis. However, Simon Gray Pty Ltd had established many contacts with professional customers, so they maintained distribution rights for Leak products in this market segment.

"As soon as we launched the TL/12 and RC/PA it started selling . . . 1949 was a very eventful year. We started to pickup up business from all over the world, and we started to appoint agencies - some of which we retained for a very long time."

Ted Ashley, 1998.

Simon Gray Pty Ltd

During 1949, Frank Hawkins wrote to Australia to gather information about suitable agencies. Soon afterwards, Simon Gray Pty Ltd of Melbourne, Australia was assigned the distribution rights for Leak products. Originally Simon Gray Pty Ltd had been an importer of jewellery and soft-goods but in the late 1940's Simon's son, Wulf Gray, decided to start a new line of business in high quality audio equipment. Wulf had heard of the Leak TL/12 through Harry Mauger at Astor Records (also known as Radio Corporation of Australia) who encouraged Wulf in his new venture, and assisted by providing part-time technical support. The first product imported by Simon Gray was the RC/PA and TL/12 amplifier.

In 1955 Joe Middleton started with Simon Gray Pty Ltd as their first Technician. He was involved in the installation, design and maintenance of sound systems sold by Simon Gray Pty Ltd. He later became Chief Engineer.

During 1955 Simon Gray Pty Ltd supplied an advanced sound system to the Princess Theatre in Melbourne. It was planned to stage a live production of the musical "Kismet" in the Princess Theatre, and an new sound system was considered crucial to the success of the show. Simon Gray Pty Ltd was engaged to provide the sound system. They decided to install six Leak TL/25A power amplifiers, with accompanying Varislope II pre-amplifiers. There were 3 Neumann M49 microphones placed along the stage in front of the foot-lights, and a sound system using twelve Wharfedale "Super 12" loudspeakers featuring cloth surrounds and aluminium voice coils. It was a three-channel system, with four Super 12's and two Leak TL/25A amplifiers per channel. The Princess Theatre was known generally to be very difficult acoustic environment, but the sound system worked very well. Fritz Langford-Smith referred to the installation as one of the best he had heard. The price charged by Simon Gray Pty Ltd only covered the cost of materials, but it won the company other business, and it was promoted in Wharfedale advertisements in Wireless World (May 1956).

Wulf Gray operated Simon-Gray Pty Ltd, distributors of Leak amplifiers in Australia. (Photo courtesy Charles Savage).

This publicity photo, taken in Australia around 1966, featured many Engineers from the broadcast industry arranged around Harold Leak in a somewhat over-emphasised pose. Left to right: Bill Vigona (Radio 3UZ), Wally Chamberlain (Radio 3XY), Noel Serpell (Radio 3AW), Harold Leak, Max Hull (East Recording Co.), Geoff Brooke (Broadcaster), Wally Shaw (GTV-Channel 9 Melbourne). (Reproduced by Permission Leak Electronics Ltd. Source: H. J. Leak (Aust) Pty Ltd.)

In 1963, Joe Middleton encouraged Wulf Gray to look into the possibility of building Sandwich loudspeakers in Australia. Simon Gray Pty Ltd arranged for Leak to ship two sets of Sandwich high frequency and low frequency drive units with crossovers to Australia, together with the cabinet design. Two pairs of prototype Sandwich speaker systems were made up for testing. It turned out that Simon Gray Pty Ltd decided not to go ahead with production, primarily because it would take the business into a new area of manufacturing, that was different from its core competencies in distribution and support for quality imported brands.

Ted Ashley recalls visits by Wulf Gray to Britain. Together with Harold Leak, Frank Hawkins, Fred Blackman and their wives, they would spend time together in the evenings on a semi-business, semi-social basis. It is unfortunate that Wulf Gray could not contribute to this book, as he died in Melbourne, Australia during October 1998.

Simon Gray
PTY. LTD.
28 ELIZABETH ST., MELBOURNE, C.1., VIC.

Logo of the Australian distributors for Leak, Simon Gray Pty Ltd.

H. J. Leak (Aust) Pty Ltd

A Sandwich loudspeaker production facility was also established in Australia. This idea was prompted by Sydney McClory – who set up an Australian company known as “H. J. Leak (Aust) Pty Ltd” located near French’s Forest, Sydney.

Syd McClory was born in 1928, and graduated in 1957, receiving a Bachelor of Science in Electronic Engineering from the University of California in Los Angeles. He had obtained a US work permit during his course and worked on Boeing B52 electronic sub-assemblies while studying. After graduation his experience in this area lead to a position with the USAF Strategic Air Command at Kindley Air Force Base, Bermuda, where B52 aircraft were kept in the air 24 hours per day as part of the “cold war”.

US military bases have a “Base Exchange” (supermarket) where goods can be bought by base personnel, tax and duty free. In 1958 Syd was approach by the Base Exchange to start a side-line business selling high fidelity equipment to 6000 base personnel and visiting air crews. In these days “British was Best” when it came to hi-fi, and Syd imported Leak amplifiers, Wharfedale speakers, and Garrard record players/changers through a Euro-American company “Bluebell International” that was based in Monaco (for tax reasons). The side-line venture proved very successful and Syd estimates that in one year he sold more product than British Industries Corporation. In return for the premises supplied by the Base Exchange, Syd paid 10% of sales to the Base Exchange.

There were usually three weeks between the Canadian and New York Audio Fairs, and a first class return ticket to North America included the side-trip down to Bermuda, giving a pleasant break from the coming English winter. In 1959 Syd received a local phone call from Harold Leak, who was visiting Bermuda for fishing, and to find out who was selling so many Leak amplifiers in Bermuda. So started a long friendship between Syd McClory and Harold Leak.

In 1960 the Sales Manager of Bluebell International visited Syd McClory, and over dinner asked him to join them in Europe – Bluebell had the franchise to sell Leak/ Wharfedale/Garrard to US military bases worldwide and to Canadian bases in

April 1960: On Syd McClory's first night in London he dined with the Leak's at The Orchard in Ruislip. Left to right: Michel Goemans (son of the President of Bluebell International), Muriel Leak and Syd McClory. A friend of Michel Goemans is seated next to Harold Leak, while Margaret Hawkins (the daughter of Frank Hawkins) is on the right. (Photo kindly supplied by Syd McClory, Australia.)

Europe. Syd accepted the offer from Bluebell and so in April 1960 he flew to London to meet with Leak staff. After arriving he phoned Harold Leak, who already knew of Syd's move to Bluebell. Muriel Leak met Syd at the airport and drove him to Holly Tree Farm. That evening he dined with the Leaks at one of Harold's favourite restaurants, The Orchard, in Ruislip.

The next day Harold drove Syd McClory to the Brunel Road factory, where for the first time he met Frank Hawkins, Ted Ashley, Fred Blackman, Don Barlow and others. He was shown prototypes of the Sandwich speaker and the Stereo 30 amplifier.

After the visit, Syd went on to Frankfurt where he established Audioson GMBH, a subsidiary company of Bluebell. During the following years Syd travelled extensively through Europe on business. Then in 1963 he moved to the London office of Bluebell International and subsequently saw Harold Leak and Frank Hawkins much more frequently (with lunchtime meetings at least once per month).

By 1965 Syd had two young children, and was spending a lot of time away from his family on European business trips, so he decided to return to Australia. At lunch one day with Harold, Syd broached the idea of an Australian Leak factory for manufacturing Sandwich speakers and Harold was immediately enthusiastic. Subsequently, Syd spent several weeks at Brunel Road (and at the speaker cabinet supplier to Leak) familiarising himself with methods for the assembly of drive-units and speaker cabinets.

In January 1966 Syd McClory established H. J. Leak (Aust) Pty Ltd. Assistance from UK-based staff was not required as Harold Leak knew of Syd's capabilities as an engineer, although such assistance would have been available if necessary. The company started, with a staff of four, in a small factory at 1 Campbell Parade, Manly Vale, New South Wales (moving later to 85 Winbourne Rd, Brookvale, NSW). Just over half the capital to start the company (51%) was provided by H. J. Leak & Co. Ltd. from the United Kingdom.

The assembly line for Sandwich speakers at H. J. Leak (Aust) Pty Ltd. To assist Syd McClory to establish the Australian operation, John Leak produced a complete illustrated manual covering all stages of speaker production. (Reproduced with permission from Australian Hi-Fi magazine. Source: AHF Vol 1, No 1, 1970)

Prior to commencing production, samples of the cabinet wall damping material (bitumen impregnated Canite) were sent to Don Barlow who approved the damping characteristics.

Speaker driver parts (magnets, baskets, cone and coil assemblies) were imported and assembled into drive units in Australia. There was no duty on these parts, versus 45% duty on finished products. All other parts were produced in Australia. Initially cabinets were bought from a local supplier but, as the quality was not satisfactory, a subsidiary company (Soundwood Products Pty Ltd) was established, giving total control of the quality. Harold Leak paid a great compliment by saying that the product was superior to the UK product.

The products produced were: the Sandwich speaker, Mini-Sandwich and the Stereo 70 amplifier. Some valve amplifiers were imported, as were samples of the Truspeed turntable. As soon as the Australian company was established, it took over distribution of Leak products in Australia (with Simon Gray Pty Ltd retaining professional distribution). Speakers were also exported from Australia to New Zealand, Fiji and Hawaii (through Ercona).

By 1968 there were 21 employees at H. J. Leak (Aust) Pty Ltd. The company was very successful, primarily as it was able to lower the cost of the Sandwich speaker and Stereo 70 amplifier by around 50%. The Australian Broadcasting Commission was a large customer as it adopted the Sandwich as a standard monitor in the 1960's, and there are still many in use today.

Harold Leak travelled to Australia many times, making his first trip around 1950 or 1951. On his business trips he stayed at the Windsor Hotel in Melbourne, and frequently went to a Collins Street restaurant owned by broadcaster Geoff Brooke. His last business trip was in 1967. By 1968/69 Harold Leak was considering selling so he did not travel to Australia.

"Made in Australia" declares this rear panel decal plate for a Sandwich Mark II loud-speaker. This unit has a serial number of 0740, and was produced in the year 1967.

In 1969 when Rank purchased H. J. Leak & Co. Ltd. in the United Kingdom, Syd McClory found that his new Managing Director was Gus Smith of Rank Industries, as they now had a 51% holding in the Australian company. Rank subsequently bought the remaining 49% of the company, although this was not the preferred outcome for Syd McClory, as the Australian company had been very successful.

The Australian operation was an important consideration when Rank purchased Leak. Rank wanted to establish a facility for making Wharfedale speakers in Australia, and also to expand it for manufacture of colour televisions. Unfortunately it had not been possible for Harold Leak to forewarn Syd of his intention to sell.

Syd McClory left the Rank operation in 1970, and established a new company (Audioson Australia) that imported speakers from KEF Electronics in the United Kingdom. KEF was operated by Raymond Cooke, who had been with Wharfedale, but left in the early 1960's after they were bought by Rank in 1959.

Harold and Muriel visited Australia once more in 1985 and met up again with Syd McClory and his wife Valda. Over dinner, Harold told them that he had emphysema, and then spent the rest of the night cadging cigarettes from Syd McClory!

Other Countries

During 1949 Leak arranged distribution to other countries, including India, South Africa, New Zealand, the Middle East, Canada and Singapore. The successes of 1949 made this year a significant one that cemented the future of the business. In France and Germany, Bluebell International distributed Leak equipment.

By 1956 Leak had distribution arrangements in many Western hemisphere countries – USA, Canada, Mexico, Cuba, Jamaica, Venezuela, Portugal and Brazil. In Brazil, Leak was negotiating with Feigenson Company of Sao Paulo, regarding manufacture of Leak products under license, however these discussions did not lead to the project proceeding.

In 1956, Harold Leak met with the agent for Leak products in Jamaica, Stanley Motta. (Photo courtesy of the Leak family.)

6. More Products

Mark 1 Moving Coil Pickup

In the early days of British hi-fi, George Wise (a precision tool-maker) became well known as a keen audio enthusiast, frequently attending various demonstrations and exhibitions. Harold Leak came to meet George Wise through a mutual interest in Hartley-Turner loudspeakers. In time George became involved with various projects at Leak, either as a consultant or on an hourly basis for doing design prototypes.

By 1948, through experience and contacts he had developed with Lowther and Voigt, George Wise had designed and built a moving coil pickup based on the Voigt patent. Initially George built the pickup for his own use, but when Harold Leak expressed interest in commercialising the design, George agreed to manage the production of parts for 500 units. An associate of George Wise was engaged to machine all the mechanical parts. George specified all the design parameters, supervised production, arranged for delivery of all parts to the Leak works, mounted diamond tips for the pickups, and trained Leak staff to assemble the pickup and arm.

The Leak moving coil pickup was released in 1948 and as it was based on Paul Voigt's patent, royalties were paid to Paul Voigt (who was living in Toronto around this time). The end-product of George's work was a high performance unit that gained much respect from owners. It had a published frequency response of +/-1dB from 40Hz to 14kHz using the Decca test record of the day. The specified high frequency resonance (of the stylus tip mass with the record compliance) was 26kHz for the 78 rpm model, and 16kHz for the LP model. (John Crabbe, who edited

"In less than a week I made a (prototype) pickup. I made drawings of the parts, found someone to make them, found a firm to mould the armature and went back to their mould-maker to get it put right, found the supplier of the ruby stylus, and found a man who could make the diamond stylus to my specification with the best finish I have ever seen for only £8. I found a firm to wind the armatures and trained their operator, checked the parts were up to standard, taught the chaps at Leak how to build the pickups, mounted all the rubies and diamonds, and cut all the rubber mountings."

George Wise, 1999.

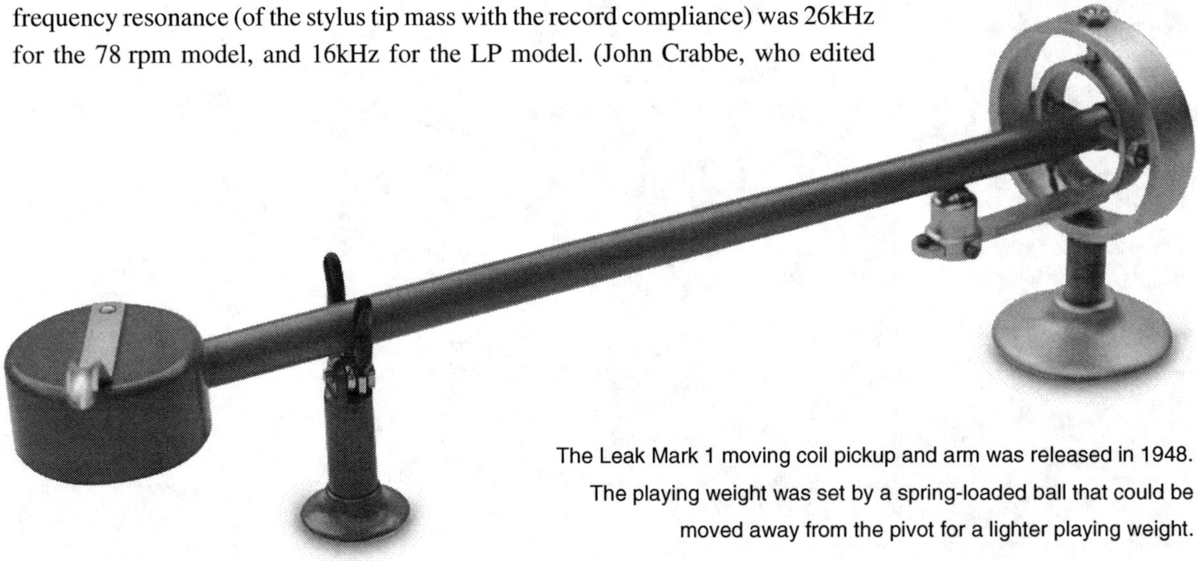

The Leak Mark 1 moving coil pickup and arm was released in 1948. The playing weight was set by a spring-loaded ball that could be moved away from the pivot for a lighter playing weight.

George Wise

In 1928, during his teenage years, George Wise became interested in high quality audio and it subsequently dominated his life. Even at 85 years of age (see photo below) he continues to build amplifiers.

George Wise was employed as a tool-maker at Evershed and Vignoles (who produced the famous "Megger" insulation tester) but during his spare time he pursued his passion for audio, through which he established contacts with Hartley-Turner, Voigt and Lowther. In 1944 Harold Leak advertised for some Hartley-Turner speakers, so George Wise visited Leak and arranged to supply them with the speakers. Through subsequent discussions and frequent visits, Harold Leak learned of George Wise's ability to design and produce high quality components such as pickups and loudspeakers – and a long association between George Wise and Harold Leak began.

In 1946 Harold Leak offered George Wise a job at the Uxbridge Road premises, but George turned the offer down as he had reservations that the job might not suit him.

During the time from 1946 to 1948, George pursued his interest in moving coil pickups. He had owned several Lowther pickups and demonstrated one at a public meeting. However they proved too sensitive to surface noise for his liking, so in time he built a pickup based on the Voigt patent of an elongated coil in a magnetic field. George was extremely pleased with the results he was getting - measurements consistently gave a very flat frequency response.

When the standard-setting TL/12 was released in 1948, his acquaintance with Leak was re-invigorated, and on a subsequent visit to the Leak works he asked Harold Leak if he might be interested in producing a pickup based on his design. Harold Leak expressed immediate interest and the Mark 1 Leak pickup was born.

George remained in contact with Leak for many years. He designed the Mark 2 pickup and built prototype components for it, built a Leak prototype ribbon speaker and assisted with component dimensions for the Truspeed turntable. He was also approached by Leak to work on a stereo pickup design but declined the offer.

His approach to high quality audio is characterised by determination and enthusiasm that has persisted for over 70 years. He is very pragmatic and a keen believer in ribbon loudspeakers (he uses Apogee Signature Calliper speakers). His philosophy on loudspeakers reflects his general approach to hi-fi: "No one can throw a pound-weight around like they can a feather."

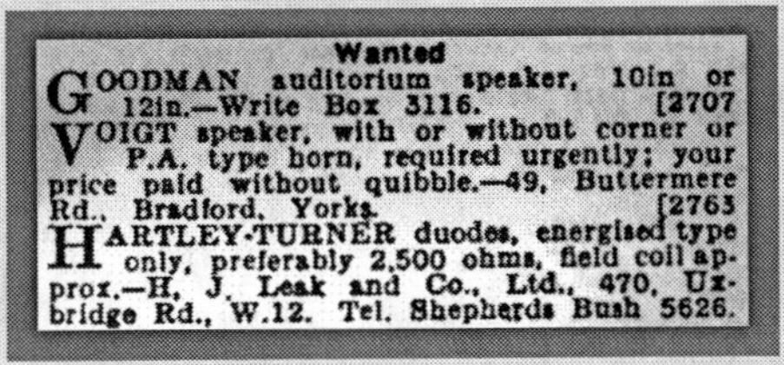

Wanted

GOODMAN auditorium speaker, 10in or 12in.—Write Box 3116. [2707

VOIGT speaker, with or without corner or P.A. type horn, required urgently; your price paid without quibble.—49, Buttermere Rd., Bradford, Yorks. [2763

HARTLEY-TURNER duodes, energised type only, preferably 2,500 ohms, field coil approx.—H. J. Leak and Co., Ltd., 470, Uxbridge Rd., W.12. Tel. Shepherds Bush 5626.

This advertisement in the June 1944 issue of Wireless World initiated the association between Leak and George Wise. (Source: WW June 1944. Reproduced by permission.)

Close up photos of the Mark 1 pickup and arm. The pickup head sat unusually close to the record surface when playing. The fine wire of the coil lead-outs can just be seen in the dismantled head. (Photo of George Wise kindly supplied by Steve Mann, of 'Slivernote', UK.)

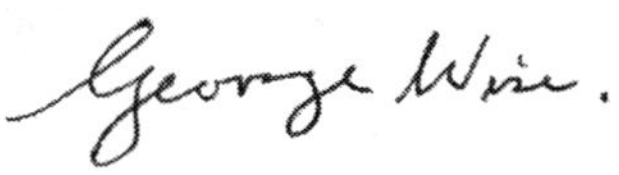

George Wise

A Leak Mark 1 Pickup in action with a Decca 78 rpm record.

Hi-Fi News magazine, recalls that when LP's were introduced, the higher compliance of the new vinyl records lowered the high-frequency resonance into the audio band giving the pickup a characteristic sound in the treble. This problem was minimised when the Mark 2 was released.)

The pickup could be supplied with a diamond or ruby stylus tip. The ruby stylus was a cheaper option, but naturally it would wear more rapidly than diamond. (To ascertain the wear characteristics of stylus tips a "shadowgraph" was used for projecting images of tip profiles onto a screen.)

Before this pickup was released, Leak were using a "Lowther-Voigt" pickup.

Mark 2 Moving Coil Pickup

A second pickup, released in 1954, was also based on the Voigt patent being a refinement of the Mark 1 design and included an inter-changeable head. Leak had commenced work on an inter-changeable head design, but at the same time George Wise had designed another pickup that also had an inter-changeable head (initially for his own use). This became the starting-point for the Leak Mark 2 pickup design. The prototype design was done by George Wise, the development was jointly done

In the Leak pickup, electrical output was generated from the long sides of a coil moving in a transverse magnetic field.

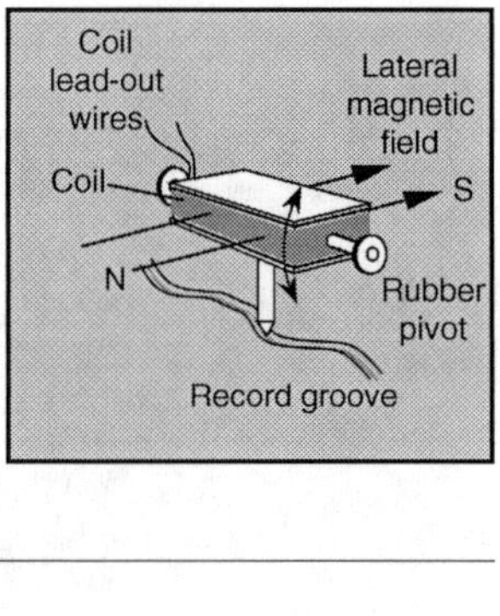

Assembly of the Leak Mark 2 Pickup

by Steve Wilkins, formerly of the Leak Pickup Department, 1960 to 1965.

The Leak Mark 2 Pickup was truly hand made. The coil was wound by hand on a watchmakers lathe. The rubber suspension for the coil assembly was made by stretching valve-rubber tubing (as used in bicycle inner tube valves) over a hot soldering iron to the required diameter - this was a hit and miss affair. The stretched rubber was then cut up into the required size with a scalpel. The smell was appalling when stretching the rubber over the soldering iron.

The rubber determined the compliance of the cartridge and was pre-tested in the pole piece sub-assembly with the tip of my finger which became very sensitive to determine the correct compliance!

The pole pieces were assembled by hand and glued together with Araldite, the gap also determining the compliance. You would become adept at choosing which particular shape and size rubber suspension to use with which particular pole piece gap to give the required compliance.

Before winding the coil each coil former and stylus was inspected under a microscope for defects (which could include poor tip polish, poor profile, glue residue or loose mounting). The major supplier towards the end of the products life was DSC, the Diamond Stylus Company in Colwyn Bay, South Wales (stylii were also supplied by W. Hodgson).

Soldering the 49swg wire onto the terminal pins was also a acquired and skilled process.

Each cartridge was tested for frequency response, output level and waveform, a manual record was kept of every cartridge's performance in a book relating the results to the individual serial number. Every cartridge was listened to for surface noise, tracking and sound quality before it was given a serial number. The serial number was scratched on by hand with a scribe.

The target for high frequency response was to keep the HF resonance above 18kHz, generally the later versions had a rising response from about 15kHz. Very early units (pre-1960) had a very flat response because the HF response was damped by putting MS4 silicone grease into the magnetic gap, but of course this migrated in the hot weather, usually down the stylus shank, but the sound of these when new was very distinctive and showed very low coloration.

The 78 rpm version went up to and was tested to 30kHz, with oversize diamond tips being available for playing older recordings

The main failing of the cartridge was the poor vertical compliance which gave rise to high needle chatter and the inability of the design to be adapted to stereo.

Repairs for worn or faulty stylus were done by replacing the cartridge as far as I can remember. The suspension rubber would eventually become stiff and dry after some time and would require replacing. Customer's units were disassembled the rubber replaced and the whole thing re-assembled. This was a tricky task which meant total re-assembly and testing, whilst keeping the stylus the same way around and trying to keep the performance the same as the original (referring to the original test figures on file).

A Leak Mark 2 Dynamic Pickup head for playing micro-groove "LP" records. The logo shown is styled from the one appearing on the front of the pickup head. British Artid Plastics were engaged to make the plastic mouldings for the pickup head. (Original photo courtesy http://www.navarrone.com)

The armature from a Leak Mark 2 pickup. The long-side of the coil former is about 6mm long. An aluminium shank mounted into the armature holds the stylus on its tip. (Pickup armature courtesy Angus Gilchrist.)

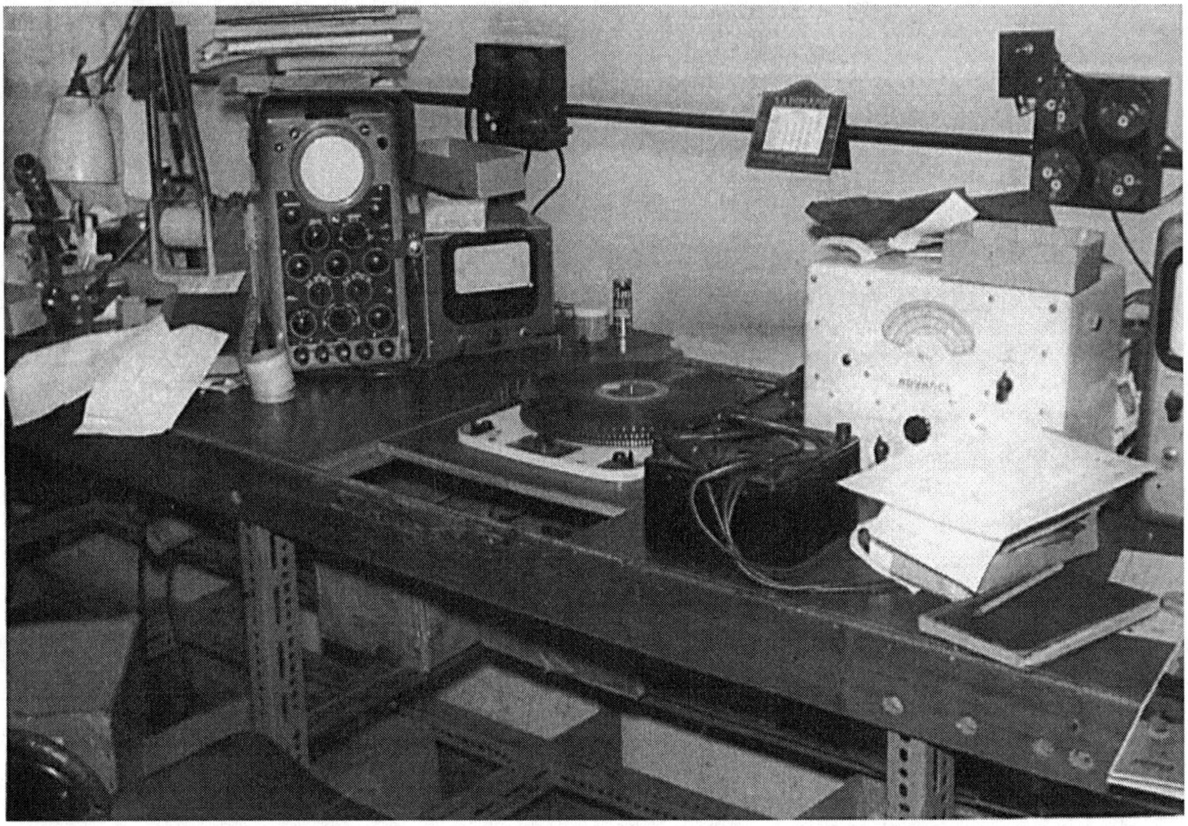

with John Bolingbroke from Leak, who then handled the transformation into quantity production.

The other main design revision was a reduction in the size of the generating system armature to two-thirds of that in the Mark 1, enabling a lowered effective tip mass and flatter high frequency response.

With the inter-changeable head of the Mark 2 pickup, owners could play either 78 rpm records (requiring a 2.5 mil stylus) or micro-groove mono records (with a 1 mil stylus). The Leak pickups had a good reputation for high performance and were regarded as being capable of tracking the heaviest groove modulations. The Leak mono pickups had very high compliance laterally but low compliance vertically and were not suited to playing stereo records. The design was conceived before stereo records were envisaged so the ability to track vertical modulations of the record groove was not a consideration. The following specification information is from Leak data sheets:

All pickups and associated transformers were measured and matched before final approval. A microscope used for parts inspection is on the left. A meter built by Leak is on the right side of the Cossor Oscilloscope and in the middle is a Garrard 301 turntable with a Leak pickup arm. The Advance oscillator on the right was used to measure transformer frequency response, and an industry-standard AVO-meter is at the front. (Photo courtesy Steve Wilkins, UK. © Steve Wilkins.)

The Mark 2 Leak pickup of 1954 with an inter-changeable head. A longer version of this arm was also available for playing 16 inch transcription discs. Lund Brothers made the pivot system and metal bushings for the arm. (Reproduced by permission Leak Electronics Ltd.)

"The moving coil impedance is 10 Ohms. No magnetic material is embodied in the moving parts, and the pickup is free from inherent distortions of moving iron types. These distortions are also present in dynamic types in which the moving-coil is wound on a magnetic core.

• Playing weight: 3 to 4 grams for LP, and 5 to 6 grams for 78's.
• Frequency response: +/- 1dB 40Hz to 20kHz. Low frequency resonance 20Hz, and high frequency resonance 21kHz for vinyl, 27kHz for shellac.
• Output: 8mV per cm/sec recorded velocity, including the associated Leak transformer with 1:80 step-up ratio."

Steve Wilkins, who was an apprentice at Leak in the early 1960's, recalls working in the pickup department (run by Angus Gilchrist). Steve wound coils for this pickup

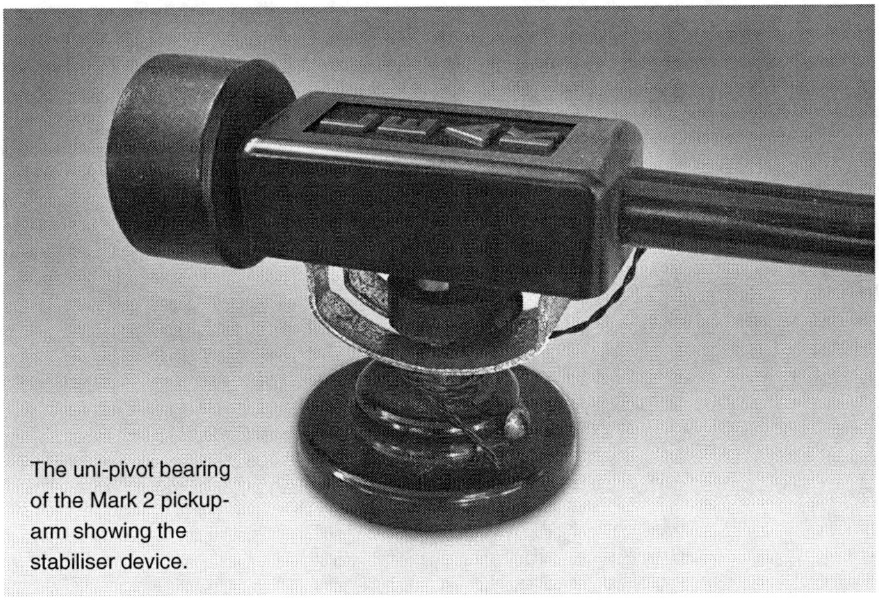

The uni-pivot bearing of the Mark 2 pickup-arm showing the stabiliser device.

Angus Gilchrist

Angus Gilchrist had been interested in electronics in his teenage years, having built several amplifiers during the late 1940's, and undertaking National Service with the Royal Signals Corps. His parents happened to meet the Leaks while on vacation, and through an exchange of interests, Harold Leak suggested that Angus contact him on completion of his National Service.

Angus started with Leak in May 1950, initially performing wiring for pre-amps alongside Norman Rowe. He recalls that:

"Practical training could be described as 'learn as you go' but you were taught to do things to a very high standard, particularly wiring and component assembly. A vigorous inspection system ensured this was maintained."

Like many others at Leak, Angus attended night school for several years with financial assistance from the firm. Angus went on to manage the Pickup Department, then in 1968 he was appointed Sales Office Manager. He remained with Leak until 1971.

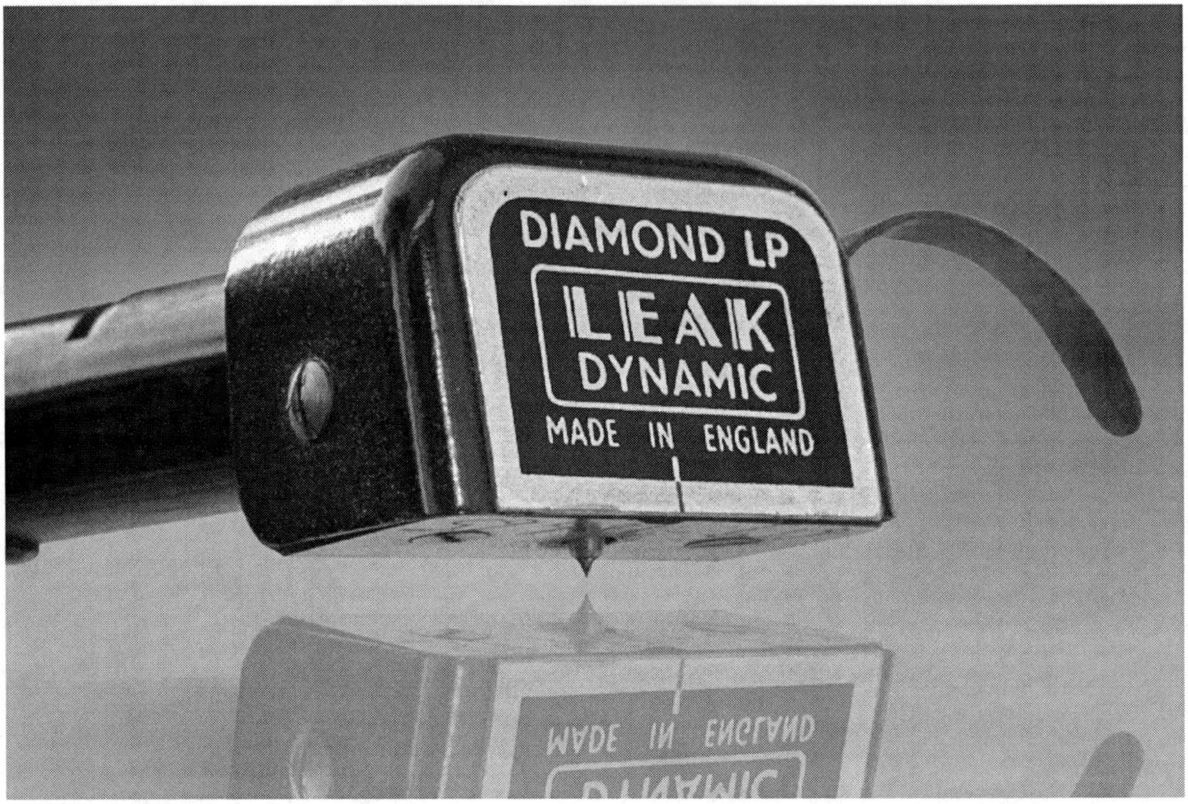

The removable-head of the Leak Mark 2 pickup. The head is a push-fit connection into the arm.

on a watchmakers lathe using 50 turns of ultra-thin 49-swg wire. It was a difficult task but on a good day 20 to 30 pickup heads could be produced.

The Mark 2 pickup was popular with radio stations and was specified by Marconi in their radio installations across Europe (along with Leak TL/12 or TL/25A amplifiers in the audio chain driving the modulator of the transmitter).

Stereo Pickup

Leak did a lot of development work on stereo pickups, but despite being promoted in company literature, none was released. Don Barlow worked on stereo pickup design in the mid 1960's after completing the Sandwich speaker project. He was assisted by Mike Deadman, who was a recent recruit having graduated from Oxford University.

George Wise was also approached to assist with the design of the stereo pickup, but at the time he was fully employed elsewhere so he turned the offer down. George knew that the development of a stereo pickup to compete with best products of the day would not be an easy task – it would require experience and good manufacturing facilities, but he also knew that Don Barlow had excellent abilities to apply to the

This Leak stereo pickup and arm was promoted in company literature during the mid-1960's, but was never launched. As well as an unresolved problem with bonding the rubber suspension to the cantilever, the pickup head had an unusual shape that was awkward and expensive to fit a mu-metal shield to. (Reproduced by permission Leak Electronics Ltd.)

stereo pickup project. George Wise had a lot of respect for Don: "I liked Don Barlow – he was a damn good scientist."

Prototype stereo pickups were demonstrated by Leak at some Audio Fairs (for instance, at the 1967 Audio Fair held at the Hotel Russell, London) and were also listed in the 1966/67 "Hi-Fi Year Book", but a product launch never materialised.

In notes by Don Barlow he wrote: "The pickup has excellent performance, but proved too difficult to make". George Wise recalled that for good reliability, the Leak stereo pickup required a special technique for bonding rubber to metal which became a technical stumbling block. Leak approached several companies both in Britain and Switzerland in search of a solution, but a suitable supplier could not be found so the pickup was never released. Another problem experienced during early development was inadequate output but this may have been solved in time through the use of powerful magnets.

The Leak stereo pickup had good overall specification, with an elliptical stylus of 0.3 x 0.7 thousandths of an inch, compliance of 10×10^{-6}dynes/cm and a tip mass of 1 milligram, allowing a playing weight of 2 grams. The pickup was of the variable-reluctance type.

The pickup-arm that was developed for use with the stereo pickup had an unusual curved shape in the vertical plane. This allowed the pivot point for the arm to be placed at the same height as the record surface, thus minimising the audible effect of record warps. An anti-skating bias device was a feature of the arm and the anticipated cost for the pickup and arm together was £26/13/8.

The Truspeed Turntable
by John D. Leak

The Truspeed turntable was modelled along the lines of the American AR turntable (and the Thorens) – the prototype used a some modified components from the AR turntable. Father had a high regard for AR products which always performed better than one might expect from their specification. When the Truspeed launch was imminent (or so we thought) we changed over to AR turntables for all our demonstrations rather than continue to promote Thorens, our potential competitor. This caused a certain amount of controversy at British Audio Fairs as exhibitors were supposed to use only equipment available on the British market, and AR were not at the time.

The turntable was the last product in which Father took a direct interest. He had lost a degree of interest in the day-to-day operation of the business (ably done by Frank Hawkins and Fred Blackman anyway) and was finding it increasingly difficult to keep up with the accelerating pace of change technically. But the turntable was something completely new to him – a new challenge. As in other things the sheer simplicity of the design owed much to his influence. One of the early prototypes came in from the designers with brass expanding bushes for mounting components and these were abandoned in favour of self-tapping screws. Another feature was a turntable bearing designed to operate without lubrication – changing viscosity would affect speed for which there was no adjustment

The test-rig for sensitivity to knocks consisted of a hammer suspended from a roof beam by a string. When it struck the bench from a specified height it gave a simple, effective and consistent testing procedure.

The design of the arm and turntable suspension was influenced by the belief that the only way the company could really expand was by getting into the integrated-systems market. This design was necessary to isolate the pickup from the cabinet and acoustic feedback. The specialised hi-fi separates market was still small, and the very limited amount we could spend on advertising was unlikely to make much difference. When companies such as Philips and Ferguson started to market audio equipment the market opened up considerably.

Truspeed Turntable

Leak didn't make a turntable ("motor") until the late 1960's. Before this, the company would source turntables from other manufacturers for customers who wanted to buy a complete system from Leak. For example, a Leak console cabinet featured in a 1950 Wireless World advertisement came fitted with a Connoisseur turntable from A. R. Sugden of Brighouse, Yorkshire. Leak had established a relationship with Arnold Sugden (who visited the Brunel Road factory on occasions to discuss the design of bearings and control of temperature rise in them).

In the 1960's the "Southdown" cabinet for housing a Leak amplifier, pre-amplifier and radio tuner was fitted with the Garrard 301 transcription turntable.

Development work on a Leak turntable commenced after Mike Deadman joined in the mid 1960's. He and Don Barlow worked on the development of a turntable which was eventually marketed as the Leak Truspeed. (In the mid 1930's the British Thomson-Houston Co Ltd had also used the name Truspeed for a turntable.)

The Leak "Truspeed" turntable from 1968 was discontinued, but the mechanical design lived on in subsequent turntables from Rank-Leak. Note that the arm is mounted on a platform attached to the sub-assembly for the turntable bearing. (Reproduced by permission Leak Electronics Ltd.)

The project began with a review of leading products in the market place – these being the Garrard 301 and 401 models, the Thorens TD124 and TD150, and the AR turntable from the USA. The Garrard was a dated design, and it was thought possible to improve on its rumble and isolation performance, along with a reduction in assembly complexity.

Leak approached a tape-recorder motor manufacturer for samples of suitable motors to drive the turntable, but after exhaustive and repetitive tests for speed stability this style of motor was discarded – much to the surprise of the manufacturer who was not aware of stability problems in the motors. (Ray Whitcombe undertook the tests by counting the slip on a stroboscope wheel!)

From this point on, the Thorens TD150 became the starting point for the Leak design – using belt drive and a synchronous clock motor. The rigid connection of the pickup-arm mount to the platter bearing was also adopted for the Leak turntable, as this approach had proven beneficial to shock-resistance and rumble performance in the Thorens design. Prototypes were built using Tufnol phenolic sheeting so that components could be easily re-positioned and holes plugged with Araldite (but eventually a prototype was destroyed by excess heat in the tea-room oven whilst the Araldite was being cured).

The first production prototype of the Truspeed cost £500 to make, and from this it was decided to produce an initial run of 50 units. George Wise recalls assisting with component specification and identification of suitable suppliers for machined parts. At the same time George Wise felt that having all machining operations under one roof, for example like Garrard, would have been enormously beneficial – but this could not be realised, and was not the way Harold Leak intended to produce the turntable.

Production of the Truspeed commenced in 1968 with the initial 50 units being produced for France, but when the Rank organisation bought Leak in 1969 they decided to discontinue the product. Leak chose to fit the Truspeed with a Connoisseur SAU2 pickup-arm, and Goldring pickup cartridge. The final choice of the Goldring pickup cartridge was based on extensive listening tests.

Whilst the Truspeed was short-lived in production, the design was not discarded. Later, in the 1970's, Rank-Leak launched the 2001 Transcription turntable which, underneath a much revised exterior, was a design based on the Truspeed.

"Like much development, the first priority was to see whet the opposition was up to. The then flagship product - the Garrard 301 or 401 was considered outdated, and of unnecessary complication. It was poor on shock isolation and rumble. The 'in' product was the Thorens TD124 or TD150. The first investigation was the supply of a suitable motor, so a manufacturer of tape recorder motors was approached and samples submitted. These were tested for speed stability by 'yours truly' using a stopwatch and counting the slip on the rim stroboscope. The results were terrible." .... "This was a shock to Leak as it now appeared that nothing would have a simple solution in this project."

Ray Whitcombe, 1999.

Fred Reynolds of Winston Electronics with Harold Leak in February 1954. The Leak VS-Tuner was designed by Fred Reynolds of Winston Electronics. Behind, at the left is Jan Prosniefskwi, an employee at Leak. (Photo courtesy of the Leak family.)

VS-Tuner

By the early 1950's, some customers were asking Leak to deliver a complete system with all-Leak components. To supply radio tuners to these customers, Leak had been making-up, virtually one at a time, tuned radio frequency (TRF) tuners, which could provide quite good quality reception if the signal strength was suitable and if the radio station maintained a good quality audio feed to its transmitter. Around several dozen of these TRF radio tuners were made.

Harold Leak decided that it was time to look at designing a radio tuner and arranged for Winston Electronics, of Park Rd, Hampton Hill, Middlesex, to design and produce them.

Jack Turner, who was Harold Leak's laboratory assistant up until the TL/12 was launched, had gone into the armed services in the late 1940's. When he left the services, Jack went to work for Fred Winston Reynolds at his company, Winston

The VS-Tuner for AM radio reception was a super-heterodyne design with continuously variable selectivity to allow for high-quality reception.

The under chassis view of the VS-Tuner is particularly neat for a radio-frequency circuit. The large dual-ganged wire-wound potentiometer at the bottom left controls the selectivity of the receiver.

Electronics. Fred Reynolds was previously an employee of STC and had come to meet Harold Leak through the various business arrangements between STC and Leak. When Leak decided to look around for a company that could undertake AM radio tuner design and production, Fred Reynolds was chosen as Harold knew of his capabilities in this area. In addition, Harold's previous lab assistant was now working for Fred Reynolds so there was a technical contact at Winston Electronics whom most people at Leak knew well.

The VS-Tuner used a modular band-selection switch facilitating quick assembly and giving easy access to alignment controls. The switch was manufactured by Wearite (Wright and Weaire Ltd), also known for the Ferrograph range of tape recorders. (Reproduced with the permission of Canford Audio, UK. Source: WW April 1949)

Winston Electronics was engaged to do a run of 100 VS-Tuners, with later production runs bringing the total number built to several hundred. As audio quality was always of prime concern to Leak, the VS-Tuner featured a continuously variable selectivity to allow for improved audio quality when signal conditions were good ("VS" being short for variable selectivity). The narrowest selectivity setting was 6kHz, and the maximum 24kHz (the latter corresponding to a nominal 12kHz audio bandwidth).

The Leak VS-Tuner required an external power supply and had three wave-bands: 700m - 2000m (long wave), 190m - 550m (broadcast); and 17m - 50m (short wave). The various coils and trimming capacitors required for each waveband were enclosed in an integrated housing supplied by Wearite (Wright and Weaire Ltd). The tuner was a super-heterodyne design, featuring the following valves: one X61M, two EF39's, and an EB34. In addition there was an EM34 dual-sensitivity magic eye tuner indicator. Whilst the design was not particularly sensitive, it did give good audio quality. It was an expensive addition to a hi-fi system as it cost around £35, being a little more than a TL/12 amplifier and RC/PA combined.

Eventually FM transmissions commenced, making possible high fidelity radio reception, and the VS-Tuner was dropped from the Leak product range.

"Fred Reynolds of Winston Electronics was previously an STC employee. Harold said to him: 'Will you do a design and make an AM tuner for us?' And so we got the variable-selectivity tuner. Up until then we made up, very nearly one at a time, a TRF tuner which was very good when used with a good signal."

Ted Ashley, 1998

FM Tuners

In the early to mid 1950's, the public was awaiting the introduction of BBC FM transmissions with much anticipation. During 1953, a subsidiary of STC had supplied 51 FM tuners to the BBC so they could undertake research and studies on practical FM reception conditions. Through contacts at the BBC, Leak were able to obtain one of these STC FM tuners, enabling the company to demonstrate FM during this BBC testing phase. Whilst this provided an FM demonstration facility for Leak, a commercial product was required in the long run.

In 1954 Leak decided to add an FM tuner to its product range. Ted Ashley was intimately involved with the design of the FM tuner series, spending much time with Stan Amos and G. C. Johnson on the development. (Amos and Johnson were well-regarded BBC Engineers, engaged to Leak under contract). Harold Leak contributed at a high level to overall design goals.

"I spent a long time with Amos and Johnson on the Trough-Line design. The problem was drift of the oscillator. With conventional coils heavy AFC was required, but then you didn't know if you were on-tune, off-tune, or where you were. Amos and Johnson decided on the shortened quarter-wavelength transmission line to reduce drift."

Ted Ashley, 1998.

Amos and Johnson had published papers in Wireless World on FM tuner design (in 1952 and 1955). Whilst Harold Leak may have noted these papers, it was through a contact Leak had at the BBC that he engaged Amos and Johnson on the design. Harold Leak arranged to introduce Amos and Johnson to staff at Leak over an evening meal.

Trough-Line FM Tuner

The original Trough-Line FM tuner broke new ground in oscillator stability for domestic VHF radio receivers. Some versions had a square-shaped aperture for the tuning indicator, and lacked the "railway line" design printed on the front panel. A special version with a white front panel and matching pre-amp was produced for the Duke of Edinburgh.

A fundamental problem with ultra high frequency radio receiver design in those days was to provide an oscillator that was stable (remain at the same frequency over a period of time). The heat from valves eventually caused the oscillator to drift because, as the coil in the oscillator's tuned circuit warmed up, it changed slightly in dimension, and hence the coil inductance changed. Thus the oscillator drifted off frequency. Early attempts to stabilise oscillators relied on the application of large amounts of "automatic frequency control" (AFC), but these were not very successful, as it could make it difficult to know exactly what frequency the radio was set to receive.

Amos and Johnson decided that a shortened quarter wavelength transmission line could act in place of the conventional coil inductance used in ordinary radio oscillators. This quarter wavelength transmission line was made in the form of an open trough, square in section, with the open side facing downwards. In the centre of the trough, a conductor ran along the length, with various taps that connected into the circuit of the oscillator valve and tuning capacitor. This transmission line is almost impervious to the effects of heat and so the oscillator frequency remains inherently stable. It was the trough-shape of the shortened transmission line the gave the Leak FM tuner its name.

Towards the end of the production run for the first model, the front panel was revised to match appearance of the new pre-amps. (Reproduced by permission Leak Electronics Ltd.)

The line was originally made of solid copper and was "a massive brute" according to Ted Ashley, making it a costly component. When the Trough-Line II was introduced, the design was revised to reduce cost. A mild steel tube, very heavily copper plated and flashed with tin was introduced in place of the copper trough (the tapped central conductor was then coaxial with the outer tube). This approach was made feasible by the fact that, at high frequencies, the conduction of current occurs predominantly along the surface of a conductor rather than in the body of the conductor (this is the "skin effect"). The quality of the plating and the brazing to the trough (or tube) influenced the "Q" of the circuit and the magnitude of the oscillator voltage, so before any alignment work was done on production units, the oscillator output voltage was measured to ensure it was satisfactory. (To minimise costs, a reduced copper thickness was tried, along with thicker tin, but this reduced the conductivity and was rejected. Aluminium tubes were also tried, but the joint conductivity at the end-plate decreased when the aluminium corroded, so this was also rejected.)

The original "Trough-Line" was released in 1955, design work being done during 1954. Originally Leak had hoped to achieve the design specification without the use of AFC. Over a period of six weeks, Ted Ashley undertook studies and tests on oscillator drift and methods for minimising it. The conclusion was that an improvement by a factor of about two or three was required to meet the design specification (of about 2kHz maximum drift). As a result of these studies, it was recommended that a small amount of AFC be applied, thus making an already quite stable circuit able to meet a very stringent specification.

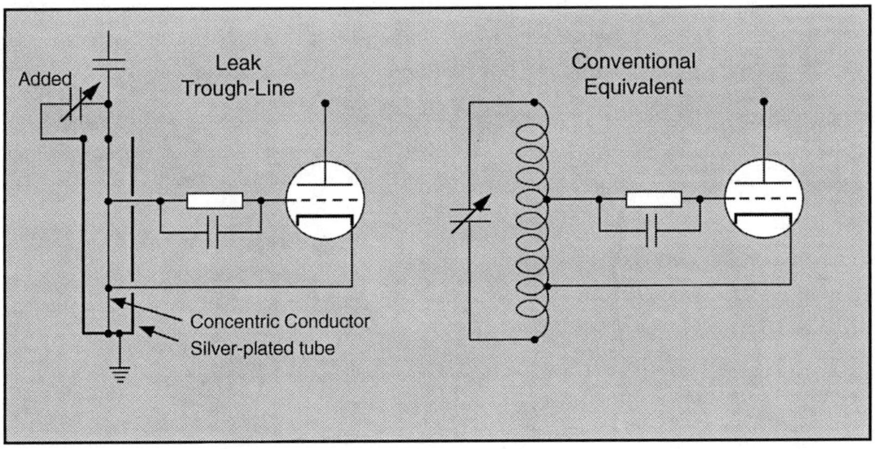

In the Leak Trough-Line, a tube with a concentric inner conductor tapped at appropriate points replaces the conventional wound coil found in most radio receivers.

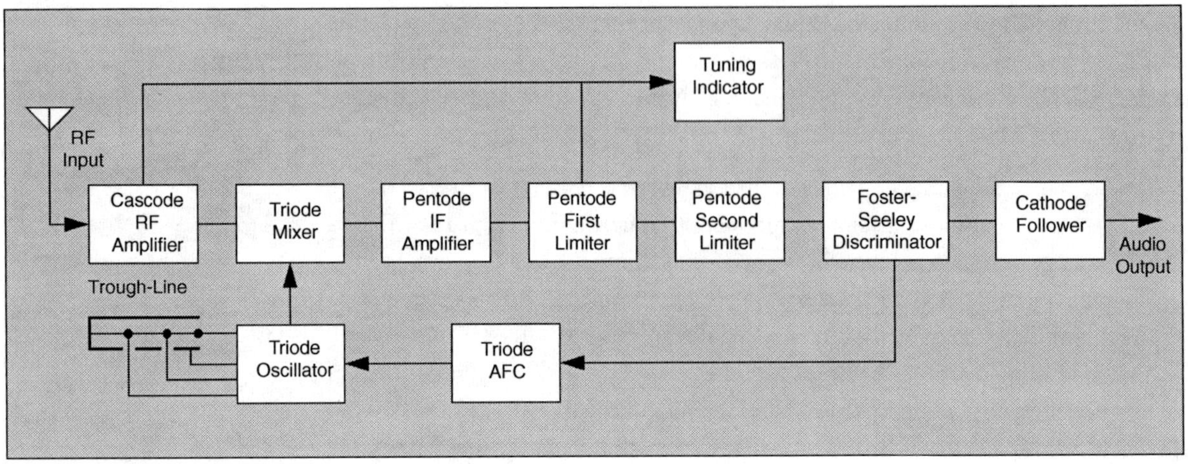

The block diagram for the Trough-Line 3 FM tuner (other models were similar). In the case of the Trough-Line Stereo the stereo decoder appeared after the cathode follower.

To convert the frequency-modulation of the received signal into audio, a discriminator or detector is used. All Trough-Line tuners use the Foster-Seeley discriminator circuit which was chosen for its good linearity, however unlike the ratio detector it does not inherently reject amplitude modulated (AM) signals such as interference. As a result, an additional intermediate frequency stage was included to clip any AM signals before reaching the discriminator (this stage has a low anode voltage to enhance the clipping function).

The tuning range of the Trough-Line was from 88 to 100MHz which suited the spectrum usage at the time, however to receive all stations these days a wider range up to 108MHz is required. The front panel controls were limited to tuning and volume only, but an EM81 indicator valve assisted with accurate tuning. At the rear of the tuner there was a "quieting" control that pre-set the level for muting the output between stations (this feature was omitted in later models). Other valves used in the design were: 2 x EF80 pentodes, 3 x ECF80 triode/pentodes, 1 x EB91 diode detector and an EZ81 rectifier.

In the 1950's British tuners required an external power supply to operate them but all Trough-Line models were self-powered. This decision was more significant than it first appears. The amount of reserve power available from typical British power amplifiers was sufficient to operate AM radio tuners, or FM tuners with mediocre performance. The Trough-Line design however used more stages and would not have been practical if the design team was constrained to a limited power budget. (The TL/12, TL/25A, TL/10 and TL/12 Plus power amplifiers had reserve HT and LT for providing power to British tuners. If a customer asked about alternative tuners, Leak would recommend Rogers or Chapman.)

The under chassis view of the original Trough-Line shows the large inverted trough that formed part of the oscillator resonant circuit. Two taps along the trough can be seen, as well as the tuning capacitor at one end.

The original Trough-Line was available from 1955 to about mid-1959. Most units were made with a gold-enamelled steel front panel similar to the pre-amps prior to 1957. When the diakon plastic front panel was introduced for the pre-amps in the late 1950's, a matching plastic front panel was designed for the Trough-Line tuner too.

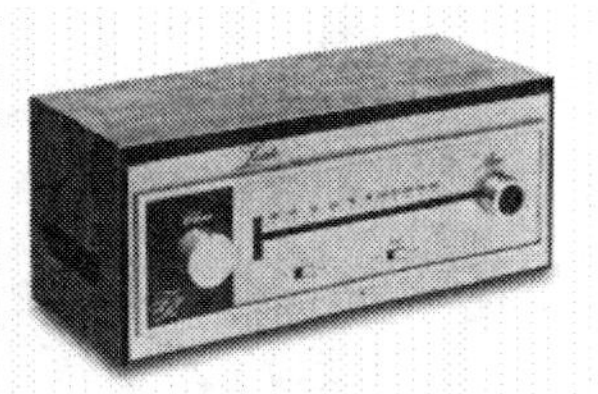

A Leak Trough-Line II with a timber case, gold-anodised aluminium fascia, and optional fascia insert (on the left side) for export to the USA. (© Hachette Fillipacchi Magazine Inc. Reproduced with permission from Audio magazine, USA. Source: Audio, Aug 1963)

Note that this original Trough–Line model doesn't incorporate an audio de-emphasis circuit, and as modern FM transmissions pre-emphasise the audio, this tuner will give excessive treble output. A simple modification for adding audio de-emphasis is shown in the Trough–Line circuit given in Chapter 12.

Trough-Line II

As subsequent Trough-Line models were designed, the majority of revisions were undertaken at Leak by Ted Ashley, but Amos and Johnson acted in an advisory role if required (these models were: Trough-Line II, Trough-Line 3, and Trough-Line Stereo).

The first stage (an RF amplifier) in this model uses a cascode connection of a dual triode valve, replacing an EF80 pentode in the original design. Other design changes were made, reflected in a different valve line up for this model: ECC84 and ECC85 dual triodes, EF80 pentode, 2 x ECF80 pentodes, EM84 indicator, 2 x OA79 diodes and an EZ80 rectifier.

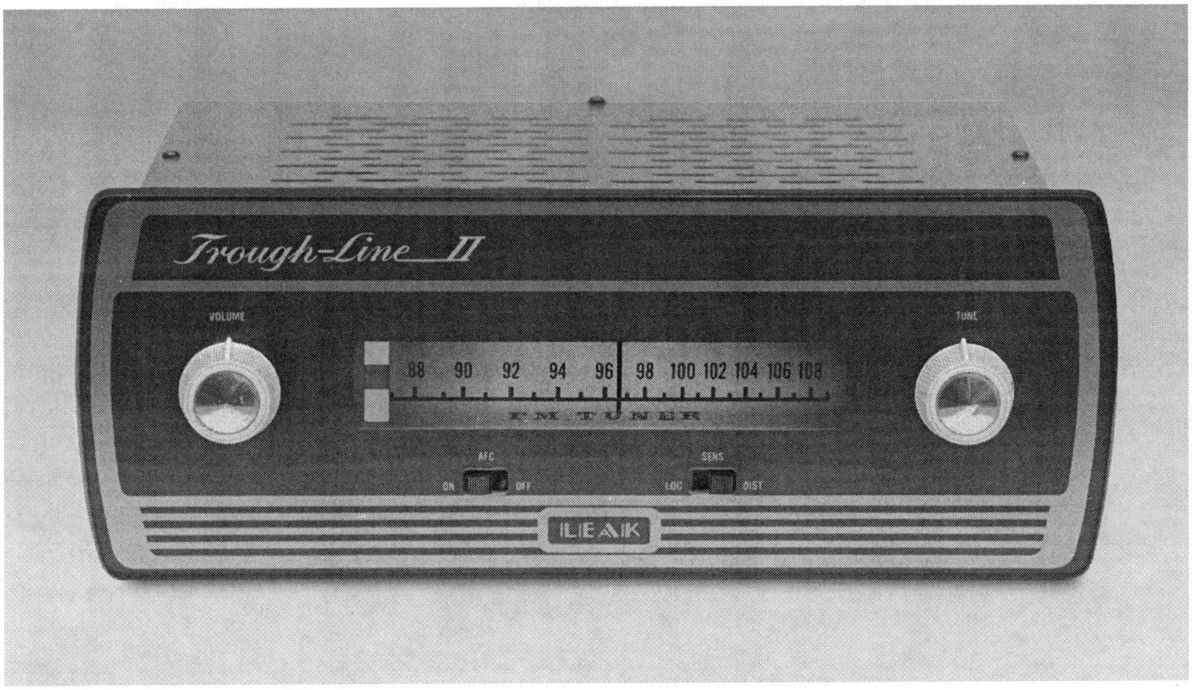

The Leak Trough-Line II had a completely new appearance, and had an extended tuning range of 88 to 108MHz.

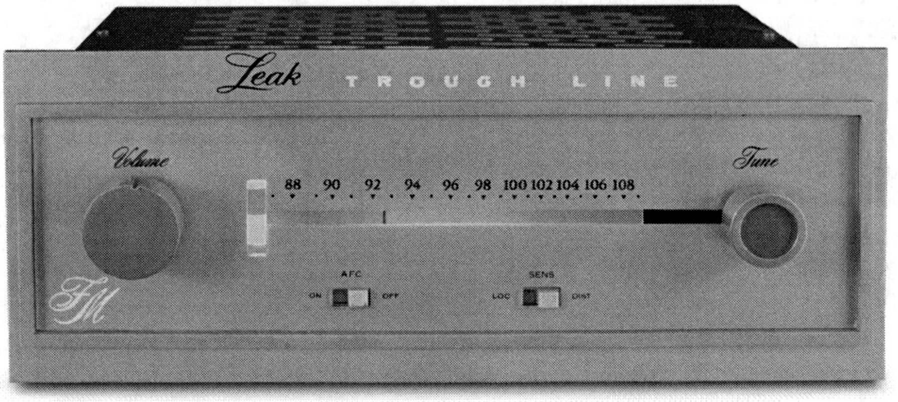

The Leak Trough-Line II as exported to the USA for Ercona Corporation, shown here without the optional decor-insert around the left hand control knob.

An automatic frequency control (AFC) on/off switch was added, which was also coupled to the power supply for the tuning indicator so that the indicator was extinguished when the AFC was on. This gives the indicator a long-life, as in normal use it need only be on while actually tuning a station, after which the AFC would be switched on, with the tuning indicator simultaneously being turned off. The "quieting" control that featured on the original Trough-Line was deleted, but an RF sensitivity control was introduced.

As a standardised method for stereo transmission had not been decided on when this model was launched, an additional output was made available on the rear panel for connection to an anticipated external stereo decoder.

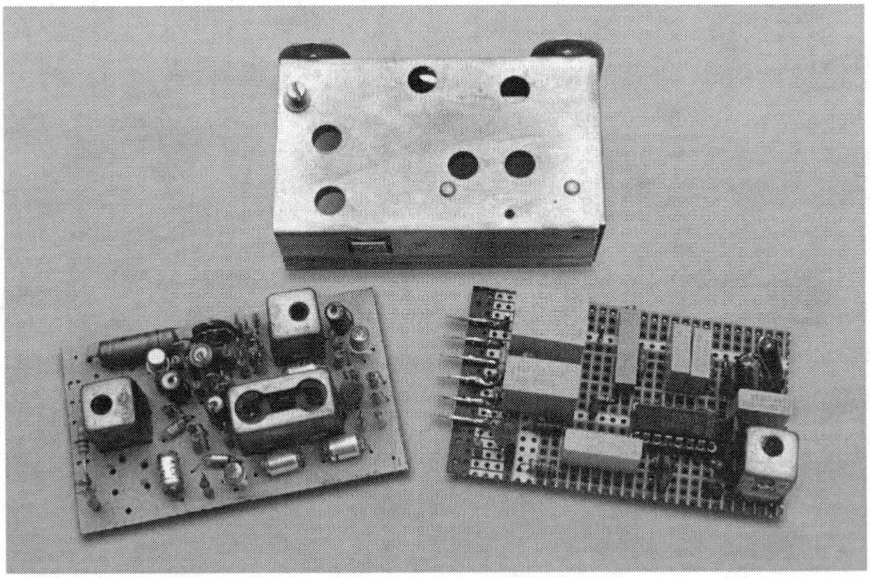

The original Leak stereo decoder used three transistors (left), was housed in a metal case (top), and is often replaced by enthusiasts with an integrated circuit decoder (right). These can provide much reduced jitter (un-wanted frequency modulation) on the recovered 38kHz sub-carrier, owing to the action of the long time constants possible with modern phase-locked-loop filtering.

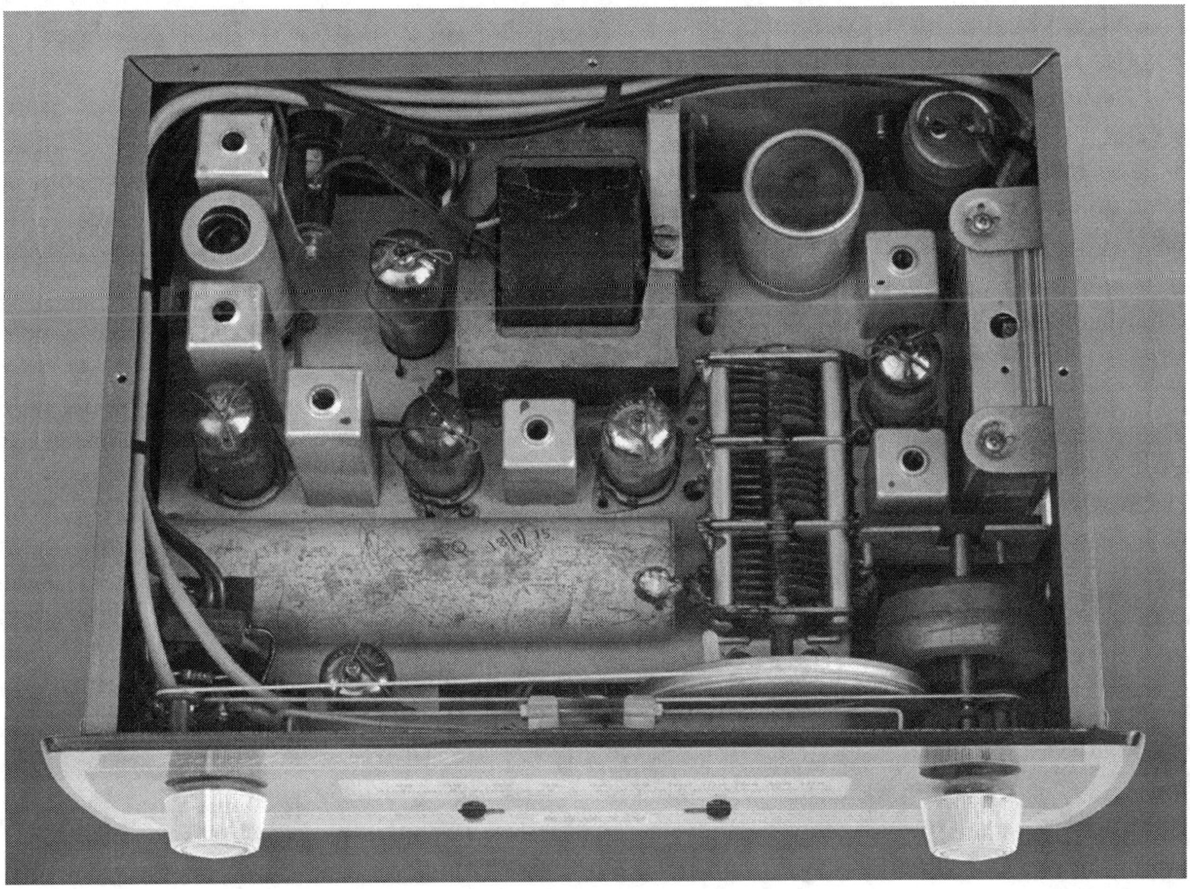

On the inside of a Trough-Line II FM tuner the tube-shaped concentric-line tuning inductance can easily be seen towards the front left of the chassis. This particular tuner has had a factory stereo decoder retro-fitted to it, along with a larger power transformer. The decoder is mounted in a metal box against the right hand side of the chassis.

These tuners have been highly acclaimed from the day they were launched. A review by A. W. Wayne (of Shirley Laboratories Ltd.) in "Hi-Fi News" was so enthusiastic that it said the tuner was suitable for "the carriage trade". The audio quality from this tuner has an excellent reputation, and makes them prized possessions, even 35 years on, despite the technical superiority of modern tuners.

This model and the ones that follow it, have the same front panel dimensions as the associated pre-amps, making for improved aesthetics.

Trough-Line 3 and Trough-Line Stereo

In 1964, the "Trough-Line 3" was introduced with a new black and silver front panel and some minor design revisions that increased sensitivity marginally, although the valve complement was identical to the Trough-Line II. The cost of the new model when introduced was £29/7/6.

Stereo FM Tuner Development

by Ray Whitcombe, IEng, FIIE(Elec).

In the beginning there was the Trough-Line, and in the end there was the Stereofetic – or as it was referred to in "factory-speak", the Stereopathetic. This latter was a bit unfair, but seemed to stem from some teething problems and the mechanical design.

Both the beginning and the end of this story were before and after my time at Leak. When I joined the firm the tuner in production was the Trough-Line II – production and sales seemed to flow steadily and reliably. The potential for further tuner developments did not exist at that time, and indeed, there seamed to be little need for anything new.

Towards the mid-1960's the promise of stereo transmissions loomed, and at first Leak saw no danger as all Trough-Line II tuners had been produced with a rear socket that gave access to the un-corrected discriminator output. It was intended, when the time came, that a suitable stereo decoder would be made available by Leak for retro-fitting to previous model tuners.

Well, the time came, and the Development Lab started work on the design of stereo decoder – at first a Mullard design was tried, but difficulties arose immediately. The capability of the Development Lab was very limited due to the lack of suitable test instruments – Cossor Telecheck signal-generators and Cossor double-beam oscilloscopes were available, but the only sweep generator for radio frequency work was an old Philips unit which used a loudspeaker drive unit coupled to a "bee-hive" capacitor to derive the sweep function! For the stereo decoder development work, Leak purchased a Fisher stereo radio test-set that could provide the necessary radio-frequency carrier, modulated with a stereo audio signal.

Work could now proceed on the Mullard decoder design, so a Trough-Line II tuner was connected to the Mullard decoder, with the new Fisher test-set feeding the aerial input with a stereo signal – but it didn't work! There was hardly any stereo effect. I don't recall who identified the cause of the problem (it may have been Mullard, or Stan Amos of the BBC), but it was simple really – not enough bandwidth. The Trough-Line tuners had been designed with performance and sensitivity in mind (at the time, long-distance reception with FM was something owners would boast about). However stereo was different, and a tuner (like the Trough-Line II) with just 200kHz bandwidth caused some losses in the stereo components of the signal.*

The remedy had two bitter pills for Leak: further development work would have to be undertaken quickly to come up with design revisions for the Trough-Line; and these revisions would need to be incorporated into old Trough-Line tuners if customers requested a stereo modification.

Better radio-frequency test equipment was now required, so a Marconi Instruments TF995 signal generator was supplied in a specially modified form for Leak that could cope with the 75kHz carrier deviation at low distortion. An American Wavetek audio oscillator was used to feed low distortion sine-waves to the Marconi signal generator.

With such equipment we set to work increasing the Trough-Line's bandwidth. The main area was the discriminator, a Foster-Seeley type. This has the advantage of good linearity but less bandwidth, and we noticed that most other manufacturers of stereo tuners used the ratio-detector instead. The idea of changing the discriminator was frowned on, so I had to try my best with the Foster-Seeley. The required bandwidth was achieved, but to maintain low distortion, very critical alignment was required. To go with the wide bandwidth discriminator the intermediate frequency amplifiers had to have similar treatment. All this achieved the wide-band tuner, but at the expense of gain – we now had a tuner lacking in sensitivity. I came up with the suggestion of using a one of the new higher gain valves developed for TV in the first stage, so the ECC84 was replaced with an ECC88.

The above measures gave us the key to the new Trough-Line Stereo production tuner, but what about those customers with older Trough-Line units? Harold Leak would not let his customers down, so apart from the cost of the decoder, anyone could have their older tuner upgraded to stereo. This included new discriminator and intermediate frequency transformers, stereo decoder and audio switching (but a revised fascia for older tuners was not made available due to cost considerations). In addition, some early Trough-Line II tuners had been fitted with a small mains transformer that couldn't cope with the extra load of the decoder, so these customers had a new mains transformer supplied with the deal – and a very good one don't you think!

FM Tuner Discriminator Alignment

The alignment of the Trough-Line Stereo was very critical. It was found that the stereo performance was very dependent

Ray Whitcombe working on the Trough-Line II FM tuner in the Development Lab at Leak. (Photo courtesy Ray Whitcombe.)

on the linearity of the discriminators's "S-curve". Initially the Lab did a final "tweak" using a distortion factor meter, having first obtained the best visual alignment with a sweep generator. This method was unacceptable for production units, so Bob Skegg and I were given permission to build some special test gear for use on the production line. The design was based on frequency modulating a radio-frequency carrier with a 1kHz sine wave that had been mixed with a three-position staircase waveform. This was fed to the tuner under test, and the tuner's audio output measured through a 2kHz band-pass filter. The output from this filter was the second-harmonic distortion at three different deviations of the radio frequency carrier. Displayed on an oscilloscope, the objective was to align the tuner so that the centre burst (at nominal carrier frequency) was as small as possible, consistent with the outer bursts being equal.

Solid State FM Tuner Development

H. J. Leak & Co. had their stereo tuner on the market by 1967, plus satisfied customers of previous models, although at a price. With the introduction of the transistor audio amplifiers, Harold Leak began to push for a solid-state FM tuner – indeed our position in the market demanded it! The excuse that valves had better performance at radio frequencies would not wear for much longer, so we set about examining products from the competition. For a number of years Leak had recommended Chapman tuners if customers required an AM/FM tuner. Chapman had introduced a simple solid state tuner – the FM1000 – and we obtained one of these for investigation. It was a straight forward design using Mullard AF11x-series transistors.

As always, HJL insisted that the performance of any future Leak product should at least match the performance of existing products, but this our Chapman test-unit did not do. So I was instructed to add a further intermediate frequency stage, plus the Mullard stereo decoder. This did produce slightly improved performance, but it was not very stable, and it was unclear how to proceed from this point.**

It was soon realised that a Design Engineer would have to be engaged to work solely on FM tuner development, but this proved more difficult than first realised. The first Design Engineer engaged produced a theoretical design for an FM tuner on paper, only to find that the design appeared flawed when prototypes were constructed. He was replaced with a second Design Engineer who had previous experience with television design at EMI, but again he proved unsuitable.

Shortly afterwards, a relationship was established between Leak and Texas Instruments (who had set up an Applications Laboratory in Bedford to promote advanced solid state design). Ted Ashley was keen for Texas Instruments to develop an FM tuner circuit that could be adapted to Leak's requirements, much as Amos and Johnson had done for the original Trough-Line. In due course a working prototype was offered, and initial scrutiny confirmed that at last the candle had been lit at the end of the tunnel! The circuit was built around some novel double-transistors configured as cascode amplifiers. The next step was to establish a cost estimate for the design, including all the work that would be necessary to give birth to the new Leak stereo tuner. There the project hit the rocks! In a turn of events similar to that encountered during the Stereo 30 development, Texas Instruments were unable to supply the special double-transistors in production quantities (apparently the devices were experimental only).

The Stereofetic

In 1968 the Texas Instruments design was scrapped, and a new head-hunt was initiated for yet another Design Engineer. This time Ian Pullen was engaged and the design for the Leak solid state tuner was now assured. The design was released as the "Stereofetic" in 1969 (see Chapter 9).

Footnotes:

* FM transmission requires receiver bandwidths in excess of 200kHz to receive audio signals with 15kHz bandwidth.

** The Chapman test-unit had alerted Leak to the underlying technical issues associated with solid-state tuner designs, and it was used in a low-key demonstration at an Audio Fair, but it wasn't used as the basis for any further development work on solid state tuners.

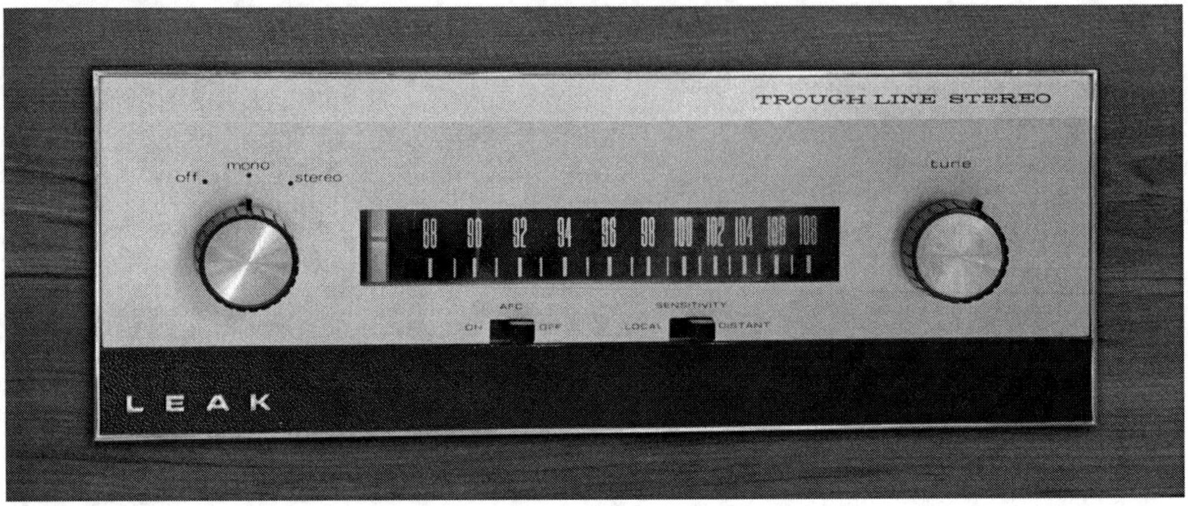

The Trough-Line Stereo had a factory-fitted stereo decoder, which could also be retrofitted to previous Trough-Line models.

By 1966, the BBC were making regular stereo broadcasts using the same GE-Zenith system that the USA had adopted in 1961, so Leak introduced the last model in the series, the "Trough-Line Stereo".

To achieve good stereo performance, the bandwidth of the tuner was increased, and the input stage was altered to a high-gain ECC88 instead of an ECC84. Mullard supplied stereo decoders (from Philips) to Leak as a pre-built module which was mounted within the tuner chassis by Leak. The decoder used three transistors and six diodes to recover the 38kHz sub-carrier for the stereo "difference" signal, and to demodulate the separate left and right signals. The design lacked a 19kHz rejection filter on the output, so the pilot tone levels in the decoded audio signal were relatively high.

For a fee of £15/0/0, owners of previous models could have the new stereo decoder retro-fitted at the factory, along with other circuit modifications to increase the bandwidth of the tuner. Leak had anticipated this requirement by fitting an output to the Trough-Line II and Trough-Line 3 that enabled a multiplex decoder to be easily retro-fitted, although the extensive modifications required to increase the bandwidth for stereo operation was an un-foreseen requirement.

The Trough-Line Stereo was introduced at a price of £47/2/1, and the Trough-Line 3 continued as a lower cost mono option at £32/3/11. The Leak company applied a smaller mark-up to the radio tuners than to amplifiers, as it was considered that this would encourage people to buy additional Leak equipment once a decision had been made to buy a Leak amplifier.

"The naming of these tuners caused me worries as many people referred to them as 'Through-Lines'. I wondered how many people did not ask for them at all for fear of mispronouncing the name. Everyone in marketing knew the story of Olivier cigarettes, which didn't sell well as people did not know how to pronounce the name."

John D. Leak, 1999.

Equipment Cabinets

The first advertised equipment cabinet from Leak was in 1950. The appearance of the cabinet was designed to match the "550" loudspeaker that was available from 1949. The cabinet housed a Connoisseur turntable fitted with a Leak moving coil pickup and arm, the RC/PA or RC/PA/U pre-amplifier, and the TL/12 power amplifier. Leak could also supply a tuner for AM reception which would have been of the "tuned radio frequency" design.

A Leak equipment cabinet from 1950 that could house a turntable, pre-amp and power amplifier. At the side are some additional controls believed to be a TRF radio tuner. (Reproduced by permission Leak Electronics Ltd. Source: WW Sept 1950)

The "Southdown" cabinet for Leak equipment was originally released for the mono amplifier range and was designed by Dennis & Robinson Ltd, of Old Portslade, Sussex, in conjunction with H. J. Leak & Co. The finish could be in either Walnut, Oak or Mahogany, with kiln-dried hardwood and birch-ply use elsewhere. Promotional pamphlets make special mention of the convection chamber cooling arrangement for the power amplifier housing. The later stereo version could house a stereo pre-amplifier, stereo power amplifier, turntable (Garrard 301) and Trough-Line tuner.

In late 1962, the Southdown cabinet was re-styled to match the Sandwich speaker system. The dimensions were reduced and it featured satin aluminium trim with chrome legs. GKD Ltd of Houghton Regis manufactured this cabinet. Another larger cabinet was also available – the "Chiltern" cabinet – which could house the same equipment as the Southdown as well as a tape recorder.

Left: The original Southdown cabinet, designed by Dennis & Robinson Ltd. in conjunction with H. J. Leak & Co Ltd. It housed a TL/10, Point One pre-amp, Trough-Line tuner and turntable. Right: A revised version of the Southdown for stereo Leak equipment. (Reproduced by permission Leak Electronics Ltd. Source: Leak pamphlet and HFN Feb 1960.)

During the 1960's the "Southdown" cabinet was a popular way to house the sound system. Shown here is the version made by GKD Ltd, with a pair of Leak Sandwich loudspeakers. (Reproduced by permission Leak Electronics Ltd.)

Portable TL/12 System

The TL/12 could be supplied with a number of options that allowed it to be used with a pre-amplifier in a portable system suitable for sound re-inforcement use, or even public address if the budget extended that far.

The RC/PA, RC/PA/U and the original Varislope pre-amplifier were all of the same width as one side of the TL/12 chassis, allowing the pre-amp and power amplifier to be bolted together via pre-drilled holes. The pre-amps could be supplied with an optional "flush" front plate for portable work, as the normal over-lapping edges would be prone to damage in such use. A top cover with a handle was available, as was a base-plate to complete the set up. A shorter top cover was another option, allowing portable use of the TL/12 without an associated pre-amp.

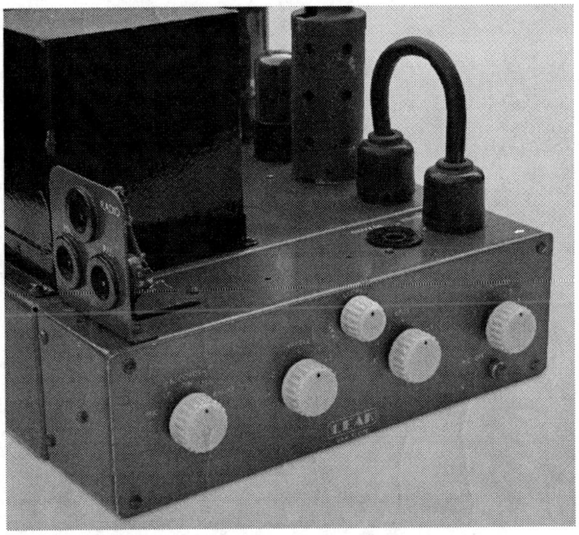

The Varislope pre-amp could be bolted to the TL/12 and a short interconnecting cable was provided by Leak for this purpose. A panel with three Bulgin 3-pin connectors extended the inputs for the pre-amp to the left side.

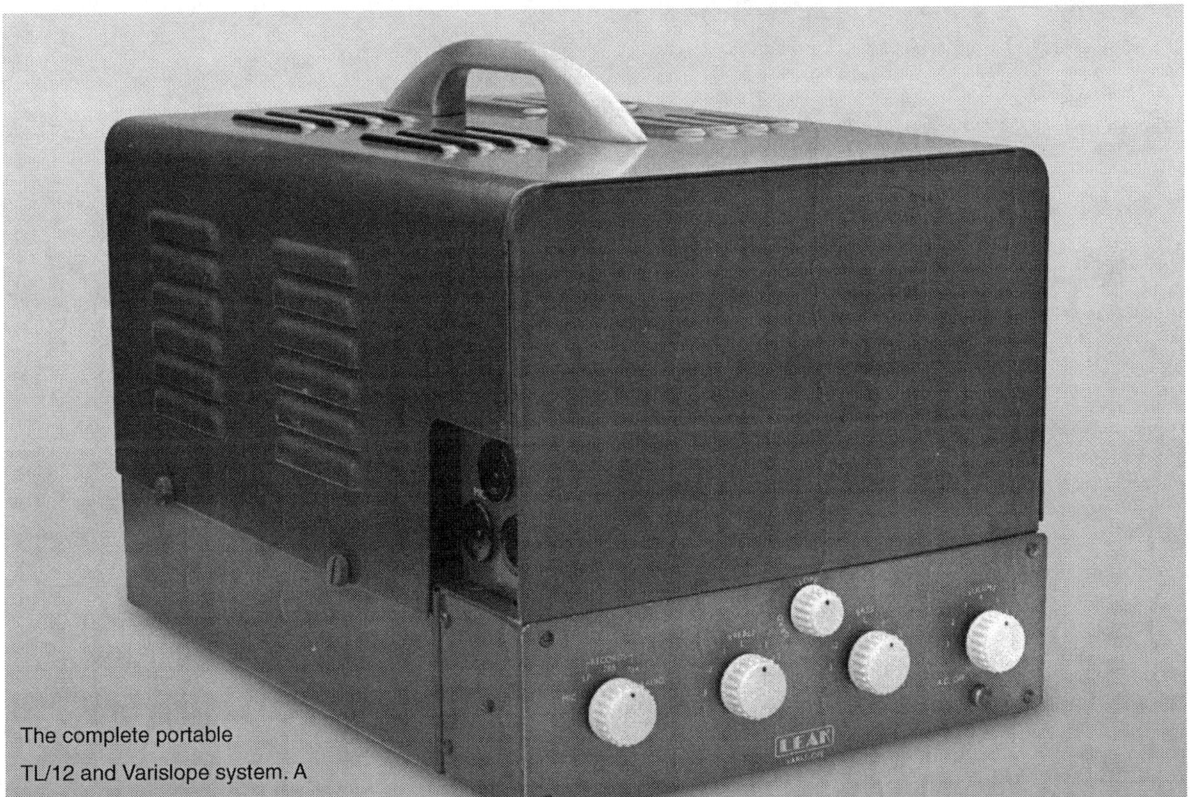

The complete portable TL/12 and Varislope system. A baseplate protected the underside of the power amplifier.

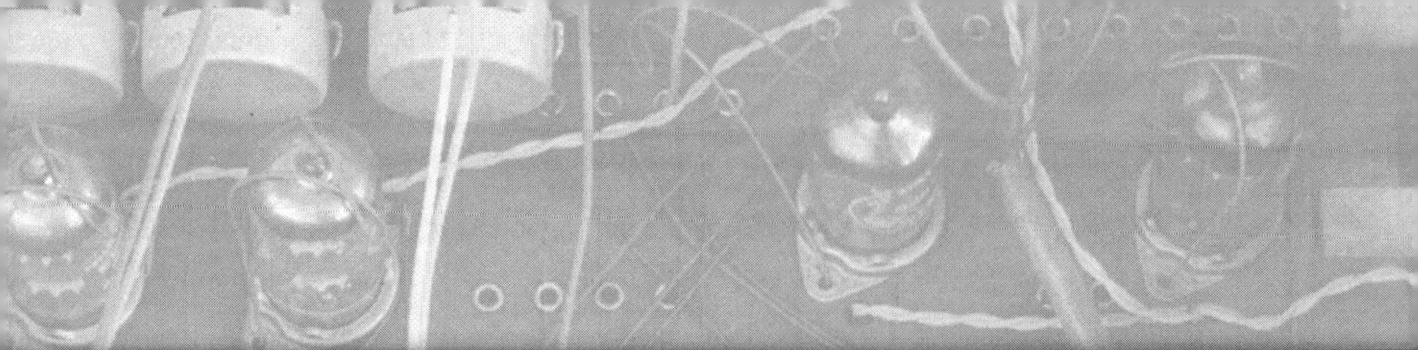

TREBLE
-3 -2 -1 1 +1 +2 +3
SLOPE
GRAD
STEEP
BASS
-3 -2 -1 1 +1 +2 +3
LEAK
VARI-SLOPE

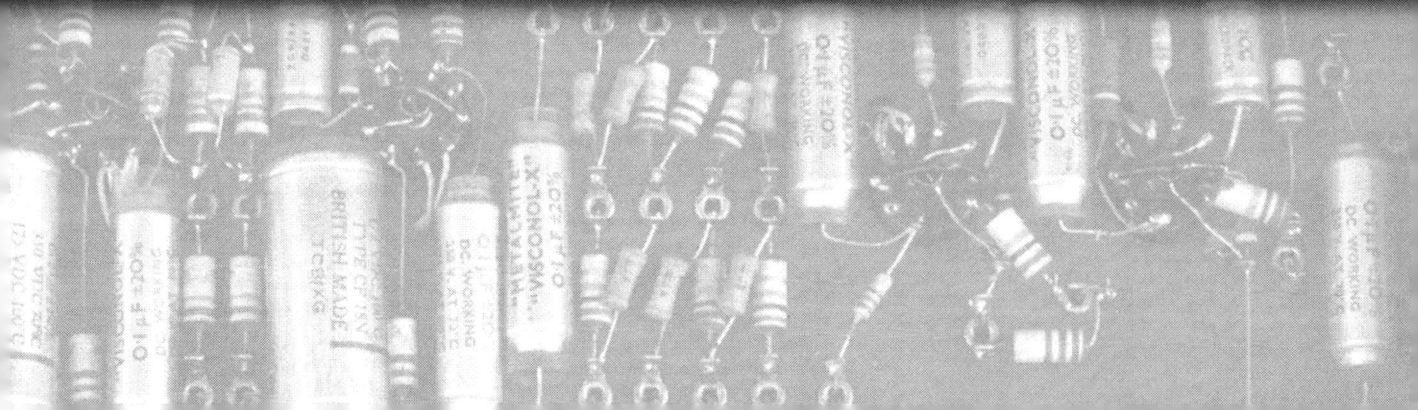

7. Pre-Amplifiers

Evolution and Design

The earliest power amplifiers from Leak did not feature separate remote control pre-amplifiers. For example, the "Standard 13 Watt Chassis" from 1936 could operate with a microphone plugged directly into it, even though it was just a three-stage amplifier.

Opposite: The original Leak Varislope pre-amp released in 1952.

Later on when significant amounts of negative feedback were introduced, the overall gain of power amplifiers reduced accordingly, so additional stages were required for low-level inputs such as microphones. With the Type 15 "Point One" power amplifier of 1945, these stages could be accommodated on the power-amplifier chassis.

But it was recognised that there were benefits in performance if these sensitive, low-level stages could be positioned away from hum-inducing components such as mains transformers – so separate remote control pre-amplifiers were introduced that contained the input selector, gain and tone controls. For the emerging post-war hi-fi market, a small pre-amp was appealing as it could be more easily mounted into a cabinet for improved appearance. As with most British pre-amplifiers, the Leak units were powered by high and low tension fed from the power-amplifier through an umbilical cord that also carried the audio signal.

The various pre-amplifier models from Leak, showing the years of production, and the increase in sensitivity that was required to match newer low-output pickup cartridges as they were developed.

Pre Amplifier Models:

Model	Years	Pickup Sensitivity	1st Stage	2nd Stage
RC/PA	Dec 1948 to 1950	50mV	Pentode	–
RC/PA/U	1950 to 1954	15mV	Triode	Triode
Varislope I	1952 to 1954	15mV	Triode	Triode
Varislope II	1954 to 1957	10mV	Pentode	Pentode
Point One	1954 to 1957	14mV	Pentode	Triode
Varislope III	1957	9.5mV	Pentode	Triode
Varislope III - Series N	to 1963	3mV	Pentode	Pentode
Point One Plus	1958 to 1963	9.5mV	Pentode	Triode
Point One Stereo	1958 to 1963	5mV	Pentode	Pentode
Varislope Stereo	1960 to 1963	3.5mV	Pentode	Pentode
Varislope 2 Stereo	1963 to 1967	3.5mV	Pentode	Pentode
Varislope Mono	1963 to 1967	3.5mV	Pentode	Pentode

A Leak Varislope Stereo pre-amp fitted with the stunning gold-anodised aluminium front panel designed primarily for the American market. The design of the knobs lacked a knurled surface finish, making the pre-amp less pleasant to operate than the conventional version of the pre-amp.

The addition of a pre-amplifier adds cost to the overall product, so it was desirable to reduce the number of stages to the minimum required to meet design specifications. This approach may seem obvious, but it had a side-effect that caused Leak to choose the rather low level of 125mV for the interface between pre-amplifier and power amplifier. Leak power amplifiers will give full power with just 125mV applied to the input, being a company standard that allowed owners to mix-and-match pre and power amplifier models. The TL/12 differed slightly in that the sensitivity was 160mV. (In the USA, it was more common to see a power-amp sensitivity of around 700mV for full output).

From the RC/PA/U onwards, all Leak pre-amps were of two-stage design, with shunt feedback around each stage. The first stage feedback was switched by the input selector so that any disc or tape-head equalisation was applied. The second stage also used shunt feedback, with a Baxandall tone control being used from the "Point One" pre-amp onwards. The Varislope twin-T circuit was available on some models.

By minimising the number of stages between signal source and loudspeaker, it reduced the cost. To do this, high-gain stages become attractive to the designer as fewer stages are required, provided these high gain stages do not have other drawbacks. In the case of the TL/12 power amplifier, the use of a high-gain pentode as the first stage made the power amplifier as sensitive as possible, consistent with other criteria such as overall stability. This then reduced the number of stages required in the pre-amplifier and so reduced cost. For this reason Leak power amplifiers have an unusually high sensitivity.

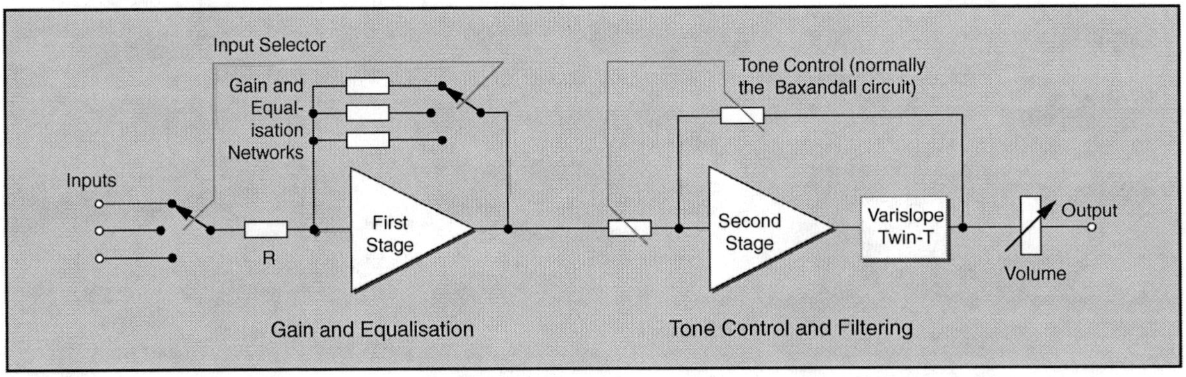

Harold Leak with an early model "Point One Plus" pre-amplifier at the display stand of an Audio Fair. (Photo courtesy of the Leak family.)

Pre-amplifiers from the 1950's and 1960's generally feature extensive flexibility – such as an ability to work with a wide range of signal levels, provide different playback equalisation characteristics, and steep-slope high-frequency filtering. The Leak pre-amplifiers are no exception, and would have been amongst the most desirable pre-amplifiers of the time in this regard.

The main contributors to pre-amplifier development were: Harold Leak, Ted Ashley, John Bolingbroke (before 1955) and Ray Whitcombe (after 1960).

RC/PA/U

During November 1950 Leak released a revised pre-amplifier – the "RC/PA/U" (the "U" standing for Universal because of improved sensitivity, and the various disc equalisations provided). This pre-amp was a complete re-design of the RC/PA (see Chapter 4). It contained a miniature twin-triode (the ECC40) with negative feedback on each stage. The input connector on this model and the RC/PA was a screw-terminal strip on top of the unit – RCA input connectors were not yet ubiquitous.

This pre-amplifier featured user-selectable equalisation for disc reproduction. At the time, the recording industry was shifting from 78 rpm records to the new LP micro-groove types, and so there was a requirement for the pre-amp to have various equalisation depending on the source of the signal (this situation was to continue for some years as the USA, UK and various companies adopted slightly differing equalisations). The cost of this pre-amp was higher than the previous model, being £8/8/- compared to £6/15/-.

The RC/PA/U pre-amp had the same appearance as the previous RC/PA, except that: the selector switch included three equalisations for playing records, the positions of the bass and treble controls were interchanged and a pilot lamp was included.

Vari-Slope

The original Leak "Vari-Slope" pre-amplifier was designed to enable the high-frequency distortion from inadequate disc reproduction of the day to be filtered out – with minimal effect on the quality of the sound. The Vari-Slope featured a low-pass filter circuit with variable turnover frequencies and variable cut-off slopes from approximately 5dB to 50dB per octave. The circuit was novel as it was implemented solely with resistors and capacitors within a negative feedback circuit. The Leak literature states:

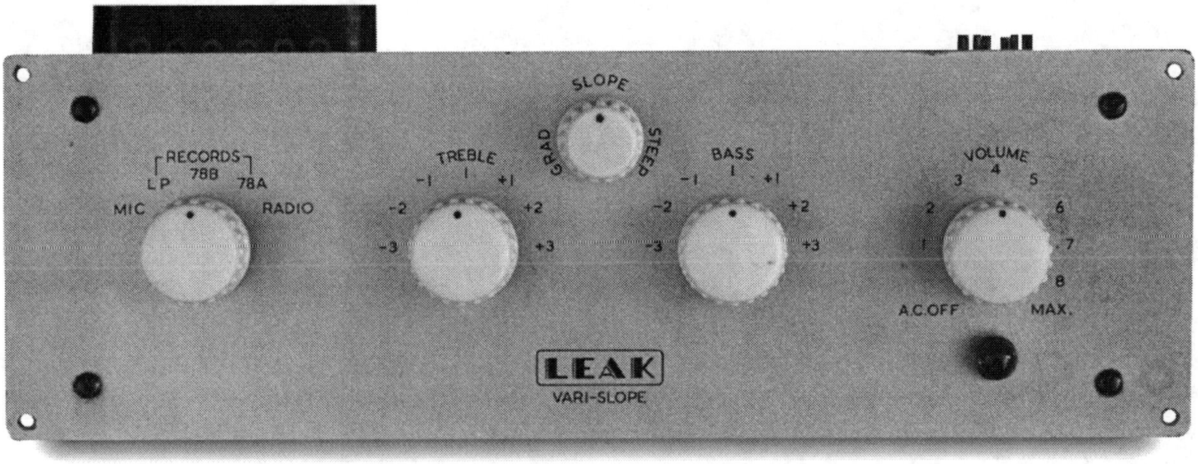

The original Varislope pre-amplifier. Some units were finished in bronze, and later ones in metallic gold.

> "The new 'Vari-Slope' pre-amplifier has a refinement which will doubtless set the pattern for future high-fidelity reproducing amplifiers. This advance consists of a variable-slope 'electronic' low-pass filter operating on negative voltage feedback principles. No inductors ('chokes') are used and all their disadvantages are completely eliminated." *

The Vari-Slope circuit is based on a twin-T R/C circuit, which was discussed in the April 1945 issue of the Journal of the Acoustical Society of America (Vol. 16, No. 4, pp 275-279). A similar circuit with fixed slope was also used by Williamson in the pre-amplifier designed for operation with the famous Williamson power amplifier (Wireless World, Oct/Nov 1949). Shown here is the basic circuit of the Vari-Slope filter, taken from the Vari-Slope Stereo pre-amplifier which was issued in 1958, however the basic operating principle was similar in other Vari-Slope circuits. In the original 1952 model Vari-Slope, there was a single ECC81 dual triode, but in later models there were two EF86 valves in each channel.

The Leak circuit uses a twin-T resistor-capacitor network in the forward path of the pre-amplifier's second gain stage. The twin-T circuit on its own is a nulling circuit, giving nominally zero output at the null-frequency. For instance the null-frequency chosen for a pre-amplifier might be 16kHz – the slope of the attenuation curve above and below the null frequency is not very high, however the Q of the overall circuit may be increased by including the circuit within a feedback loop. Thus a steep slope of attenuation between say, 8kHz and 16kHz can be achieved with little or no loss below 8kHz. Above the null-frequency the loss decreases, but simple capacitive loading can reduce any residual output here.

* Extract reproduced by permission Leak Electronics Ltd.

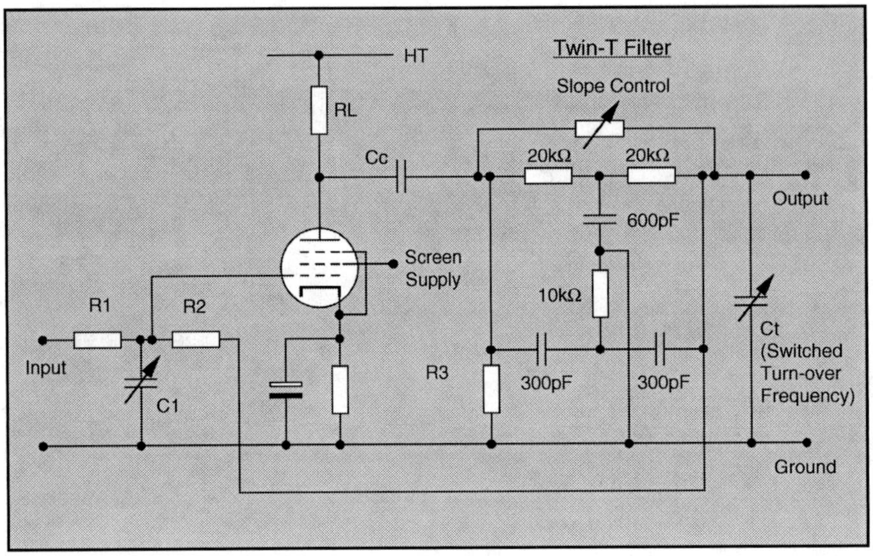

The Leak Vari-Slope circuit was based on a twin-T configuration of resistors and capacitors that gave a sharp null near the high end of the audio band. The inclusion of the twin-T in a feedback stage enabled steep roll-off rates to be achieved. The Q of the circuit, and hence the roll-off rate, could be varied with a single resistor.

To provide a variable slope characteristic, a variable resistor is placed in parallel with the twin-T circuit. When this resistor is open circuit the filter has the highest Q and steepest slope. To provide different turn-over frequencies a variable capacitive load is placed on the output and for the lowest turnover frequency an additional capacitance C1 is switched into circuit (unlike the Williamson circuit in which the actual capacitors in the twin-T were switched in value). In later circuits the resistors R1 and R2 were replaced with the famous Baxandall tone control circuit (developed by Peter Baxandall), thus giving the ability to have treble boost simultaneously

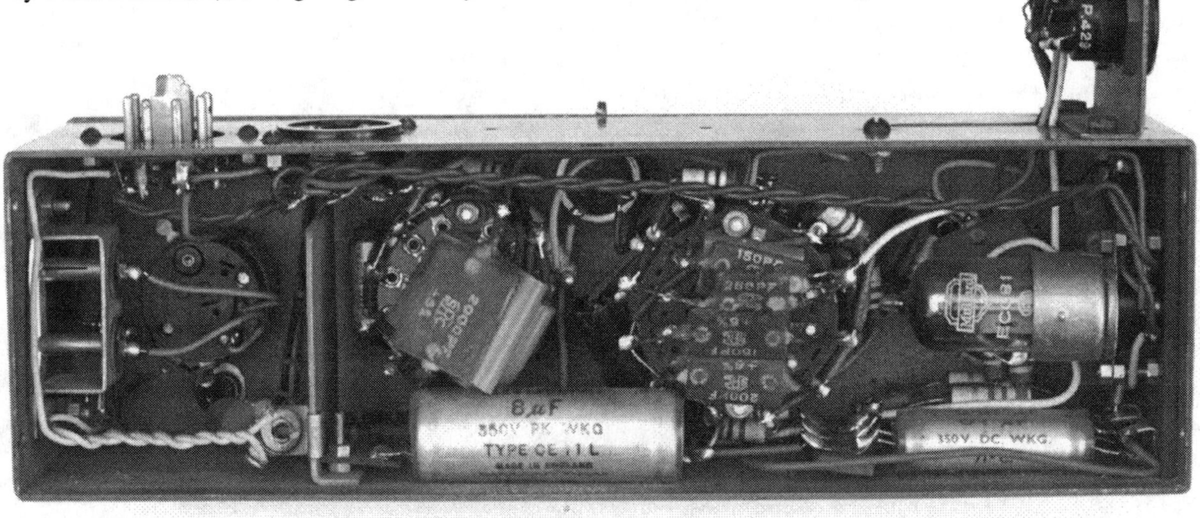

Inside the Vari-Slope pre-amp: some components are fitted to a terminal strip but the majority are attached to the control switches. The single ECC81 valve is on the right of the photo. This particular pre-amp is fitted with the optional input connectors (Bulgin 3-pin) that were used with the portable TL/12 set.

with high-frequency filtering. (Ironically the solid state Leak amplifiers of the late 1960's had a high-frequency filter implemented with inductors and capacitors. However, by the 1960's distortion from records and pickups had been reduced, so less flexible filtering was acceptable.)

The front panel appearance of the original Vari-Slope is unusual as the slope control is placed in an unorthodox position, above the central line of the four other controls. This position allowed minimal alteration to the design of the pre-amp chassis. It also allowed the pre-amplifier to retain the same width as the side-dimension of the TL/12 power amplifier so it could be coupled via four screws to the TL/12, and then assembled with a cover into a single portable unit (as with the RC/PA and RC/PA/U pre-amps). The cost of the pre-amp alone was £12/12/-.

The standard input connector on the RC/PA, RC/PA/U, and the Vari-Slope pre-amps was a screw terminal block.

Steep Cutting Filter

One of the options available with the TL/12 in 1951 was a "steep slope" low-pass filter that was housed in a mu-metal case, and featured cut-off frequencies of 10kHz, 8kHz and 6kHz. When the Vari-Slope pre-amp was introduced in 1952 with an improved filter circuit, Leak offered a service for owners to have the RC/PA/U model re-built to incorporate the Vari-Slope filter circuit, at a cost of £8/8/-.

"The switched disc equalisation was a valuable feature in the early 1950's, particularly on 78's, which varied enormously. The sound of some 78's could be transformed significantly by actually ignoring the equalisation markings on the controls and selecting by ear. The theoretical equalisation was often completely subverted by the choices of various people in the process of recording and mastering, particularly where labels covered a multitude of sources for recorded material."

Peter Stinson,
Audio Enthusiast, Australia.

The Leak "Steep Cutting Filter" was aimed at owners of previous model pre-amps that did not feature the Varislope circuit.

But a lower-cost cost option was maintained for owners of previous Leak pre-amps, where a "filter-only" could be interposed in the cable between the RC/PA/U and the TL/12. The product was called a "Steep Cutting Filter" and had three cut-off frequencies of 10kHz, 8kHz and 6kHz, with variable slope attenuation. The product was first advertised in Wireless World in the October 1953 issue – the price was £5/10/- compared to £12/12/- for a complete Vari-Slope pre-amp.

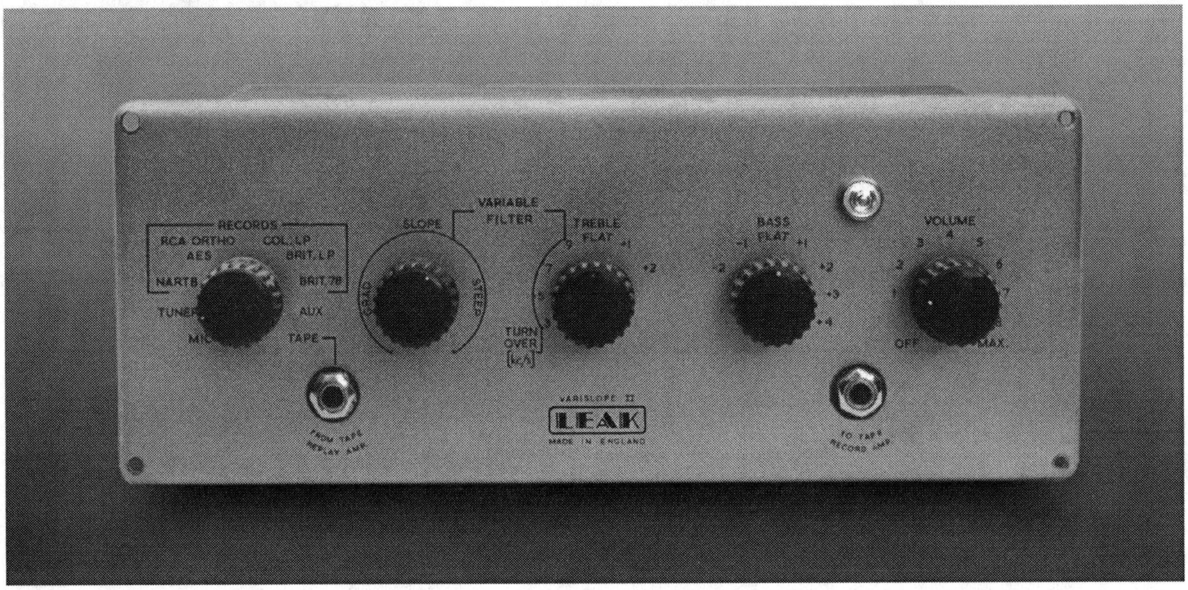

Varislope II

The Varislope II pre-amp has six different equalisation settings for replay of 78 rpm and LP records.

This pre-amp was introduced in 1954, and features higher input sensitivities than previous pre-amps, with improved appearance. Two EF86 valves, both operating in pentode mode, are used to obtain the higher sensitivity and each stage has similar shunt feedback circuits to that used in the previous RC/PA/U.

There are six disc equalisations provided (the American distributors specified this high degree of flexibility): BRIT-78, BRIT-LP, COL-LP, RCA-Ortho, AES and NARTB. Tuner, Tape, Mic and Auxiliary inputs are also provided.

Bass and treble controls are adjusted by switches for boost and cut, with the variable slope facility available only for treble-cut (later model pre-amps allowed treble boost and cut simultaneously with variable slope filtering).

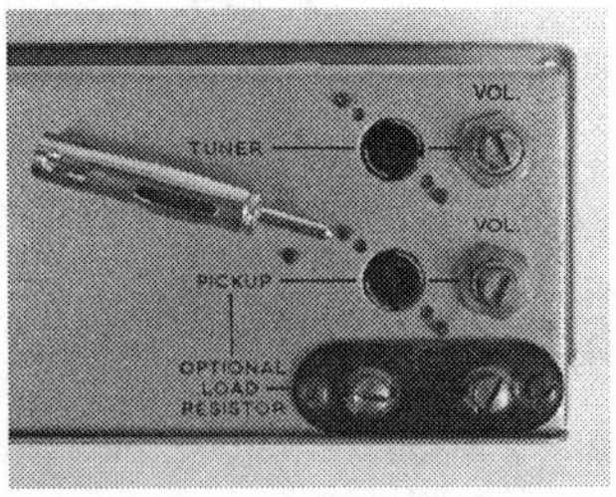

The input connectors on the Varislope II and Point One pre-amps used an unusual connector similar to that used for car-radio aerial connectors. In the case of the Point One pre-amp there was provision for an external load resistor for the pick-up input (as shown here).

Inside the top of the Varislope II. Many components are fixed to the switches, and others are located on the underside of the terminal board.

The physical appearance of the pre-amp was designed in-house at Leak, with the steel front panel being finished in gold enamel. The wide range of inputs and extensive tone controls would have made the Varislope II one of the best and most flexible pre-amps available in its day. These features came at some cost though, as the price had risen around 30% to £16/6/-.

Point One

This 1954 pre-amplifier was the accompanying unit for the TL/10 power amplifier, although it could be used with the TL/12 if required. It featured the new "Baxandall" feedback tone control circuit published in the October 1952 edition of Wireless World, and so has continuous control of boost and cut, instead of stepped switches. Variable slope high frequency filtering is not provided on this cheaper model.

There are two gain stages in this pre-amp, each with shunt feedback. The first stage performs the required equalisation, and the second stage performs tone control functions. This unit has two EF86 pentodes, but the second stage is configured as a triode. The sensitivity of the pickup input was 14mV, which was satisfactory at the time of its production, but would have been inadequate for reproduction from the high-quality low-output pickups that would appear in later years. Four equalisation

The Leak "Point One" pre-amp. Some units were fitted with black knobs, while late in the production knobs of the same style as those on the Varislope III pre-amp were used.

Interior layout of the Point One pre-amp – a second EF86 valve is located beneath the top one.

settings are available for disc (AES/RCA-Ortho, Col-LP/FFRR-LP, NARTB/HMV-LP, and BRIT-78), plus Tape and Tuner inputs. The steel front panel was finished in gold enamel and the cost was £10/10/-, representing a considerable saving compared to the Varislope II pre-amp.

The Baxandall Tone Control

The "Baxandall" negative feedback tone control circuit is the most famous and popular circuit for providing the user a means to adjust the frequency response of an amplifier. It has been adopted by nearly every amplifier manufacturer at some stage, either in its original form or with slight modifications.

The circuit was developed by Peter Baxandall and published in Wireless World in October 1952. Desirable features of the circuit are: independent and continuous variation of bass and treble without switches; boost and cut available from the one control without the need for a change-over switch; and flat frequency response is obtained with the controls at mid-position.

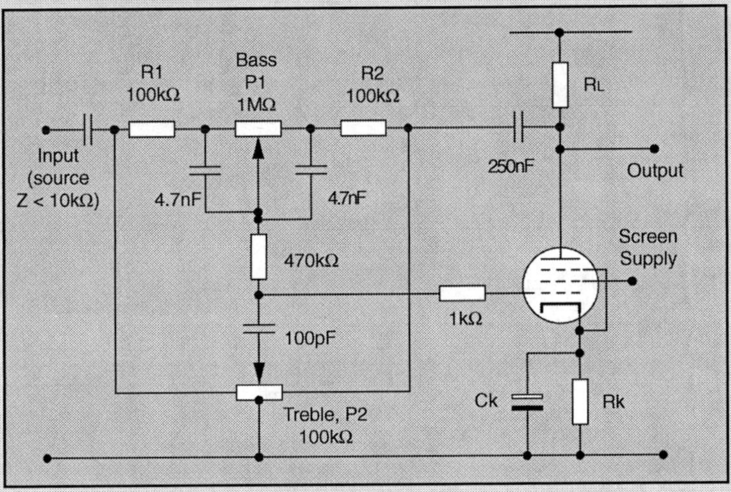

The circuit shown is based on the original published in 1952. The overall topology applies shunt feedback by sensing the output voltage and summing the input current with the feedback current at the virtual earth formed at the grid of the valve. A desirable effect of this is a lowered output resistance. One immediately noticeable feature of the circuit is the symmetry of the design, from which is can be correctly assumed that the frequency responses obtained with boost and cut closely mirror each other. The Baxandall design will give around 20dB of boost or cut (this being set by the ratio of P1 to R1 and R2) and turnover frequencies being set by the choice of various capacitor and resistor values.

One common variation often adopted is to omit the tap on the treble potentiometer. Baxandall investigated the characteristics of this option and found that the treble boost and cut curves differ slightly by levelling off at the frequency extremes, compared to a response that continues to rise or fall at the frequency extremes if the tap is included.

Leak first adopted a Baxandall-style tone control circuit in the Point One pre-amp released in 1954 and continued its use in subsequent models.

"An important characteristic of the Baxandall circuit is the sliding turnover point, which differentiates it from the fixed hinge point of conventional systems. However, Peter Walker would say, and I would agree with him, that while the Baxandall characteristic is almost ideal in the bass, it is not so well suited to treble, particularly treble boost, where it can result in excessive boost at the extreme top when only a little mid-top lift is wanted."

Peter Stinson,
Audio Enthusiast, Australia.

Varislope III

This was the first Leak pre-amp where the appearance was designed by an external industrial design house (Richard Lonsdale-Hands Associates). The pre-amp has a diakon plastic front panel with a mild three-dimensional in the lettering and graphic design. Originally the front panel had a rectangular outline with rounded corners, but later units had the left and right hand ends rounded with a large radius curve (as also seen on the Point One Stereo and Varislope Stereo front panels).

For the first time, two pickup inputs were provided on a Leak pre-amp each with adjustable sensitivities. Four different equalisation curves (RIAA, LP, NARTB and 78E) were included for reproduction from records. Two EF86 valves were used, with a Baxandall-style tone control circuit, incorporating the Varislope twin-T circuit.

During the production run the circuitry was revised to increase the sensitivity of the pickup input. Initially the second stage EF86 was configured as a triode and the sensitivity was 9.5mV. Around the time Leak released stereo pre-amps (with improved sensitivity) the second stage EF86 in the Varislope III was altered to operate as a pentode for additional gain and the feedback modified (this revision was designated "Series N" by Leak).

Introduced around May 1957, at a cost of £15/15/-, and continuing in production until 1963, it was replaced by the Varislope Mono in 1964.

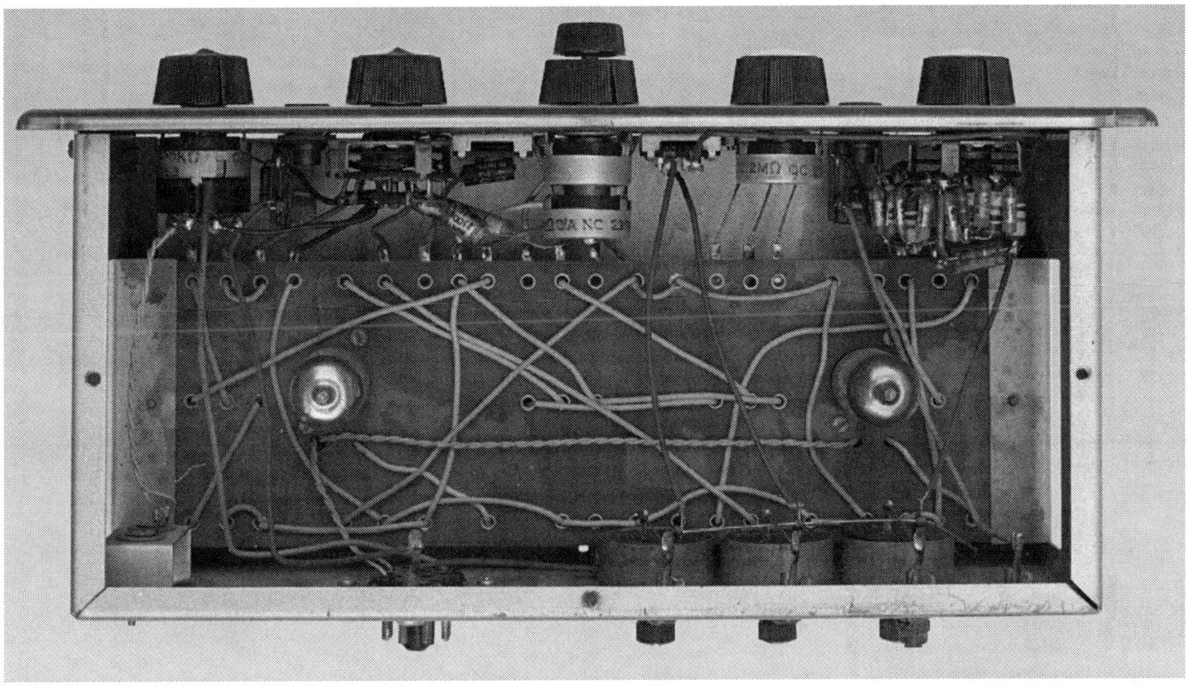

The chassis layout of the Varislope III pre-amplifier showing two EF86 valves. The input stage is on the right hand side and potentiometers on each input (shown at the bottom) allow the levels to be equalised between various sources.

With the Varislope III pre-amp, construction technique became closely aligned with the practices used on Leak power amplifiers. Unusually, this unit has Hunts coupling capacitors instead of TCC.

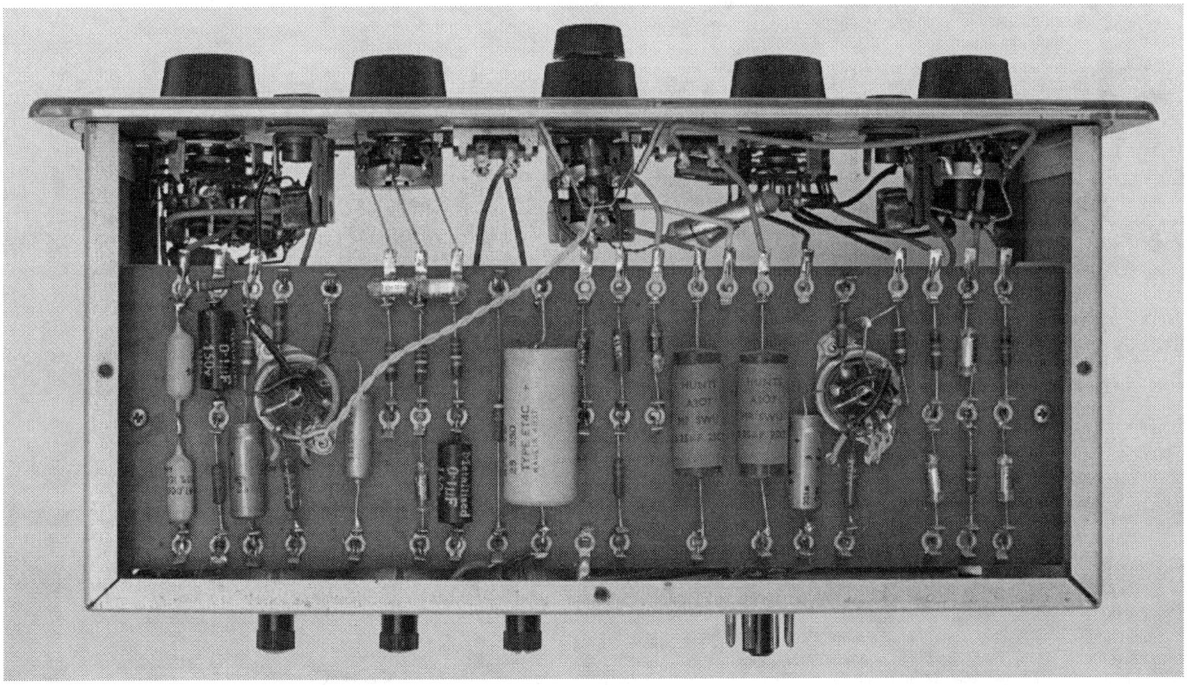

The later version of the Point One Plus pre-amp with curved-ends on the front panel and cream knobs. Early versions followed the appearance of the Varislope III pre-amp.

Point One Plus

The circuit of the Point One Plus was almost identical to the original Varislope III, except that the slope control was omitted from the high-frequency filter. The user could select one of three turnover frequencies (9kHz, 6kHz and 4kHz), or by-pass the filter, but the slope was pre-set for a steep rate of attenuation above the turnover frequency. Equalisation for RIAA, NARTB, LP and 78OE records was provided. This pre-amp was released early in 1958, with the first advertisements appearing around April. The cost was £12/12/-.

The American version of the Point One Stereo introduced in 1961. Various coloured decor-inserts were available, which could also be removed completely if required.

Point One Stereo

This model was the first Leak stereo pre-amp, and was announced at the 1958 London Audio Fair, however full production did not start until late 1958. As the model name implies, the overall functionality was similar to the Point One pre-amp, except that this was stereo.

While the overall architecture of this stereo pre-amp followed previous designs (two-stages, comprising EF86 pentodes with feedback), circuit revisions increased the sensitivity for pickup inputs to 5mV. The additional gain was achieved by raising the anode load resistor on the first stage, and increasing the impedance of the Baxandall tone control circuit.

The first stereo pre-amp from Leak was the "Point One Stereo" released in 1958.

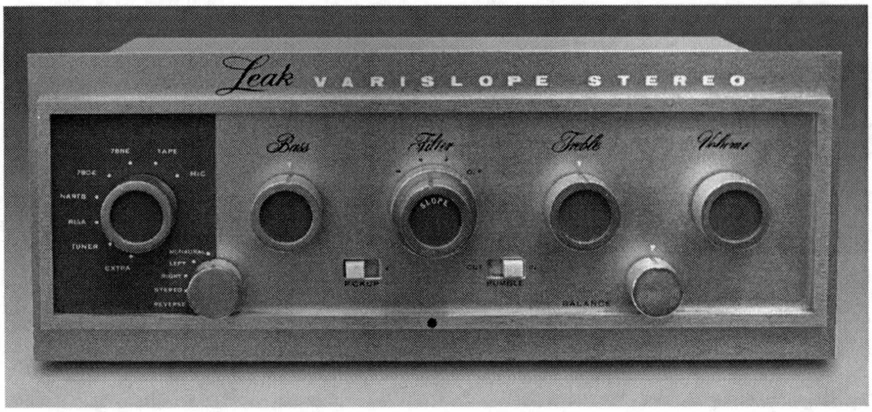

Like the Point One Stereo pre-amp, the Varislope Stereo was also produced with a gold-anodised front panel for the USA. The unit shown here is fitted with the optional decor inserts which are placed in the middle of the knobs and behind the input selector knob on the left hand side.

The normal front panel design followed the brown/gold diakon plastic of the Varislope III, but as with the Varislope Stereo pre-amp a specially designed front panel featured on pre-amps sold in the American market after 1961 (see also "Varislope Stereo"). The cost of the pre-amp was £21/0/- and it was discontinued in 1963.

Varislope Stereo

This pre-amp is effectively a stereo version of the Varislope III, and was issued in mid-1960 with the same gold and brown livery as the later versions of the Varislope III.

The circuit uses two EF86 valves in each channel, with facilities similar to the Varislope III. In this pre-amplifier, Leak had to obtain very high gain from the pair of EF86 valves in each channel – this was the first Leak pre-amp to have a pickup sensitivity of 3.5mV, making it useable with all the low-output pickups at the time, and also with most pickups that followed in later years. The cost was £25/0/-.

In 1961 the American agents for Leak requested a special gold finished front panel for the American market (see also "Industrial Design"). The panel was made from

Advertising graphics used by Leak. (Reproduced by permission Leak Electronics Ltd. Source: HFN Aug 1960)

The Varislope Stereo pre-amp was introduced in 1960.

The classic design lines of the Leak Varislope Stereo are a striking feature when seen from close-up.

3mm thick aluminium, and had a square-section escutcheon (attached from behind) that formed a visual frame for the controls. Different coloured inserts were supplied for the knob-centres, and for a backing plate around the selector switch, making it possible to match the pre-amp to an existing decor.

One manufacturing variation for this model was the production of 30 special units fitted with dual-concentric potentiometers for the bass, treble and gain controls. These were delivered fitted into Southdown cabinets and were made at the request of Angus McKenzie, a well known figure in the audio industry who often reviewed equipment for the hi-fi press.

Stereo Adaptor

In early 1959 Leak announced a Stereo Adaptor that had a ganged volume control and a function switch, and was designed for use with two mono power amplifiers (and presumably two mono pre-amplifiers). The cost was £5/10/-.

The adaptor was produced at the request of the American distributors, but three or four years later nearly all of them were still gathering dust in the factory. Phil Biggs salvaged a few of the parts, and the rest was disposed of.

A Varislope 2 Stereo pre-amp with its original packing case.

Varislope 2 Stereo and Varislope Mono

With the introduction of Leak transistor amplifiers, a revised black-and-silver "Leica" appearance was designed by Noel Haring Associates. Late in 1963 the new look was adapted for the mono and stereo valve pre-amps, and some minor circuit revisions were made. The Varislope filter was reduced in complexity to a choice of just two slopes (steep and gradual), but 'Tape Monitor' facilities were added. These design revisions led to the release of the Varislope 2 Stereo and the Varislope Mono pre-amps.

In common with other Leak pre-amplifiers the signal wiring between the input connectors and the input selector switch was un-shielded. Normally this can result in crosstalk, particularly from a high-level input such as 'Radio' to a low-level input such as 'Pickup'. To eliminate crosstalk, Leak used a shorting ring on the input selector switch that connected high-level inputs to ground when they were not in use. This approach made assembly time quicker as individual earth connections for each signal shield were not required.

A mono version – the Varislope Mono – was also available, although these are less common as most people were purchasing stereo sound systems by the early and mid 1960's. A rather minor point relating to these models is that the front panel screen-printing has a reputation for wearing off with use, so restoration to original appearance may require the front panel lettering to be re-printed.

The cost for the Varislope 2 Stereo was £25/0/-.

The Varislope 2 Stereo had a two-position slope switch, and was the last stereo valve pre-amp made by Leak, being discontinued in 1967. This particular pre-amp has had the front panel screen printing restored in a slightly different form than the original printing.

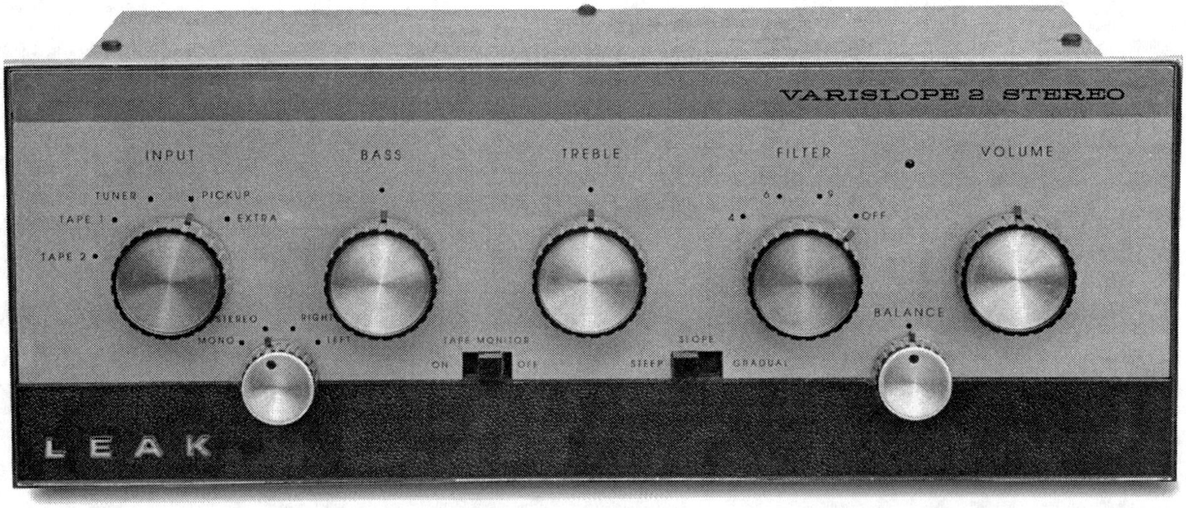

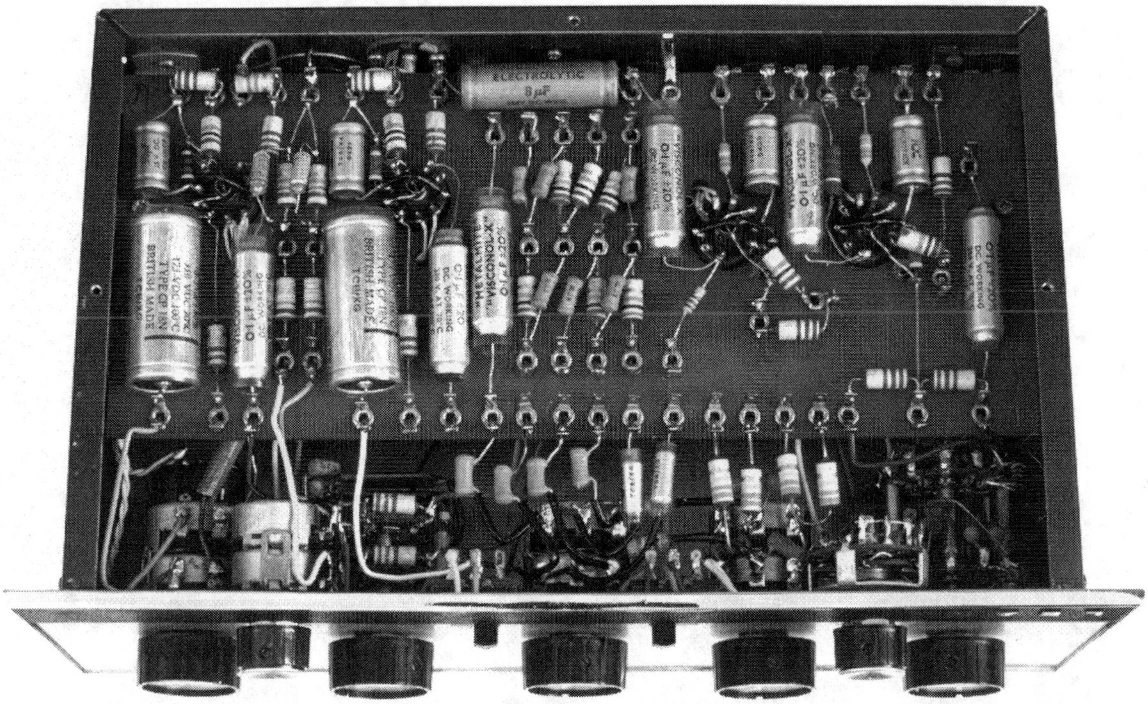

Most of the Varislope 2 Stereo pre-amplifier circuitry was mounted on a main component board which was hard-wired on the other side, but some components were mounted on the input and filter switch terminals.

Four EF86 valves are seen inside the Varislope 2 Stereo pre-amp, with the first stages on the left side. Input attenuators can been seen at the rear, and the twisted-pair wiring for switching the AC supply to the power amplifier is at the right.

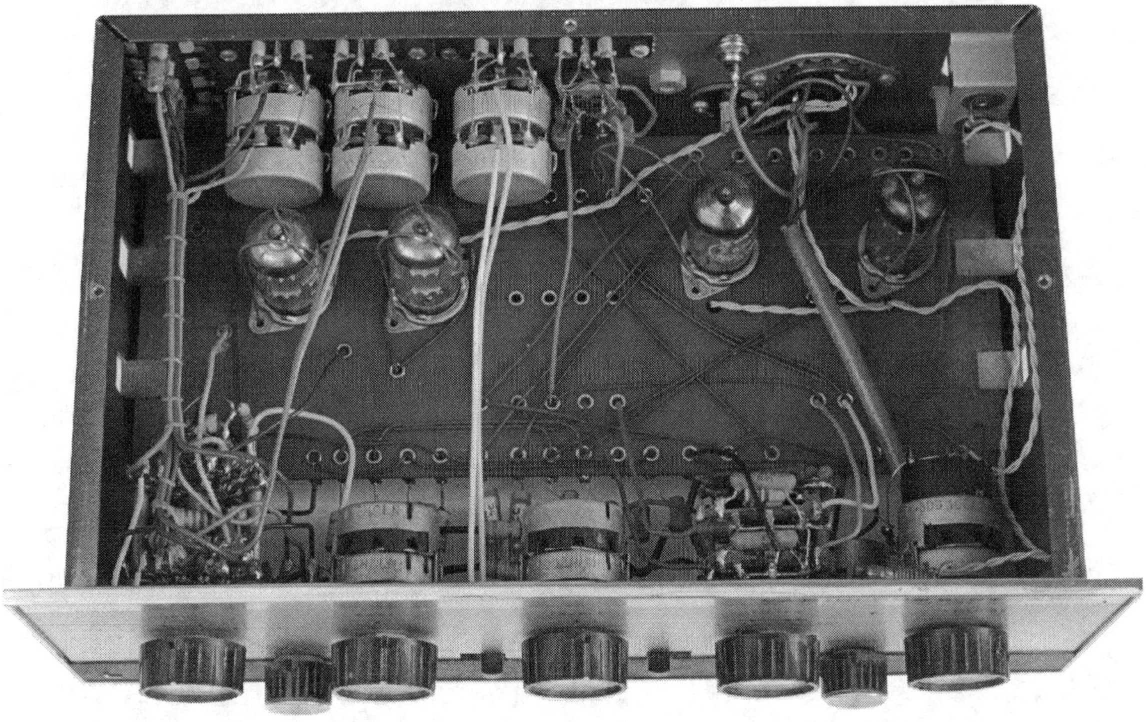

EL34
STEREO 50
LEAK
MADE IN ENGLAND

8. Power Amplifiers

Evolution and Design

In one form or another, the TL/12 circuit topology developed in 1947/48 was used for nearly 20 years in all Leak valve power amplifiers. The topology can be summarised briefly as: a common cathode first stage for voltage gain; a long-tailed pair double triode phase splitter, and push-pull cathode bias output stage, with feedback of about 26dB inclusive of the output transformer. It was really the new developments in output stage valves, high gain double triodes (12AX7), and the introduction of the ultra-linear output stage that enabled a string of power amplifier models to be made by Leak, all based on the original TL/12 design from 1948. The fact that the TL/12 circuit was the mainstay of Leak power amplifiers for nearly 20 years testifies to its design excellence.

Model	Power Output	Output Stage	Years
TL/12	12 Watts	KT66 Triode	1949 to 1957
TL/25	25 Watts	KT66 Tetrode	1949 to 1951
TL/25A	25 Watts	KT66 Tetrode	1951 to 1957
TL/10	10 Watts	KT61 Ultra-linear	1954 to 1957
TL/12 Plus	12 Watts	EL84 Ultra-linear	1957 to 1967
TL/25 Plus	25 Watts	EL34 Ultra-linear	1957 to 1967
TL/50 Plus	50 Watts	KT88 Ultra-linear	1957 to 1968
Stereo 20	10 Watts/channel	EL84 Ultra-linear	1957 to 1967
Stereo 50	25 Watts/channel	EL34 Ultra-linear	1957 to 1963
Stereo 60	30 Watts/channel	EL34 Ultra-linear	1963 to 1967

A tabulation of Leak valve power amplifiers from 1949 to 1968.

First Stage

The first stage is always a common-cathode triode or pentode. The input is applied to the grid and the feedback from the output stage is returned to the cathode by injecting it across a 100Ω resistor, which had a small capacitor in parallel with it to reduce radio-frequency injection into the feedback loop (from loudspeaker leads acting as an antenna). The anode load was sometimes purely resistive, but in some models a resistor-capacitor network in parallel with the first stage anode resistor was used to improve the stability of the amplifier. These networks are commonly

Opposite: The Leak Stereo 50 power amplifier fitted here with EL34 valves, although KT66 or 5881 valves could be used if required.

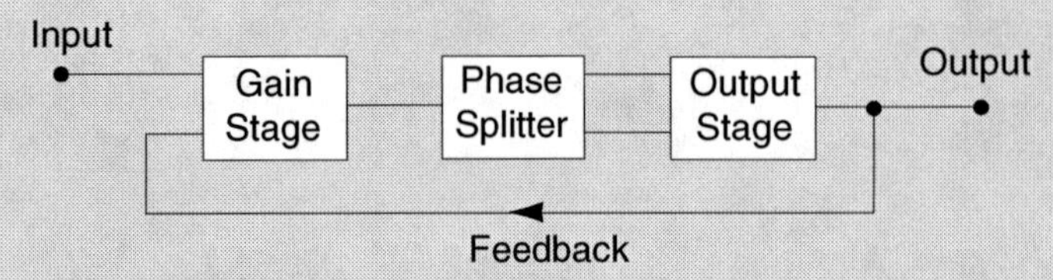

Amplifier Model	Gain Stage	Phase Splitter	Push Pull Output Stage
TL/12	EF37 Pentode	ECC33 Double Triode Long Tailed Pair	KT66 Triode
TL/25, TL/25A	EF37 Pentode	" "	KT66 Tetrode
TL/10	EF86 Pentode	" "	KT61 Ultra-Linear
TL/12 Plus TL/25 Plus, TL/50 Plus	EF86 Pentode	ECC81 Double Triode Long Tailed Pair	Ultra-Linear
Stereo 20 Stereo 50, Stereo 60	ECC83 Triode	ECC83 Double Triode Long Tailed Pair	Ultra-Linear

Leak power amplifiers from the TL/12 onwards all used a similar circuit topology, as shown in this diagram. The table summarises the differences between models. All the power amplifiers listed had 26dB negative feedback, except for the TL/50 Plus which had 23dB negative feedback.

used in many amplifier designs because the stability of the amplifier can be improved by altering the gain and phase characteristics. However, as such networks reduce the gain of the amplifier at high frequencies, feedback is reduced and there is a resultant increase in high frequency distortion.

The first stage is capacitively coupled to the second stage. Whilst it may have been possible to arrange the DC voltage between these two stages so that direct coupling would be possible, the DC voltages varied too much, with and without the pre-amplifier connected, to allow this.

Second Stage

The second stage also provides gain and acts as a phase splitter. A double triode valve is configured as a long-tailed-pair to obtain the phase splitting function, with two output signals of equal magnitude but of opposite phase.

To obtain a good amplitude balance between output signals, the "tail" resistor from the coupled-cathodes to ground needs to be much higher in value than the impedance looking up into the cathodes of the valve (which is typically is a few thousand ohms). If we assume this impedance is 2kΩ, then a tail resistor of 20kΩ will give an imbalance of about 10% in the output signals (this is because some of the signal current flowing out of the first triode's cathode flows into the tail resistor instead of into the cathode of the other triode. The proportion lost into the tail resistor is about 2/20'ths in this case, giving a 10% imbalance). To offset this problem, the anode resistor of the second triode in the pair is raised in value by 10%, thus giving satisfactory balance in the drive to the output stage.

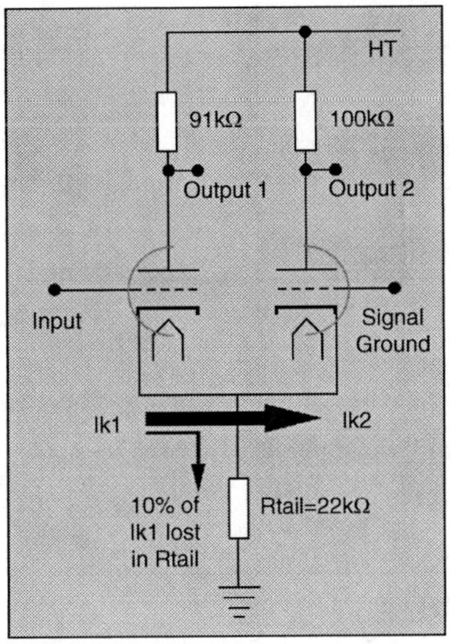

In the Leak phase splitter a long-tailed-pair is used, but as the tail-resistor is not high enough in value for perfect balance in the output signals, the second triode's load resistor is made 10% higher, thus giving adequate signal balance (ie: so Output1 equals Output2, but is out of phase by 180 degrees).

Output Stage

The output stage in the original TL/12 circuit was push-pull and triode connected but later models used the ultra-linear arrangement. Individual cathode resistors made the amplifier tolerant of un-matched valves, as the bias tended to automatically adjust for variations in valve characteristics and act to minimise any DC imbalance in the output transformer.

However individual cathode resistors require a large capacitor to by-pass the cathode. At low frequencies this capacitor becomes ineffective and so the gain of the output stage reduces. Typically this occurs at around 30Hz, with the gain reducing at still lower frequencies until a limit is reached (usually around 3Hz) where the gain levels off to a value equal to the DC gain. This "step" in the low frequency response has a beneficial effect, making the amplifier easier to stabilise when the feedback is applied (in the example given, it introduces little phase shift at frequencies below 3Hz, with some mild phase shift between 3Hz and 30Hz, and asymptotes to zero phase shift above 30Hz). Other components in the amplifier, such as the output transformer and coupling capacitors, start to introduce phase shift for frequencies below around 5Hz, so it is beneficial to reduce the gain of the amplifier before

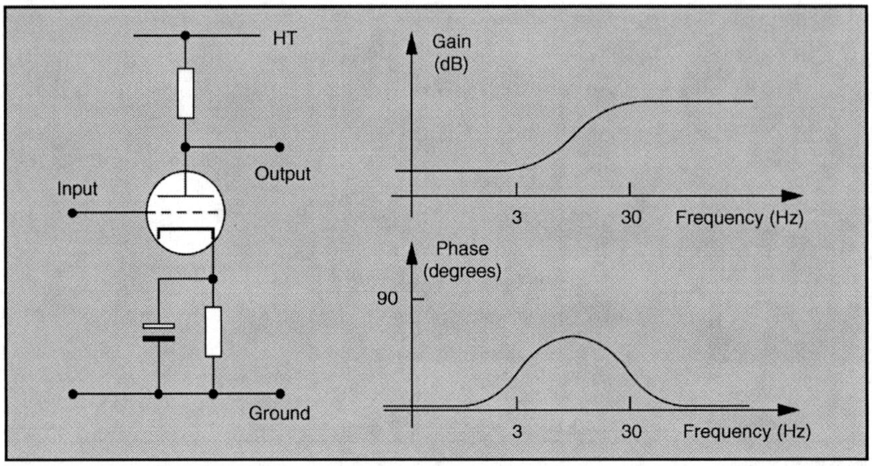

The Leak output stage uses individual cathode bias resistors with by-pass capacitors. This introduces a "step" in the low frequency response that reduces the gain at low frequencies, thus improving the stability of the amplifier because the feedback is reduced before other phase shifts from coupling capacitors and the output transformer accumulate below 3Hz. (Similar "step" networks are used by amplifier designers for high frequency compensation too).

these other phase shifts can affect stability. A drawback from the reduction in gain (and hence feedback) is a rise in distortion from 3Hz to 30Hz, although most music does not demand much power output here so the increased distortion is usually of little practical significance.

Of course, when the feedback is applied, the "step" in frequency response and the attendant phase shift (either from the output stage cathodes, or from the first stage compensation network) is not apparent as the feedback corrects for this, giving the amplifier a flat frequency response from 20Hz to 20kHz.

Development

Harold Leak was involved in the introduction of all these power amplifier models. Other engineers who worked on development of these products were: Ted Ashley, John Bolingbroke (prior to 1955) and Ray Whitcombe (after 1960). During the later part of the 1950's Harold's contribution at the detailed design level lessened but he continued to set overall design goals and any constraints within which designers had to work.

Lund Brothers in South-West London (near Balham) supplied all the chassis for Leak power amplifiers. Quotations from other suppliers were obtained but Lund Brothers usually provided the best products at a satisfactory price.

Customised versions of most power amplifiers were produced over the years for many specialised purposes. A common modification was the fitting of 100 Volt line output transformers for companies such as Marconi and Planned Music.

Ray Whitcombe recalls that Harold Leak loved to talk to customers, whether at the Brunel Road factory, or at Audio Fairs as shown in this promotional photo. The logos for British Industries Corporation on the wall, and the Garrard record changers in the background, suggest that the location is at an Audio Fair in the USA. A Stereo 50 power-amplifier is just visible behind the customer, and Harold Leak is holding an EL34 power pentode in his hand. (Photo courtesy of the Leak family.)

LEAK
AMPLIFIERS AND
PREAMPLIFIERS
PROFESSIONAL COMPONENTS AND WORKMANSHIP
IN AN AMPLIFIER FOR THE HOME
STEREO AND MONAURAL MODELS
"POINT ONE STEREO"
STEREO AND MONAURAL SYSTEMS
quiet!
it's
RC88 DELUXE

TL/25 and TL/25A

The TL/25A had a massive transformer, and a KT61 "Partridge control valve" to regulate the screen supply for the tetrode output stage. In this photo, the KT61 is just visible behind the right-hand KT66.

When the TL/12 was developed, it was recognised that a higher power output could be obtained by operating the output stage in tetrode mode: with the same voltage on the anodes, about double the power output could be obtained from the KT66's. An amplifier with 25 Watts output and with a similar distortion rating to the TL/12 would find application in commercial sound systems (e.g. cinemas), so Leak developed a 25 Watt version of the TL/12. Initially this amplifier was designated the TL/25 and was released in December 1948 with the TL/12. As the output stage was operating in tetrode mode, it was important to have a reasonably stable screen supply. To achieve this, the TL/25 used a gas regulator valve to derive a DC supply of about 300 Volts for the screen grids (by dropping about 130 Volts across an S130 gas regulator which was fed from the main HT of 440 Volts). The amplifier used the same chassis and driver circuit as the TL/12 with the exception of a lower gain input stage.

Later on, a revised version of the TL/25 – the TL/25A – was introduced. In this amplifier the screen regulation was performed by a normal "hard valve" configured as a DC cathode follower (a KT61). This configuration is also known as a "Partridge control valve". The regulating valve requires a separate heater winding as its cathode is operating at around 300 Volts.

The TL/25A also had a larger output transformer than the TL/25, and so the filtering choke was moved closer to the output valves. Rather than have special chassis made for this variation, the TL/12 chassis was retained, and additional cut-outs for the filter choke terminals were done by hand. This was acceptable as production numbers for this amplifier were low. Leak also used a tertiary feedback winding on the TL/25A output transformer. This is common practice, particularly with amplifiers operating over a wide range of load conditions, as it helps to improve the stability of the circuit.

As the TL/25A was not really aimed at the domestic market it sold in much fewer numbers. It is very rare today and usually in very poor physical condition as a result of being in a harsh commercial environment. (Six of these amplifiers were installed in the Princess Theatre, Melbourne, Australia for the stage show Kismet in 1953.)

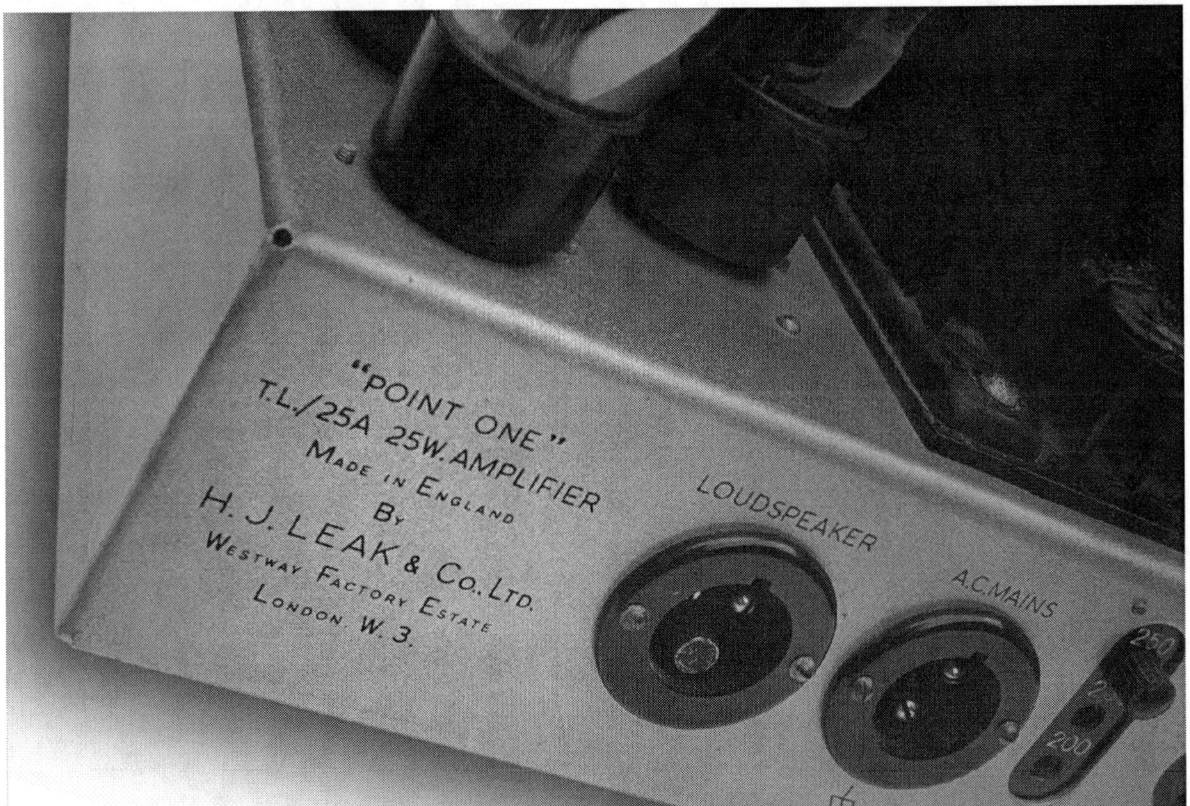

This end-on view of a TL/25A amplifier shows the base of the GZ32 rectifier and KT61 screen regulator, the screen printing, and a modified speaker outlet.

TL/10

The TL/10 was introduced in 1954 as a lower cost amplifier, with fewer parts and lower production costs, but maintained the distortion rating of 0.1%. Examining the amplifier, it is apparent that Leak spent some time assessing which parts of the TL/12 could be dispensed with, without sacrificing performance significantly. For instance, expensive connector fittings were dispensed with and replaced with screw-terminal connections for A.C. power, A.C. switch, and loudspeaker connectors. However the output transformer used the same sized core as that on the TL/12. The aim here was to bring "Point One" hi-fi performance to a wider audience. The TL/10 amplifier was sold at about two-thirds of the cost of the TL/12 and it was very successful commercially. During the time the TL/10 and TL/12 were available on the market simultaneously, the TL/10 model out-sold the TL/12 by a ratio of three to one.

An overhead view of the TL/10 shows the chassis layout, and the electrolytic capacitor pack made especially for the TL/10.

The power rating relied to some extent on so called "music power", so the amplifier could not quite deliver 10 Watts continuous average (distortion was rated at 0.1% for 7.5 Watts output). From this point of view the design was a compromise but in practice the audible difference between 7.5 Watts and 10 Watts was not significant and many consider the TL/10 to be much under-rated these days.

This amplifier was the first Leak amplifier to incorporate the "ultra-linear" output stage configuration (patented by A. D. Blumlein of EMI Ltd in 1937 and popularised by Hafler and Keroes from the USA in 1951). The output stage uses KT61 kinkless tetrodes (introduced by Osram in 1939), although the American 6L6 could be used if minor circuit modifications were made. (See Chapter 12).

Some Leak circuit diagrams for this amplifier show that it used a single common cathode resistor in the output stage (all other Leak valve circuits used separate cathode resistors for each output valve). However, the amplifiers with a single cathode resistor appear to be uncommon as most TL/10's encountered have separate cathode resistors that are individually by-passed.

Despite its lower cost, the TL/10 was impressively constructed. This particular amplifier is a fine example with completely original components. Note the large 1kΩ power supply resistor at the bottom right - it was used to filter ripple in place of a choke. (Photo and amplifier courtesy of Richard White, UK.)

An interesting feature of the amplifier is the electrolytic capacitor pack made especially for the TL/10. This enables the amplifier to retain its good appearance. Presumably Leak anticipated that sales would be high enough to offset this addition to the overall cost.

The output transformer on the TL/10 was supplied by Hinchley Engineering of Devizes, Wiltshire, signalling Leak's move away from Haddon Transformers (as used on all TL/12's). Once again cost considerations were the driving force here, although it would not have been possible to sacrifice quality, as the level of feedback used made high performance from the output transformer mandatory.

One of the engineers who contributed to the development of the TL/10 was John Bolingbroke. After the design was finalised he also did some of the pre-production work and laid out the design of the main component board in the amplifier.

To accompany the TL/10 a new pre-amplifier was released, called the "Point One". Sold together, the TL/10 with its pre-amp cost just £27.

TL/12 Plus

The TL/12 Plus came about primarily from new output valve developments from Mullard. When the EL84 output pentode was released, George Cooksey of Mullard made sure that Leak was fully briefed on the capabilities of this small but efficient 9-pin output pentode. As a result, Leak became aware of the possibilities for a new twelve watt power amplifier design that would be smaller, more efficient, cost less and yet retain the same performance as the TL/12. At the time of its release in 1957, the TL/12 Plus was priced at just £18/18/0, where as the TL/12 had previously been selling at £28/7/0. Orders for the amplifier were excellent, with the USA distributor placing an initial order for 1150 units immediately after its debut at the 1957 New York Audio Fair.

The TL/12 Plus uses an ultra-linear output stage like the TL/10, with the first and second stages based directly on the topology of the TL/12, although more modern valves are used (EF86 and ECC81 respectively). The EL84 output valve requires considerably less voltage drive than the KT66 used in the original TL/12 (about 22 Volts peak to peak, compared to 72 Volts peak to peak). This gives the TL/12 Plus the potential to perform better than the TL/12, as the driver stage will produce less distortion and the first stage is not required to produce as much gain.

The amplifier was produced for around 10 years and was issued in three colours during its life: champagne (originally referred to as golden-beige at the factory), bronze, and metallic grey (the latter two paints being supplied by Carson-Paripan

A Leak TL/12 Plus, in this case fitted with a Mullard 5AR4 rectifier featuring a metal base ring. With the "Plus" series of amplifiers a black plastic finish on the transformers was introduced to cover the lamination stack.

at Runnymede). Around the time of the last colour change (circa 1963), the output transformer design was revised and the designation number changed from 3921 to 8778 (see also Chapter 12).

When the TL/12 Plus and TL/25 Plus were released in 1957, this signalled a move away from the Osram beam power tetrodes to the new pentodes from Mullard. Leak would continue to use Mullards as the prime supplier of valves for the stereo amplifiers that would follow in 1958. The exception here was the TL/50 Plus mono amplifier that used Osram's fabulous KT88.

The TL/12 Plus had reserve high-tension and filament supplies available to power British tuners but the two other amplifiers in the "Plus" series did not.

Ultra-Linear Design Revisions

When the TL/12 Plus amplifier was being designed, one of the prime requirements set down by Harold Leak was that the power output and distortion should be at least as good as the TL/12 amplifier it would replace. Another design constraint was that the output stage was to be operated in ultra-linear mode, with screen taps at 50% of primary turns. This would mitigate the low-fi image that tetrode or pentode operation had in the audio industry.

Initial tests revealed that the EL84 valves in the TL/12 Plus prototype were a little stretched to achieve the power required. So Harold Leak decided to raise the supply voltage until a generous 12 Watts was obtained, regardless of the maximum voltage ratings for the EL84 (14 Watts at clipping was quoted in some Leak literature). Previous experience with "proper" valves such as the KT66 had shown that this practice was safe, as the ratings were somewhat conservative. To compound the situation, Harold had decided on a high grid resistance of 1MΩ, which increased the stability margin at low frequencies.

With the design finalised, TL/12 Plus production commenced, and a year or so later the Stereo 20 was introduced with a similar output stage. However, in time a number of TL/12 Plus and Stereo 20 amplifiers were being returned to Leak with output stages that had "burnt-up". When Ray Whitcombe joined Leak in 1960, one of his first tasks was to investigate the cause. Soon after making initial investigations, George Cooksey of Mullard suggested that the EL84 valves were being operated beyond the maximum rated voltage and maximum grid resistance. This practice can lead to thermal run-away: additional dissipation in the valve results in the grid heating up and in turn emitting some electrons, which results in a reduction in grid bias, more plate current, more heat, until the valve destroys itself.

It was important that this problem be resolved quickly. So the output stage was re-designed using a lower supply voltage. To regain some of the power lost through a lower supply voltage, the position of the ultra-linear screen-grid tap was moved from 50% to 25%, the combined effect of which still gave 0.1% distortion at rated power. In addition grid-resistor values were lowered on some models. The other outcome from this experience was the realisation that the newer valves were less conservatively rated than the older style types and it was important to follow manufacturers ratings. In line with Leak policy any amplifier that was returned for service was modified.

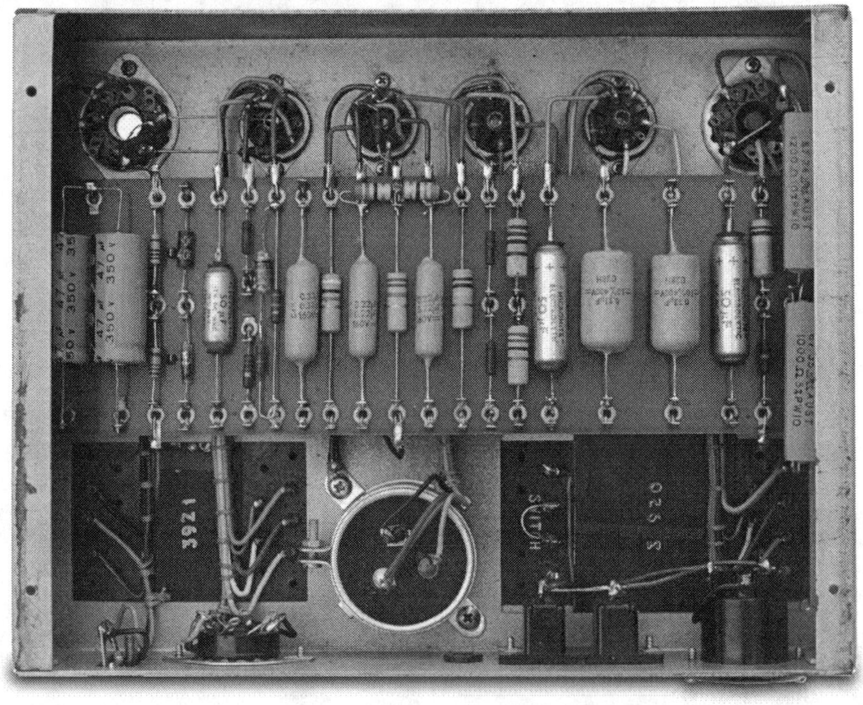

The under-chassis layout and construction of the TL/12 Plus is of a similar standard to the TL/12. Some capacitors have been replaced on this unit.

The TL/25 Plus, showing the heat-shield to protect the power transformer from radiant heat.

TL/25 Plus

After the introduction of the ultra-linear output stage, there was little benefit to be had with tetrode-connected output stages in hi-fi amplifiers. They would remain popular where the main requirement was maximum power but this is not the case in hi-fi amplifiers. As a result, the Leak TL/25A, which was a tetrode design, was replaced by an ultra-linear design – the Leak TL/25 Plus.

Mullard's new EL34 output pentode was chosen for the design, although there is sufficient latitude in the circuit to allow the owner to substitute KT66's or 5881's without circuit modifications. As with all amplifier in this series, the "Plus" designation is an indication of the amplifiers ability to deliver more power than it was actually rated at – usually around 20% more. With around 440 V of HT on the output stage, this mono amplifier will give around 30 Watts at clipping.

The first and second stages of this amplifier are similar to the TL/12 Plus – with an EF86 and ECC81 being used again, and the circuit being based on the same "TL/12 topology" as the original 1948 circuit.

Surprisingly, the TL/25 Plus is on the same sized chassis as the TL/12 Plus (only the height differs for the taller power transformer on the TL/25 Plus). Early American advertisements for the TL/25 Plus show a chassis without a heat shield, however these were fitted to later units to protect the power transformer and capacitors from radiant heat. The TL/25 Plus was issued in the same three colours as the TL/12 Plus – champagne initially, then bronze and lastly metallic grey.

The TL/25 Plus from above.

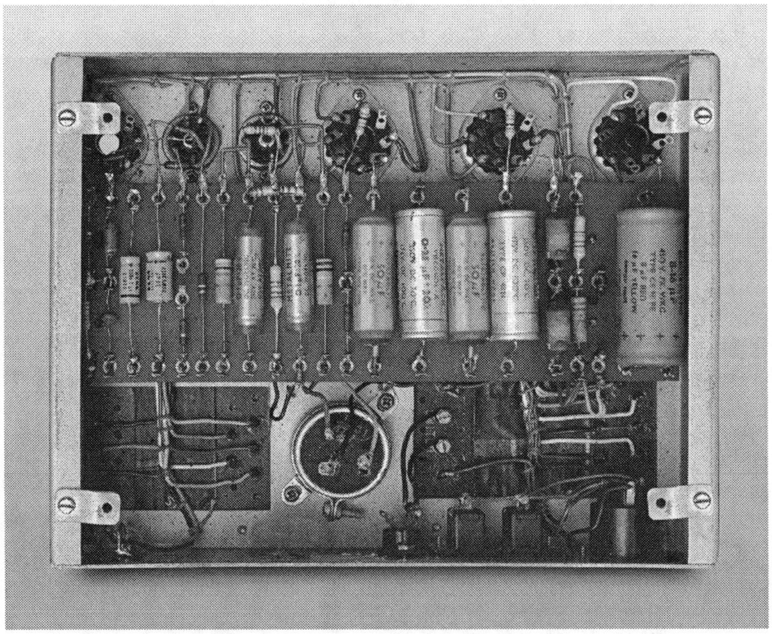

The under-chassis view of the TL/25 Plus shows some dis-colouration of the component board near the cathode resistors (on the right hand side) which is normal, and is caused by the heat dissipated from them.

Opposite: Crowds at a Hi-Fi fair entranced at the quality and performance of Leak products. A TL/12 Plus, Stereo 20 and a TL/50 Plus amplifier can be seen on display with a Point One Stereo pre-amp, along with products from other manufacturers. (Photo courtesy of the Leak family.)

LEAK
HI-FI

TL/50 Plus

The Leak TL/50 Plus design was initiated at the request of the American company Ercona who would later become distributors of Leak products in the USA. (© Sunrise Press, Devon. Photography by Jonathan Hill. Reproduced by permission. Source: Audio! Audio!)

Leak had not considered the possibility of using the KT88 valve, but Charles Frank Jnr. (from the American company Ercona) was keen for Leak to start using this valve – "you've got to use the KT88 – it looks sexy!". The staff at Leak may have worded this differently, but the appeal of the KT88 is undeniable and so work was commenced on the design of the TL/50 Plus primarily as a result of this suggestion from Charles Frank Jnr. (Some years later in 1961, Ercona would replace British Industries Corporation as distributors for Leak in the USA). The TL/50 Plus was launched in 1957 and remained in production for around 11 years, although in much smaller quantities than other Leak amplifiers as most customers did not see a need for such power levels with the efficient speakers of the day.

The KT88 ultra-linear output stage in the TL/50 Plus requires around 100 Volts peak to peak drive, which is about 50% more than that in the original TL/12 circuit.

As with all Leak power amps from TL/12 Plus onwards, the sensitivity had to remain high – at 125mV for full output (so all power-amps could be used with any model Leak pre-amp). Together this meant getting the most gain possible, with an ability to drive the KT88's properly, from the original 3-stage TL/12 circuit. One concession made in the design by Leak was an alteration in the feedback level from 26dB down to 23dB for the TL/50 Plus, which would have eased the above gain and drive constraints slightly.

The TL/50 Plus can deliver around 60 Watts at clipping, although the circuit is rated at 50 Watts. The distortion rating was 0.1% at 45 Watts (+/- 1dB).

The under chassis layout of the TL/50 Plus shows a spacious layout of components. (Photo courtesy of Stone Audio, UK).

On the TL/25 Plus and TL/50 Plus a heat shield was added early on in the production run. Originally these models did not have this, but it was found necessary to protect the power transformer and electrolytic capacitors from the radiant heat of the output stage valves. Another source of above-chassis heat on the TL/50 Plus are the cathode resistors made by "Standees" dissipating around 9 Watts total.

Information from the owner's manual for the TL/50 Plus amplifier.

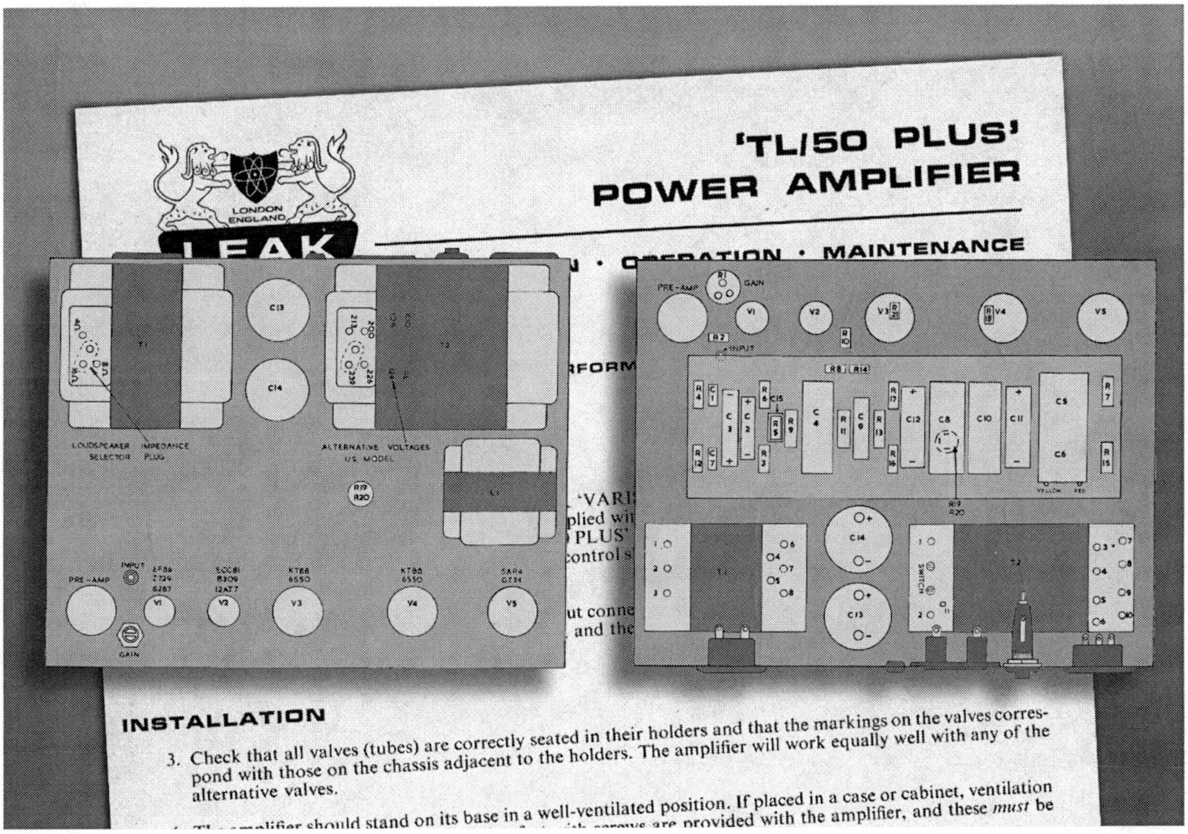

Stereo 20

This amplifier is basically a stereo version of the TL/12 Plus – the output stage and output transformers are identical, but the first and second stages use high-gain ECC83 double triodes. In the TL/12 Plus, overall voltage gain was provided by a high-gain pentode and medium gain ECC81 triode, whereas in the Stereo 20 two high-gain triode stages are used to get the same overall sensitivity of 125mV. The ECC83, with a high-μ (internal amplification factor) of 100, allowed Leak to design this circuit entirely with triode gain stages.

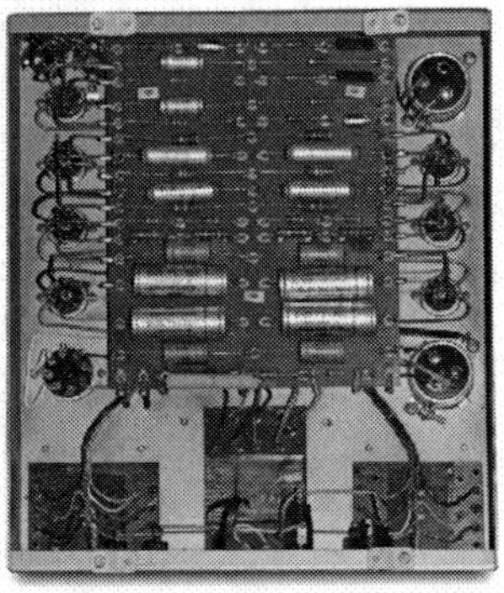

The fabulous symmetry made possible when stereo was introduced was exploited well in the Stereo 20 amplifier.

The power transformer on this model has a reputation for running warmer than some other Leak amps. A safety mechanism used by Leak on this amplifier, and other models, was the use of a close-rated 100Ω resistor in the power supply. If excessive current was drawn from the supply, the resistor overheats and de-solders itself from the connection to the component board, thus protecting other components in the amplifier from further damage.

Initially issued in champagne (golden-beige), the colour was changed to gold-bronze and then to dark metallic grey when the matching "Varislope 2 Stereo" pre-amp was introduced around 1964 . Like the TL/12 Plus, output transformers with different designations were used during the production run. A type 3921 output transformer was used initially, with a type 8778 from around 1963 onwards.

Two ECC83 twin triodes, two EL84 pentodes and a 5AR4 rectifier on the Stereo 20 chassis.

The Stereo 20 amplifier is highly acclaimed forty years on for its sound quality, reputed to be one of the best Leak amplifiers in this regard.

One of the best visual features of stereo amplifiers from Leak is the use of symmetry in the component layout. Whilst the shared input stage driver rather spoils perfect symmetry from above, the under-chassis view is quite stunning – particularly if it has not been modified over the years.

The amplifier is the most common Leak amplifier on the second hand market, and the audio quality is rated very highly by many. The amplifier sounds good and doesn't become harsh if the signal waveform is clipped during overload conditions. When it is used with reasonably efficient speakers it gives excellent results.

Stunning wiring practices continued in the Stereo 20 amplifier. The central earth point is located on a lug adjacent to the electrolytic capacitor on the left.

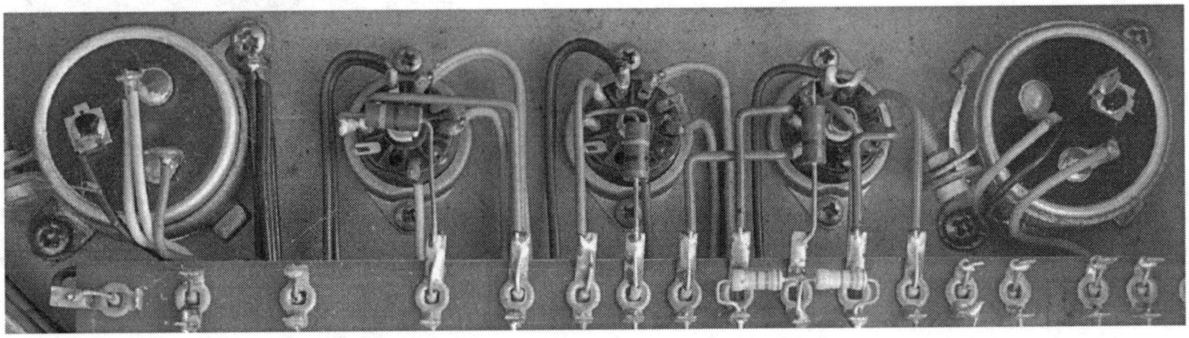

Stereo 50

Above: Leak Stereo 50 top chassis layout. Below: Some Stereo 50's were fitted with input attenuators like that on the Stereo 60.

Just as the Stereo 20 was a development from the TL/12 Plus, the Stereo 50 used the TL/25 Plus design as the starting point. Once again, ECC83 twin triodes were able to provide sufficient gain to replace the pentode and medium-μ triodes in the TL/25 Plus, whilst maintaining the 125mV sensitivity. The circuit shows that the anode resistors on the first and second stages are quite high in value, as are the grid-resistors on the power valves. This indicates that as much gain as possible was being squeezed from each stage – with the designers being constrained here to meet a 125mV sensitivity and a 0.1% distortion rating that required 26dB feedback.

The power supply voltage on this amplifier is just 390 Volts which results in a conservatively run output stage in which valve life is extended. The output valves are the EL34 from Mullard which can deliver more power than 25 Watts in ultra-linear mode, but in this design power is limited to 25 Watts per channel. The use of

The Leak Stereo 50, fitted here with EL34 valves although KT66's can be used without changes to the circuit. The speaker connectors on this unit have been modified from the original screw terminal. Models for the Canadian market had the voltage tapping adjustment (shown here on top of the power transformer) moved beneath the chassis to comply with the newly introduced Canadian safety standards.

resistor-capacitor ripple filtering in the power supply may have also been a limiting factor leading to a lower power supply voltage, and resultant lower power output than that obtainable with inductor-capacitor filtering.

The Stereo 50 used wire-wound cathode resistors that were mounted above the chassis. These were special resistors that Leak had constant trouble obtaining. On one occasion 50 amplifiers were shipped from the factory fitted with resistors from Radiospares which were annular in shape and had required special brackets to mount them. The power supply ripple filtering resistor of 100 Ohms was also mounted above the chassis. Unfortunately, these resistors have generally failed over the years, and many amplifiers found second hand have had replacements rather poorly installed above or below the chassis.

The Stereo 50 was issued in both champagne and bronze enamel. Late in the production run, some other variations were introduced:

- some units featured a "Stereo 60" chassis – complete with attenuator switch, and mounting holes for a non-existent filter choke (presumably this would have been close to the time the Stereo 60 was introduced).
- the original output transformer, designated type 3925, was re-designed and replaced by the type 8615 as used on the Stereo 60.

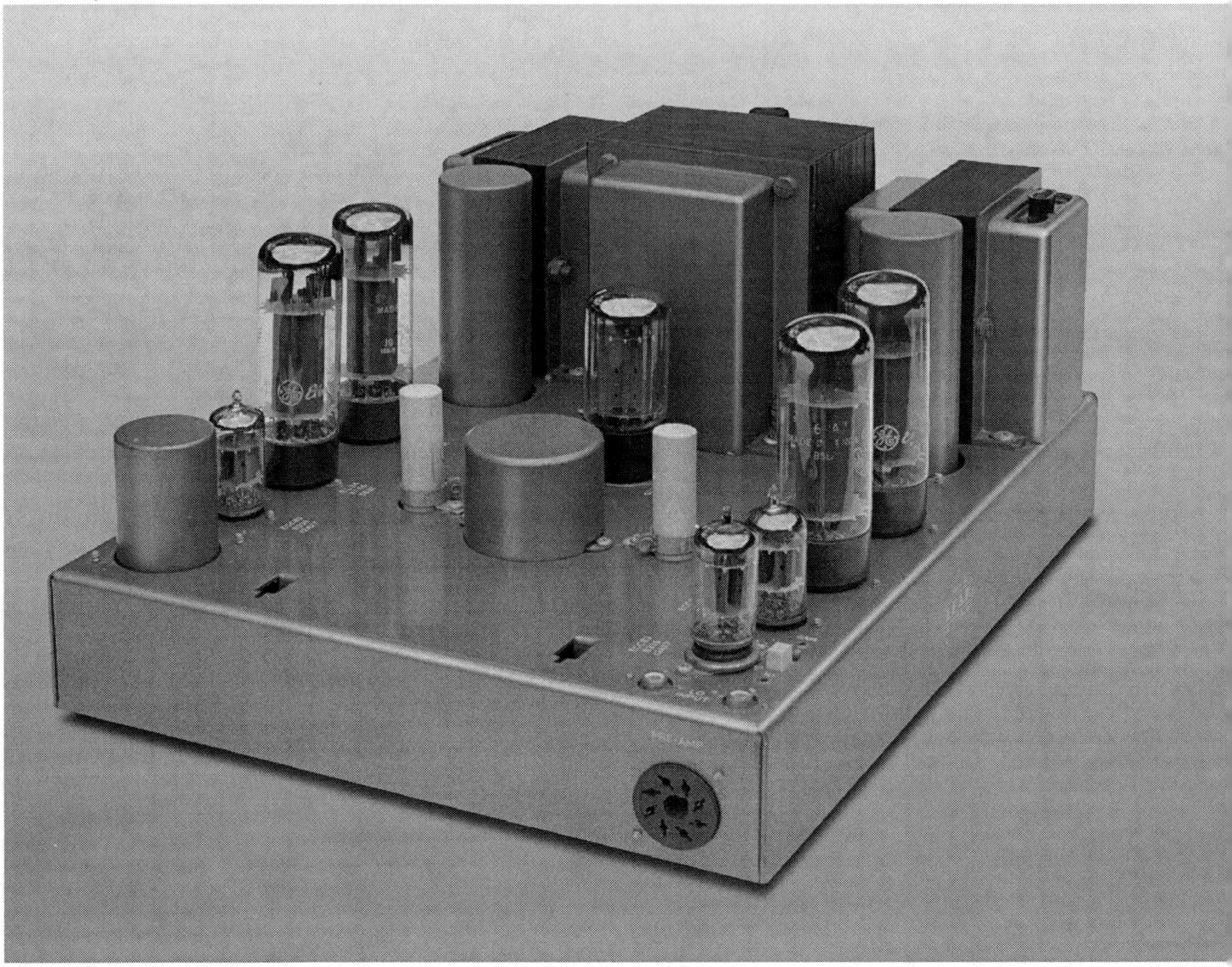

Stereo 60

The Leak Stereo 60, shown here finished in bronze, but most units are in dark metallic grey. Note the cylindrical ripple-filter choke in the centre of the chassis: under the cover is a conventional core with "E-I" laminations.

Introduced around mid-1963, this amplifier is closely related to the Stereo 50 model, which at first glance looks quite similar. External differences compared to the Stereo 50 are:

- the Stereo 60 has an input attenuator switch
- the first stage valve socket is mounted above the chassis instead of from below, and is repositioned in-board by about 3mm to make room for the attenuator switch
- a power supply filter choke in a round can was included in the Stereo 60, being mounted in the middle of the chassis

Circuit-wise, the Stereo 60 is almost identical to the Stereo 50 except that the Stereo 60 includes a step-network in the first stage anode to compensate the feedback loop. The power supply voltage is higher in the Stereo 60 at 420 Volts – the power transformer outputs more voltage and the filter choke results in less voltage drop in the filtering circuit. As a result the Stereo 60 delivers slightly more power than the Stereo 50.

The output transformers on the Stereo 60 were designated type 8615 and differed considerably from those used on most Stereo 50's. The 8615 had the screen taps at 25% whereas the 3925 used on the Stereo 50 had taps at 50% (which gave an output stage closer in nature to triodes). In addition, the 8615 primary impedance was 5600Ω compared to 7000Ω for the 3925. As a result some difference in performance between the two amplifiers can be expected, given the different loading conditions on the output stage.

Initially these amplifiers were finished in bronze, but when the pre-amp appearance was revised in 1964 (with the Varislope 2 Stereo) they were then finished in dark metallic grey.

The under chassis view of the Stereo 60.

The Last Leak Valve Amplifier

With the introduction of the Stereo 30, and then the Stereo 70, the stereo valve amplifiers from Leak were the first to be discontinued, with the higher-powered mono amplifiers remaining in production for a short time. At the 1967 Audio Fair in London, Leak valve amplifiers were conspicuous by their absence, as they were about to be deleted from the product range. The TL/50 Plus, being a specialist amplifier by then, was the last Leak valve amplifier to be discontinued, around 1968. Some distributors were still listing valve amplifiers as late as 1968/69, possibly from remaining stock reserves.

The rectifier socket was mounted from above the chassis to allow easier servicing.

9. Germanium and Silicon

Stereo 30

Until 1961, Leak had a range of amplifiers, none of which featured printed circuit boards or transistors. In addition, the company was looking for a way to produce an integrated amplifier small enough to be put into a cabinet, eliminating the need for the pre-amplifier to be in a separate unit to the power amplifier.

Early work was begun on re-designing the Stereo 20 and the Varislope Stereo pre-amplifier so that they could be produced on one chassis. To minimise the size, the output stage used Mullard's ECL86 (which has a triode and power-pentode within the same envelope). The possibility of incorporating a tuner into the design was also on the drawing board at this stage – which was an influence from the American market.

However external events overtook the integrated valve amplifier project. In the November 1961 issue of Wireless World, R. Tobey and J. Dinsdale published a paper titled "Transistor Audio Power Amplifier". This amplifier was capable of 10 Watts output with distortion of about 0.25% and drew on the work of H. C. Lin – using a quasi-complementary output stage. Harold Leak approached Tobey and Dinsdale who subsequently submitted a prototype for evaluation. Harold was also considering engaging either of them on development work, but this did not eventuate. (Another prospective employee was T. D. Towers of Pye who was also well acquainted with transistors.) The publication of the Tobey-Dinsdale design resulted in the development work at Leak changing focus – away from the integrated valve amplifier project to a transistor-based design.

In prototype form, this amplifier was first dubbed the "Transtereo 30", but the name reverted to the more easily marketed "Stereo 30", mainly because the word "Transtereo" had been copyrighted by a speculator. During a five-year production run 50,000 units were produced. Early models had an attractive steel slip-cover finished in grey vinyl. When teak cases were introduced, the steel cover was discontinued, being replaced by cheaper metallised fibre-board (as shown here) that attracted a lot of criticism.

Ray Whitcombe

Ray Whitcombe, photographed in 1997.

Ray Whitcombe was born during the War in 1942 and "toughed out" most of the Blitz with his family in London. After education at a semi-private secondary school he obtained nine "O-Levels" and went on to study Telecommunications at a Technical College as he was keen to establish a career in electronics. Ray was an avid reader of Wireless World and in 1960 he noticed an advertisement for a position at H. J. Leak & Co. Ltd. which he applied for:

"I got a reply inviting me to attend an interview with E. H. Ashley! I duly turned up with my school and technical Certificates and a request that I be allowed to continue my technical education. At the time, a Government scheme subsidised an employer if they allowed an employee a day off per week to attend college. Ted Ashley was keen on this idea, considering it far better than the Night School system he had to endure. Anyway, Leak offered me a position as a Lab Technician with day release during term until I achieved an advanced Telecom Certificate.

"On starting at Leak in August 1960, initially I was given simple tasks until my capabilities were assessed. One of the first things I remember was being given a heap of circuit diagrams to study, not only to learn the products, but also to learn the way of drawing the circuit diagrams. This was based on the guidelines of L. Bainbridge-Bell, a Senior Engineer at the Admiralty.

"I was soon accepted as more than a mere 'bottle-washer' and did investigation work on the amplifiers then in production, as well as helping Don Barlow with the Sandwich loudspeakers. Many of the amplifiers that the Lab processed were special designs and it was our task to determine stability margins etc. In fact Leak produced many specials for purposes as diverse as: PA, store-casting, servo-motor drives and ultra-sonic transducer testing. They never seemed to refuse a commission, even though many must have been loss-makers, however it all added to the Company's prestige!

"Working with Don Barlow gave me a grounding in acoustics, a subject that was completely foreign to me. My main task was testing in the roof-hut. This entailed carrying trays of tweeters up a ladder from the Lab to the roof and then doing a manual frequency response on each. A little known fact is that most, if not all, of the production tweeters in the Sandwich speakers were thus tested. I did quite a few!"

Ray remained with Leak until November 1968. During his time there he was intimately involved in the design and development of the Leak Stereo 30 and Stereo 70 amplifiers. He gained his Full Technical Certificate while with Leak and subsequently obtained a Fellowship in the Institute of Incorporated Engineers. "All this thanks to the foresight of Ted Ashley and the 'little' Leak company."

The development staff had been monitoring transistor applications for audio through published papers and technical data from transistor suppliers. Leak was aware of the classic 1956 paper by H. C. Lin (of the RCA Labs in New Jersey) that described a transistor amplifier using a quasi-complementary output stage with a distortion of 1% at 6 Watts output. Later on, in 1963, Mullard published several high quality transistor amplifier designs in a series of Wireless World articles by P. Tharma, one of which was a Class B design with distortion of about 0.1%. However it was noted that performance would vary if accurate biasing was not maintained.

Suppliers of transistors (such as RCA and Mullard) were providing circuit design information and transistor specification data to Leak, to assist early development work. Often Leak had to retort to these suppliers that distortion in transistor circuits was of prime concern – and that the company could only move towards transistors if distortion could be held below 0.1%. ("After all, Leak didn't have that trade mark for nothing" recalls Ted Ashley.)

The frequent papers appearing in the technical press regarding transistor audio amplifier design, and design data from transistor manufacturers all contributed to the knowledge base at Leak. This enabled Leak to have prototype transistor amplifiers under development by 1962. (A prototype of the Stereo 30 was demonstrated to the BSRA during late 1962 and reported on in the May 1963 edition of Wireless World.)

To ensure high performance from transistor circuits, Leak approached Mullard for the best possible transistors – asking them to provide transistors with parameters falling within a specified range. These transistors were of the germanium p-n-p variety (type AD140), which required a negative power supply voltage of -42 Volts

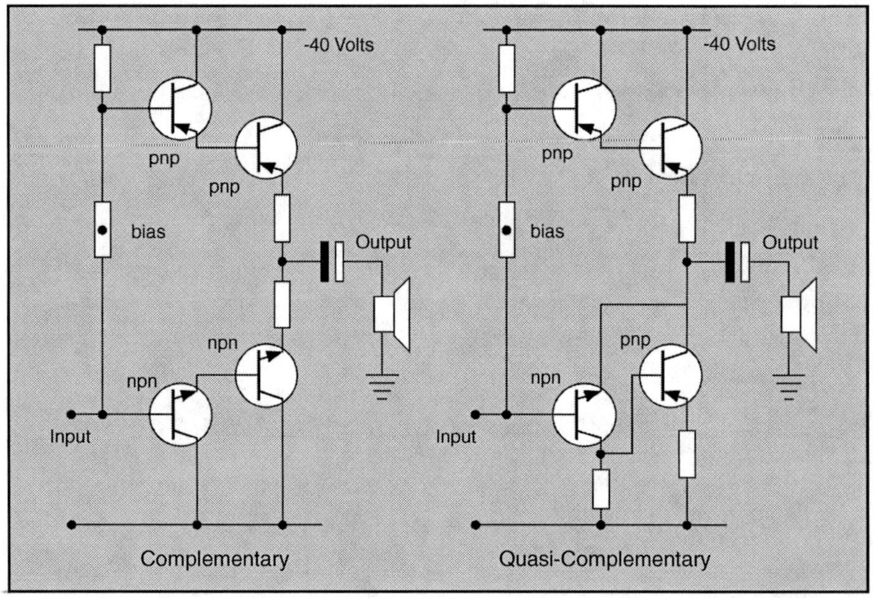

A transistor push-pull output stage ideally uses complementary devices - a high power npn transistor and a high power pnp transistor, each in turn driven by smaller complementary transistors (left). Early transistors were most easily fabricated as pnp devices, with high-powered npn devices being difficult to make or expensive. So the quasi-complementary stage was developed (shown at right), in which a low powered npn device and a high powered pnp device are coupled to emulate the circuit on the left. The bias device is bonded to the output transistor heat sink, so the bias is reduced if the temperature rises.

for the required power output. The output stage chosen by Leak was of the push-pull quasi-complementary design similar to that proposed by H. C. Lin, whilst the driver circuits were of similar topology to that proposed by Tobey and Dinsdale. The amplifier was capacitively coupled to the loudspeaker, and the majority of signal feedback was taken from the loudspeaker side of this coupling capacitor. A smaller amount of signal and "DC" feedback was taken directly from the output transistors to set the DC operating conditions of the amplifier.

Two major issues that were tackled by Leak engineers during the design phase of the Stereo 30 (and later amplifiers) were thermal run-away and cross-over distortion. As is well known, cross-over distortion can be alleviated by raising the zero-signal current but only to increase the likelihood of thermal run-away from the additional heat dissipation. This made early transistor designs a balancing act between low distortion and self destruction. The Stereo 30 design used a thermistor device (temperature sensitive resistor) that sensed the temperature of the output stage heat-sink to regulate the zero-signal current in the output transistors to between 15 and

The Stereo 30 had a single printed circuit board across the base of the chassis. One of four AD140 output transistors is shown on the left. The Canada Standards Association (CSA) requested that Leak use a multi-stranded wire with a varnished textile sheaf. This would have made the neat wiring Leak was known for impossible, or too time consuming to achieve. After discussions with suppliers, another CSA approved wire was submitted and then adopted by Leak.

Reflections on the Stereo 30: by Ray Whitcombe, IENG, FIIE(Elec), Design Engineer, formerly of H. J. Leak & Co. Ltd.

Harold Leak was always very aware of export markets. He kept a close eye on the competition by arranging for amplifiers from competing companies to be investigated in the laboratory at Leak (e.g. H. H. Scott, Fisher, Marantz and Harmon-Kardon). Around 1960 it became apparent that the market trend was towards integrated amplifiers, so the development team embarked on an integrated valve amplifier design that would combine the and Stereo 20 and Varislope Stereo onto one chassis. It is important to recognise the close links that existed between the design lab at Leak and the technical departments of various suppliers (as is common throughout industry). This practice effectively enables larger technical backing for projects than would otherwise be possible.

The valve supplier to Leak was Mullard and their representative, George Cooksey, suggested the use of the triode/power-pentode ECL86 valve for the output stage of the integrated amplifier. Hinchley Engineering of Devizes was engaged to design a suitable output transformer for ultra-linear operation. Harold Leak had some reservations about this project, even though very good results were obtained on test. The issue was size – the integrated amplifier was a problem when it came to fitting it into the Southdown cabinet. The design was made as compact as possible, and the project was about to enter a production-design phase when Tobey and Dinsdale published their transistor amplifier design in Wireless World. Upon reviewing the article Harold Leak approached his development team and decided to suspend work on the ECL86-based amplifier so that an investigation into a transistor-based design could be commenced.

The design aims for the transistor amplifier were clear: it must achieve 0.1% distortion at rated output; its output should compare well with the Stereo 20; it should feature similar flexibility in controls and function to the Varislope Stereo; it should be of compact design, similar in size to the Leak pre-amps, and it should sell at a lower price than the Stereo 20 with a Point One Stereo pre-amp. As the Tobey-Dinsdale design did not achieve these objectives a significant amount of development work was envisaged.

Mullard were approached to supply suitable transistors for the development work to be based on, however they were initially reluctant to encourage this type of development – this was surprising, but they seemed unconvinced that Leak were serious about transistor amplifier development. Ted Ashley subsequently contacted Texas Instruments who were able to supply transistors and technical support for the new project.

Harold Leak's solution to achieving 0.1% distortion from the prototype designs was "more feedback" – after all there were no transformers in the loop. But it soon became apparent that large amounts of feedback caused problems with the restricted frequency response of the available transistors –

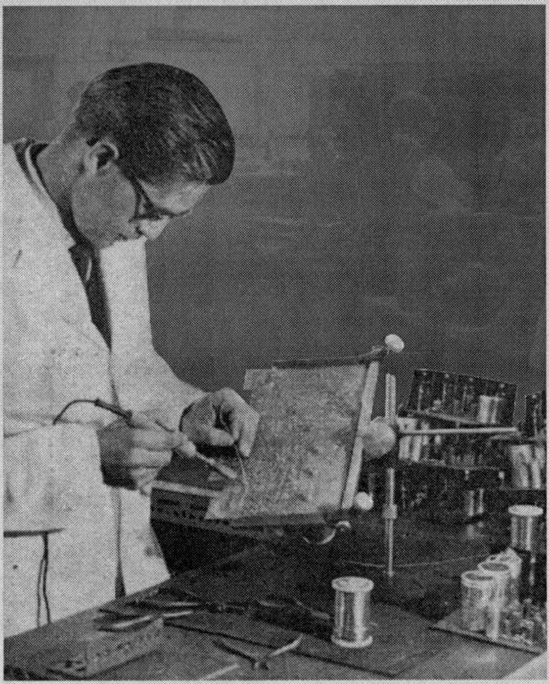

Les Kirkpatrick demonstrates assembly of a Stereo 30 circuit board for an Ersin solder advertisement. (Reproduced with the permission of Multicore Solders, UK. Source: WW, April 1964)

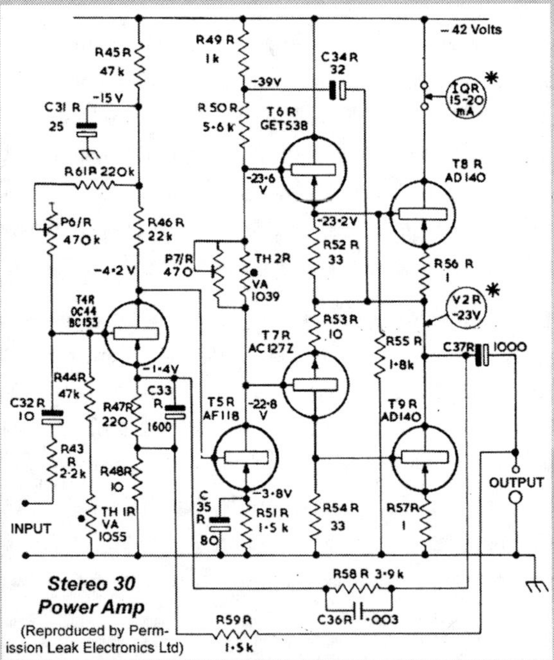

Stereo 30 Power Amp
(Reproduced by Permission Leak Electronics Ltd)

germanium types are not that brilliant. An added difficulty was Harold's insistence on 12 Watts output, which required a 40 Volt rail, necessitating "high voltage" transistors by 1960's standards. Such transistors tended to be fragile, or of poor bandwidth which compounded the stability problems when feedback was applied.

The requirement to obtain 12 Watts from the amplifier resulted in Leak adopting the IHFM method for power output measurement (where the output is measured with a power supply assumed to have perfect regulation) although some of the development staff didn't approve of this approach. Power output was quoted at 15 Watts IHFM into 4 Ohms, but distortion was quoted at 0.1% for 8 Watts output into 15 Ohms. John Leak recalls that Frank Hawkins was very reluctant to have the unit tested by the hi-fi press – his usual excuse was that demand was so high that none could be spared.

Particular attention was paid to the method for heat-sinking the output transistors and stabilising the bias current with variations in temperature. Tobey and Dinsdale had used diode-stabilisation for the output stage bias current, but Leak chose to use a thermistor bonded to the heat sink. To stabilise the mid-point voltage on the output stage another thermistor was introduced at the input stage.

The IHFM rating for power output also had the desirable effect of allowing smaller heat sinks on the output transistors, which in turn reduces the size of the chassis. Ray Whitcombe considers that the design team took a gamble on the heat sink size, but hindsight shows that the approach worked.

After a lot of experimenting, much of it very empirical, a workable and robust design emerged which was demonstrated in 1962 by Harold Leak.

The project then entered a production-design phase but immediately encountered difficulties. Mullard approached Leak with revised proposals to supply transistors to Leak – the possibility of Mullard obtaining transistors from Texas Instruments was even considered. At the same time, Texas Instruments advised Leak that production quantities of the types used in the Stereo 30 were not available, so there was little choice but to adapt the design to Mullard transistors. The new AD140 proved a good replacement for the output stage and GET113's and AC107's were used in the pre-amp stages.

Much more difficulty was found in selecting transistors for the gain and driver stages of the high-feedback power amplifier design. Initially the OC81 was considered for the PNP-driver but it proved unsatisfactory on test, so Mullard came up with the GET538 which was as good as anything Leak had tested so far. The NPN driver suggested by Mullard was the AC127, but unfortunately the V_{ce} rating was too low, so Mullard offered specially selected versions with a higher V_{ce} rating (designated AC127Z). For the class-A pre-driver stage, a wide bandwidth transistor was desirable, which also needed to tolerate the 40 Volt rail. Fortunately Mullard could supply the AF118 which had been developed for video output stages in early solid-state televisions. Finally, the first stage was replaced with the OC44 radio-frequency transistor.

The requirements for the pre-amp section of the Stereo 30 were similar to the specification and function of the Varislope Stereo valve pre-amp. A super-alpha pair was used in the first stage to obtain sufficient gain for equalisation. Also the first transistor (an AC107) was operated in starvation mode to minimise noise. The familiar Baxandall tone control, coupled with the Varislope filter circuit, was adapted for transistor circuit impedances.

As a result of the co-operative relationship with Mullard, the transistors for the Stereo 30 were painstakingly selected – as they were critical to the performance, stability and reliability of the design.

The Stereo 30 was the first Leak product to use a printed circuit board. The layout was done by Ted Ashley (Ray Whitcombe recalls that it took Ted a few days and evenings and that "it was not a bad job!"). Development printed circuit boards were made using a new "deposition process" as the turn-around time was quick for this technique. However better reliability was obtained from the traditional etched process, so this was used for production boards (supplied by Tectonic).

The introduction of the Stereo 30 caused a minor-level of euphoria amongst the industry and suppliers to Leak. In the April 1964 issue of Wireless World an advertisement placed by Leak for the Stereo 30 is followed up by 10 pages of advertisements from suppliers to Leak. Amongst the list are: Tectonic printed circuits, Mullard transistors, Lunds chassis, Dubilier resistors and potentiometers, Plessey accessories, Belling Lee connectors, Hinchley transformers, TCC capacitors, Ersin solder, and AB rotary switches.

The design of products using new technologies evolves from many different directions and scientific effort over a period of time. The Stereo 30 was inspired by Tobey and Dinsdale's work, and their work in turn was inspired by that of H. C. Lin who had been employed by RCA Laboratories to develop transistor applications. In reality the step from an initial circuit design to a product that sells is the greatest step. Although the Stereo 30 appears similar to the H. C. Lin amplifier, the number of man-hours required for development of a commercial product far exceeds that for the original conception. In this regard the Stereo 30 was designed by Ted Ashley and Ray Whitcombe, with assistance from Texas Instruments and later Mullard. Harold Leak was the driving force for the project with his company's prestige and technical expertise enabling the commercial success of the Stereo 30.

Acknowledgement: This text was kindly contributed by Ray Whitcombe, IEng, FIIE(Elec), and was edited by Stephen Spicer.

20mA. Another thermistor associated with the first stage of the power amplifier regulated the DC voltages in the amplifier.

Other manufacturers were keenly developing transistor amplifiers as well. Two models from around 1963 were the Pye HFS30T of 15 Watts/channel, and the Radford SC4-20 of 20 Watts/channel and 0.1% distortion. By late 1963, a review of transistor applications in audio (by Hi-Fi News magazine) showed about a dozen manufacturers in Britain with amplifiers, controls units, or tuners using transistors.

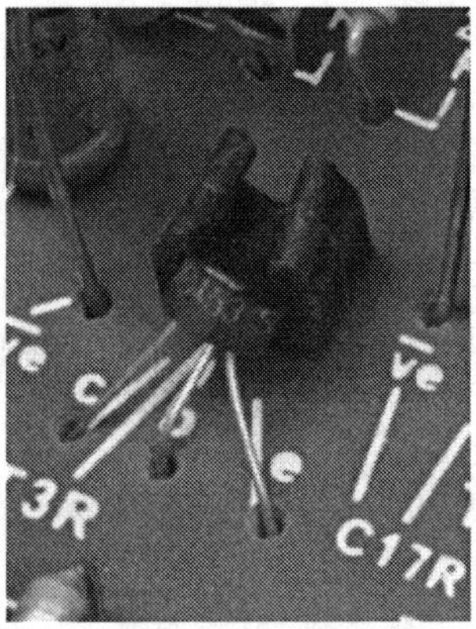

Plessey designed an 8mm version of its transistor clip for use in the Stereo 30.

A particular difficulty with the design of the pre-amplifier stages was maintaining a high input impedance for the pickup input. This was a natural characteristic for valve circuits but it required more skill to obtain this with transistor design. The first demonstration of the Stereo 30 was done with a pre-amplifier stage that was a pre-production design and which was considerably altered before the product was finally launched. For a time, Leak were considering various plug-in modules for the pre-amplifier that would require the owner to set the amplifier up for the pickup that was being used. However Harold Leak rejected this approach, as he felt it was not in line with the way the product should be used by the owner. This increased the time required for the development of the pre-amplifier section of the Stereo 30.

The Stereo 30 pre-amp incorporated the variable slope filter circuit used in Leak valve pre-amplifiers, but adapted for use with transistors. A minor revision to this part of the Stereo 30 was made during the production run – a compensation capacitor was added to the pre-amp circuit, presumably to increase stability. The Stereo 30 development took around two years and it was launched in December 1963 whilst its Audio Fair debut was in 1964.

The Stereo 30 also signalled Leak's use of printed circuit boards – on this model there is a single large board across the bottom of the chassis that has all the pre-amplifier and power amplifier components mounted on it (with the exception of the output transistors). Ted Ashley recalls that the process of laying out this board was a very intensive period of work. The number of above-board wire links had to minimised, earthing arrangements had to prevent any hum loops, and the interaction between components had to be minimised.

An appealing combination with chalk and cheese in the same cabinet: the valve-based Trough-Line Stereo could be mated with a Stereo 30 (shown) or Stereo 70 amplifier in this attractive packaging option from Leak in the late 1960's. These cabinets were sold empty to avoid attracting the 27.5% tax rate applicable to tuners, and which would have applied to the whole combination if sold complete! (Reproduced by permission Leak Electronics. Source: HFN April 1968)

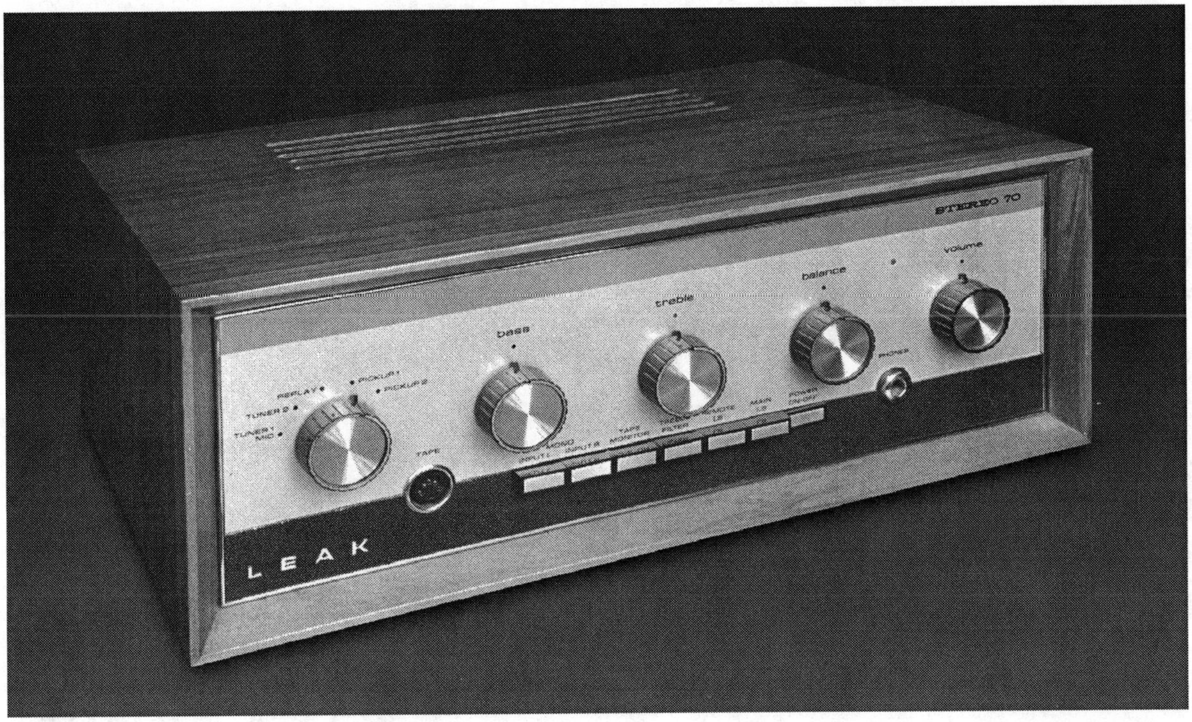

Stereo 70 and Stereo 30 Plus

The Stereo 70 amplifier could be purchased as a free standing unit for mounting into a cabinet, or complete with case as shown here. (Photo courtesy Philip Spicer.)

After silicon transistors became more common, RCA came up with selected versions of the high-powered 2N3055 n-p-n transistor which was used by Leak in the Stereo 70 amplifier. The output stage was of the quasi-complementary type, and was capacitively coupled to the loudspeaker. To obtain the higher power output, the 2N3055 transistors were operated from a positive supply of 75 Volts.

The Stereo 70 was released in 1967 and Ted Ashley remembers being very pleased with the design, on which Design Engineer Ray Whitcombe contributed a great deal. The physical layout of components in the Stereo 70 was completely new – the amplifier used a pair of plug-in boards for the pre-amplifier stages and another pair

Harold Leak with a Southdown system comprising: Stereo 30, Troughline Stereo tuner and Sandwich loudspeaker. (Photo courtesy of the Leak family.)

of plug-in boards for the power amplifier (but the power transistors were mounted separately on a heat sink that wrapped around the power transformer). The output stage biasing is more conventional, using a pair of diodes to sense the output transistor temperature, instead of a thermistor as used on the Stereo 30. The amplifier's performance is a considerable improvement over the Stereo 30.

Ray Whitcombe recalls: "Up to the time of the introduction of the Stereo 30, final production testing was undertaken by skilled personnel. The production rate was relatively low, and a few Testers who had an understanding of the circuitry involved, had no problem keeping up. Enter the Stereo 30 and 'all hell broke out'. Everyone wanted one, thus production leapt skywards. Together with the new production technology that printed circuits and transistors presented, Leak had trouble meeting targets. This problem was approached in two main ways. Leak had to sub-contract some Stereo 30 production off-site (to Elcom of Northampton). At first this made testing matters worse, as the build-quality was bad, involving re-work and re-testing. The Test Departments's load rose to an unacceptable high, so a decision to employ un-skilled personnel to do the testing was taken."

Ted Ashley adds: "Staff who were not able to diagnose and rectify faults were employed on testing to a strictly laid down procedure, and products which failed these tests were set aside for the technical staff to rectify. This remained our testing philosophy throughout Leak manufacture and into Rank days."

The Stereo 30 Plus followed soon afterwards, effectively being a revised version of the Stereo 70 with a lower voltage power supply of 48 Volts, and an output stage using either 2N3055 or 2N5036 types.

In both the Stereo 30 Plus and the Stereo 70 designs, the original twin-T variable slope filter was discarded and a fixed-frequency filter that used an inductor and capacitor was used instead. The roll-off rate was 12dB per octave.

The Stereo 30 Plus, which had a Headphone connector covered by a knob that was removed to reveal the connector (lower right small knob).

Opposite Above: Chassis layout of the Stereo 30. Opposite Below: The later Stereo 30 Plus used a chassis layout based on the Stereo 70.

Unlike the Stereo 30, the Stereo 70 used plug-in boards for the pre-amplifier and power amplifier driver stages (see left side of chassis). Interconnecting wiring was done with cable looms.

Hi-Fi News had the following caption for this photo of Harold's daughter at the 1968 Audio Fair: "Hannah Leak was the Melinex clad nymph that caught many an eye, here she operates the new Stereo 70." (Reproduced by permission Leak Electronics Ltd. Source: HFN June 1968)

Stereofetic FM Tuner

By 1967, Ted Ashley was heavily involved in the pre-production work on the Stereo 70 and Stereo 30 Plus amplifiers and Leak was planning to design a solid-state replacement for the valve-based Trough-Line tuner. Harold Leak placed advertisements in Wireless World and Hi-Fi News for a "Technical Assistant" position, with a view to locating a Design Engineer for the new FM tuner. There were approximately one hundred applicants for the position, and the last one of the applicants accepted for interview was Ian Pullen (who was employed by Tectonic Printed Circuits in Wokingham at the time). During the interview, Harold Leak asked Ian Pullen if he could design an FM tuner, to which Ian replied "he would give it a go". Ian Pullen had previously worked for Rank-Cintel, where he designed the first head-up display for the Fleet Arm Arm's Buccaneer aircraft. Ian pointed

"You can assume that most of the opposition to the Stereofetic name came from me! I had to market them. Despite the name, and it was Ian Pullen who coined it, having him there was like a breath of fresh air. After only a week he was bringing bits of hardware into management meetings, and his enthusiasm was infectious."

John D. Leak, 1999.

out to Harold Leak that the techniques used in the head-up display project would have applications in audio. Harold employed Ian, perhaps because he was impressed with his "cheek" apart from his technical competence. Apparently Harold felt that Ian Pullen reminded him of himself when he was Ian's age.

On commencing with Leak, Ian Pullen was briefed by Harold who required him to produce the finest FM tuner possible. The work environment at Leak meant that Ian had a free hand to direct the technical project work as he saw fit. He could come and go as he pleased from his development laboratory, provided he kept Harold informed on the progress of the project. In a passionate voice, Ian recalls his time at Leak prior to the 1969 take-over "as the happiest six months of my life".

All the development work on the FM tuner was undertaken at Brunel Road. The work was progressed in conjunction with suppliers that could provide the necessary components. RCA supplied the discrete transistors, Murata provided the ceramic filters for the intermediate frequency stages and Toko provided coils in the tuner. To maximise the tuner's suppression of spurious RF signals, the transistors were soldered into position against the plated metal chassis of the tuner (and so were subjected to significant heat during assembly). Norman Rowe of the Test and Service department recalls suggesting to Ian Pullen that the tuner could be difficult to service, and how Ian Pullen had looked aghast and replied "what do you mean 'when it needs servicing'?" – implying that the Stereofetic was most unlikely to fail.

In 1969 Leak discontinued the valve Trough-Line FM tuner, and replaced it with the Stereofetic FM Tuner which was of an advanced design. This tuner was the first application of Field Effect Transistors (FET) in the input stages to the tuner, and Integrated Circuits (IC), of which there are two in the intermediate frequency stages (although Truvox had used IC's in a tuner design too). The name for the tuner was coined by Ian Pullen. When he suggested "Stereofetic" to the staff at Leak it was initially met with negative responses like "what sort of word is that?". It was however, a "stereo" tuner, with "fets" and "IC's", so the name was adopted.

The tuner was very small and light compared to the normal valve tuners, and drift was not a problem as heat dissipation is low with solid state designs. However there was a drift problem with very early units, because a capacitor chosen by Leak was not drifting in accordance with its published temperature co-efficient. Subsequent production used a different capacitor.

A new feature introduced with the Stereofetic was the so-called "quasi-stereo" mode, where the treble range of the stereo audio signal was blended into mono. This provided a useful reduction in background noise while receiving weak signals, but still gave stereo effect in the bass and mid-range where background noise is not as apparent.

Ian Pullen

Ian Pullen was born in 1931 and was educated at Imperial College in London. He undertook studies there with his friend Angus McKenzie who would later become well known in the British audio industry. Ian had always been interested in audio and had assembled Williamson amplifiers during his youth. During his time at Imperial College, Ian Pullen and Angus McKenzie gave joint presentations and demonstrations to staff and students interested in hi-fi, which sometimes showed how good results could be obtained from quite different approaches. He worked with Leak from 1968 to late 1970, and moved to Rank's Yorkshire facility after the take-over of Leak. Before leaving the Rank organisation he designed the Leak Delta range of electronic products. In subsequent years he worked for Lustraphone designing an amplifier and tuner, and also operated his own consulting business. He retired to Eastbourne in East Sussex.

Leak also intended to produce another version of the Stereofetic that would have included an AM tuner within the same case. Some Stereofetic tuners even have the tuning capacitor for an AM section present inside. The design work on the AM section was under-way in late 1969, but when the company was bought by Rank this project work stopped.

The Stereofetic was recognised by the industry as a high performance tuner. It received glowing reviews, and won a design award. It was also used as the basis for the Leak Delta FM products that followed under the Rank organisation during the 1970's. According to Ted Ashley "Ian was a very talented Engineer" and this is borne out by the design approach to this tuner.

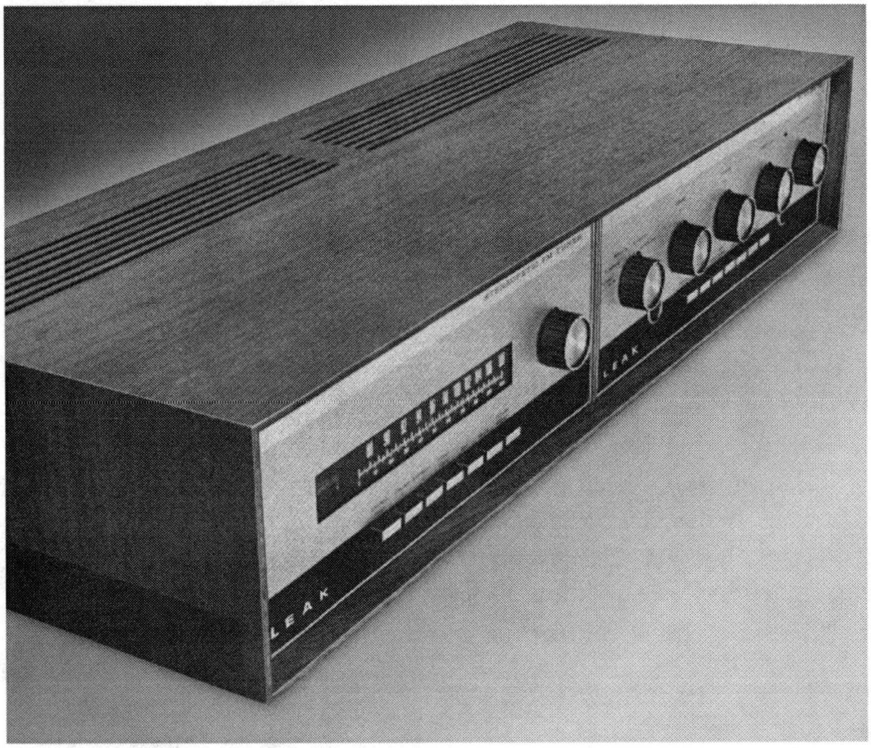

The Stereofetic FM tuner broke with the Leak tradition of symmetry in front panel styling - it is shown here with a Stereo 70 amplifier. At least one prototype of the Stereofetic had a circular tuning indicator where the left hand knob would have been. (Reproduced by permission Leak Electronics Ltd. Source: HFN Feb 1970)

"The factory-parlance referred to the Stereofetic as "Stereopathetic". This came about because of manufacturing difficulties caused by the mechanical design. The set was constructed on a chassis plate similar to the Trough-Line tuners. The I.F. and decoder printed circuits were mounted to this, but the tuner "front-end" section was actually an integral part of the whole plate. The trouble was that the most likely part to fail on test for drift or low sensitivity was the front-end. This meant re-building whole tuner!"

Ray Whitcombe, 1999.

Personal Recollections

by Ian Pullen

The short time Ian Pullen spent at Acton with H. J. Leak & Co. Ltd. was full of incidents – there was never a dull moment. Ian's recollections give an insight into the work environment at Leak and the relationships between staff:

"I had the use of any test gear equipment I asked for with no questions asked. I was *trusted* by the management to the best job possible and not to be unnecessarily wasteful; not something one finds a lot of these days. The only equipment I did not care to ask for was hot and cold environmental boxes. They were expensive and cumbersome. I got round the problem of testing between 5 and 40 degrees Celsius by opening all of the outside windows of the laboratory to let in the cold (it was still late winter) and by turning on all the radiators plus imported electric heaters for the hot tests.

"One time I had to go with Harold Leak to Bradford, to look at the Rank factory, and it was foggy as we belted up the motorway with him trying to show how powerful and fast his car could be. Apart from being scared to death at the speed, I was worried about a red light winking on the dashboard. 'Don't worry,' he said, 'it's the brake system warning light, but the garage tell me it's just the light that is faulty. There's nothing wrong with the brakes, and anyway the car has three independent systems, so you're quite safe.'

"The day after our return to London, Harold turned up at the factory in a Hillman Minx several hours late. All three braking systems of his car had failed simultaneously and he was lucky to have been able to stop by scraping along a kerb on a slight uphill. The service agents had collected their faulty machine in a surreptitious way by loading it into a windowless pantechnicon. It appeared that the three 'independent' systems were all dependent on a single micro-switch which had gradually failed on the journey to Bradford and back. I was never keen to go with him after that.

"Harold had a habit of suddenly wanting to do things – slightly different things – and one day I was working away in my laboratory when he poked his head round the door and said that we were all going off to lunch together. The 'we' was Ted Ashley, Fred Blackman, Phil Biggs, Frank Hawkins, Harold and me. So we piled into a couple of cars (I made sure I was in Frank's this time) and off we went. To the Post Office tower no less which, in those days, had a revolving restaurant with a panoramic view of London that repeated every hour or so. We had a leisurely and highly calorific meal as we watched the city sliding by and in the end, Phil had to point out that there was a factory somewhere down there which still needed running, and all the senior staff were lording it in the clouds. So Harold reluctantly let us return to our labours. I noticed that Frank, when he got behind the wheel of his Mk 10 Jaguar, held a bit of string which disappeared beside his right leg. When I asked about it he told me that the return spring on his accelerator pedal had broken, so he had to pull it back up all the time. This was with a hand which was controlling our cornering, etc. It almost made me want to be back in Harold's car!

Ian Pullen, designer of the Stereofetic FM tuner, photographed in 1991. (© I. Pullen, 1991. Reproduced by permission.)

10. Loudspeakers

The 550 Loudspeaker

During the time that the TL/12 was being designed, Harold didn't let his interest in loudspeakers wane, so other staff at Leak were assigned to investigations on possible speakers for a commercial product. Some early activities in loudspeakers (or "motors" as Leak referred to them) focused on suspension design and experiments on various types of cone surrounds with drive units from other manufacturers. (As early as 1936, Leak were listing an electro-dynamic speaker, the D.3, in company literature.)

"I joined the company in November 1945, a matter of weeks after finishing service with the RAF. I was employed in loudspeaker development because at the age of 16 I had been an articled pupil to P. K. Turner, and had gained experience in development, testing and construction of loudspeakers."
John Bolingbroke, 1999

In 1949 Leak released the 550 speaker. The company developed the 550 because they required a loudspeaker to conduct "live sound versus reproduced sound" demonstrations but couldn't find a commercially available speaker that they were completely happy with. The 550 was never produced in any quantity – it was made primarily for Leak's own purpose and product demonstrations, although it was advertised in Wireless World during 1950. The 550 was supplied to customers requiring a complete Leak system and so was made on a one-off basis. In total, less than a hundred units would have been made.

The cabinet for the 550 speaker was manufactured for Leak by an external supplier. The cabinet was lined with beaver board and had sheets of cotton-wool stuck to timber frames as baffles. The frames were about 50mm thick and there were about four or five of them within the cabinet. Other details are:

- separate bass and treble loudspeakers (possibly Goodmans)
- cross-over filter included
- available in Walnut, Mahogany, Oak, Elm or Sycamore

This model was not advertised beyond the early 1950's, possibly being deleted from Leak's product line around this time.

The Leak 550 Loudspeaker. (Reproduced by permission Leak Electronics Ltd. Source: WW Sept 1950.)

The Ionophone, Plessey and the "Duad"

Harold Leak got involved in talking to Plessey around 1953 because the latter had obtained the United Kingdom rights to a novel loudspeaker called the Ionophone (developed by the French company Audax). Leak was interested in this loudspeaker because, even in the early 1950's, Harold was searching for practical methods to achieve piston-action of the loudspeaker diaphragm.

With the Ionophone speaker, a very high voltage, radio frequency signal was applied to an electrode that caused the air to ionise, and so expand and contract in the vicinity of the electrode, in sympathy with the magnitude of the applied signal. By amplitude-modulating the input voltage with audio, the speaker would reproduce the sound of the modulating signal. A horn was attached to the speaker to increase the efficiency.

Major Falkus was employed at Plessey at this time and, when Harold Leak made inquiries about the Ionophone, Major Falkus suggested that they send some units to Leak for assessment. As well as conducting tests on the Ionophone, Leak used it in demonstrations at an Audio Fair.

From many points of view, the Ionophone was under-developed and had some practical problems. The speaker was driven by a pair of 807 transmitting valves and if the driving circuits became de-tuned the 807's could be destroyed. In addition, there were problems with spurious radio frequency radiation. While in use at the Audio Fair, Donald Chave from Lowther came across to Ted Ashley at the Leak stand and asked Ted to come upstairs to the Lowther stand where they were demonstrating FM radio. As Donald tuned the Lowther FM radio off-station, he could receive the sound from the Leak demonstration. Another problem with the Ionophone was that it produced excessive levels of ozone gas in a confined space. As a result of these problems Leak didn't pursue any further interest in the Ionophone.

The Ionophone was however eventually launched on to the market and Major Falkus became Fane Acoustics of Batley (from FAlkus and NEwbold, who was his partner). Fane Acoustics became dominant in the field of heavy-duty drive units for discotheques and high output sound systems.

An early version of the Ionophone that was developed by Plessey. The electrode structure at the throat of the horn is shown below. (Reproduced with the permission of Electronics World magazine, UK. Source: WW July 1956)

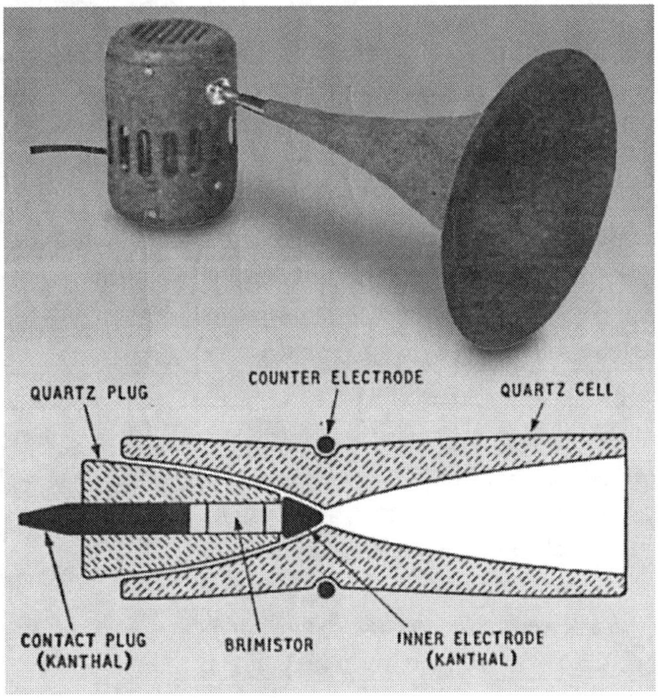

During the time that Harold Leak was in contact with Plessey, they showed him a 15-inch drive unit which had a pair of 2-inch high frequency units mounted on a pressed-steel frame across the face of the low frequency diaphragm. The unit was in turn modified slightly either by Leak or Plessey, whilst at the same time marketing plans were developed based on the name "Duad" for the loudspeaker (pronounced dew-add). This name became known to Peter Walker of Quad, who contacted Leak by letter, concerned that the "Duad" name resembled the Quad name far too closely. As it turned out the speaker wasn't put into full production – either because the commercial aspects were not viable or possibly the performance wasn't satisfactory. However Joe Middleton recalls a handful of units being shipped to Australia – they had a huge magnet structure and the cabinet carried the insignias of both Leak and Plessey. (Often Harold Leak could apply quite a lot of resources to a project and yet decide not to proceed, if results were not of sufficient quality.)

Ribbon Speakers

Another loudspeaker which exhibited piston action was the ribbon speaker. Harold Leak started to investigate these speakers by using a ribbon microphone from the BBC, acting as a loudspeaker – it was quite viable to drive these in reverse through a suitable transformer, particularly if the frequency range was restricted to the treble.

Following this initial investigation, prototypes using very large magnets taken from magnetron valves (used in high frequency applications like radar) were developed. George Wise was also active with ribbon loudspeaker development and had built his own prototype which was demonstrated to staff from Leak. It was an impressive demonstration and Harold Leak was very enthusiastic – so he visited George one weekend and offered to purchase the design. As a result George Wise made the prototype Leak ribbon speaker.

The prototype Leak ribbon speaker developed as part of a research programme into piston-action transducers. (Reproduced by permission Leak Electronics Ltd.)

The aim of the ribbon speaker project was to produce a high-frequency transducer to accompany a bass speaker, however the concept was never put into production. The ribbon did make it to prototype stage and as such it played a role in a research programme on suitable transducers for a high quality loudspeaker. This research was conducted around 1954 to 1955, and some of the results were published by Leak in a paper published in "The Journal of the Brit. IRE", Dec 1955, and in The Gramophone. The paper focuses heavily on the search for a piston-action loudspeaker. The ribbon loudspeaker was a contender as the actuating force is applied evenly across the surface of the ribbon diaphragm, thus preventing diaphragm "break-up" found with cone speakers at high frequencies.

George Wise in 1997 with the prototype ribbon loudspeaker he developed for Leak in the early 1950's. (Reproduced with permission from Steve Mann, of 'Silvernote', UK. Source: HFN June 1997)

The Leak ribbon speaker features a very light aluminium alloy diaphragm 2 inches long and 3/8 inch wide, and only a fraction of a thousandth of an inch thick. A powerful magnetic field is applied across the face of the ribbon diaphragm, which has the signal voltage applied along its length. No mention is made of an associated matching transformer but one was probably required to match the very low impedance of the ribbon (typically less than one-quarter of an Ohm) to the amplifier.

A strong magnetic field was required to obtain sufficient efficiency. The cost of such magnets limited the size of the unit (since a wide range unit needed a long ribbon and hence more magnets resulting in higher cost). The small size restricted the frequency range to 3kHz and above. A horn was also added to improve efficiency.

The measured performance of the ribbon showed a very smooth frequency response being +/-3dB from 2kHz to 20kHz, both on-axis and 30 degrees off-axis. Leak noted the excellent transient response of the unit and stated that "the ribbon loudspeaker has been the most faithful reproducer available in the past".

It seems that Leak rejected further work on this prototype on the basis of cost. Leak were searching for a suitable high-frequency unit to operate with a 15 inch bass unit as a two-way system. The cross-over frequency required was around 1kHz for the 15 inch bass unit but the ribbon could not quite make it down to 1kHz with reasonable cost. These days, of course, a three-way system would be used and the ribbon would perform very well indeed from 3kHz upwards. However H. J. Leak regarded this as "rather costly and cumbersome" in the mid-1950's, so the development work went to a point but the results didn't warrant any further investigations.

Electrostatic Speakers

Harold Leak also undertook research into electrostatic loudspeaker types. He had met staff from Pickering at an Audio Fair in New York, where Pickering had an electrostatic speaker that was of interest to Harold. One of these drive units was brought back to England for assessment by Leak. In these early drive units the vibrating diaphragm of the speaker was held in position (and electrically insulated) by a coating of flock or plush-pile material deposited on the diaphragm surface. This flock restricted the motion of the diaphragm but it was still viable to use the speaker for high frequencies. However it was found that the flock wore away with use owing to the high voltage gradient it was subjected to, so the speaker could not be made reliable. The limited viability of such a design was underlined when an electrostatic driver demonstrated at an exhibition had worn the flock material completely away. As the speaker had attracted a lot of enthusiastic interest, the reaction at Leak following the failure was pessimistic.

An improved prototype was then made, this time using a central diaphragm with two acoustically transparent plates either side, making the speaker suitable for push-pull drive (the previous prototypes were non-push-pull). Ananta Sarkar was employed by Leak and during his eighteen month stay conducted work on this type of speaker. This speaker was described by Harold Leak in a paper published in "The Journal of the Brit. IRE", Dec 1955, and also in The Gramophone. The company appears to have come close to production, as Leak stated in company sales literature that such a speaker would be available soon (around 1956).

The prototype was of the balanced push-pull variety, and featured very low distortion. In push-pull types of electrostatic speakers, the distortion is reduced by charging the central diaphragm via a large series resistor (as originally proposed by F. V. Hunt). Leak considered that it was possible to achieve distortions as low as 0.1%, which would have been excellent acoustically as well as from a marketing point of view (matching the distortion of the Point One amplifiers). The unit showed ultra-smooth frequency response from 1.5kHz to 20kHz, +/- 2dB, both on-axis and 30 degrees off-axis. In conclusion H. J. Leak stated that (in December 1955):

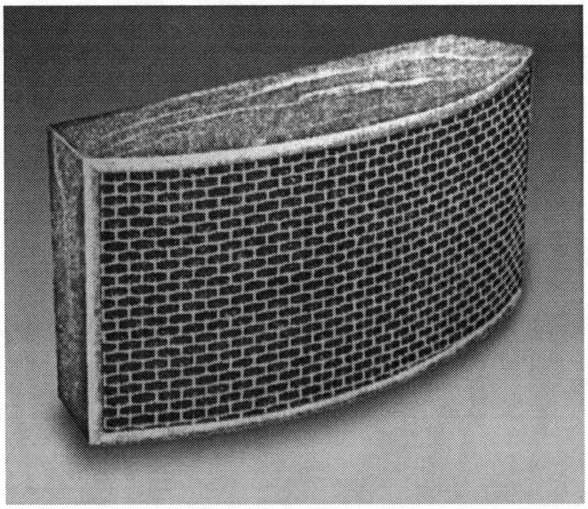

This is an early prototype electrostatic loudspeaker for frequencies above 1kHz. (Reproduced by permission Leak Electronics Ltd.)

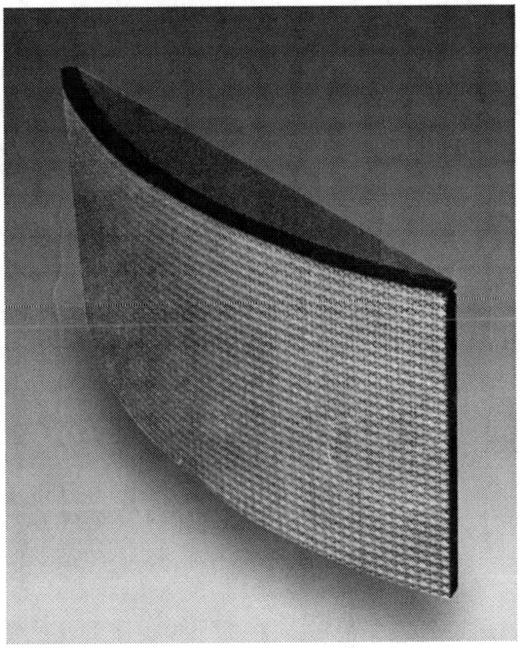

This loudspeaker was promoted in company literature as a forthcoming product in 1957, but it was never launched. (Reproduced by permission Leak Electronics Ltd.)

> "At present the author cannot produce an electrostatic bass loudspeaker having both the necessities of reasonable size and great reliability. This is because materials of suitable form and physical properties have not yet been made. The problems are technological and they will be solved in time – probably within a year."

This was a prophetic statement – but for Quad, not Leak. The research team at Quad were able to overcome such difficulties and launched an electrostatic speaker in 1956 (the first Wireless World placed advertisement appeared in March 1956). This speaker is still very highly regarded 40 years on, being one of the best loudspeakers ever made. Leak had knowledge of the work Quad were doing on electrostatics but there was no liaison between the two companies regarding the project work.

Harold Leak had envisaged using the electrostatic speaker in conjunction with a 15-inch bass unit mounted in a cabinet of about five cubic feet. But full production of the electrostatic did not eventuate, despite advertisements appearing for an anticipated launch in December 1956 of a revised design christened the "FREL":

> "FREL is the trade name of the Leak Full Range Electrostatic Loudspeaker which will be available to the public in 1957. The design is original and has great theoretical and practical advantages over previously-described electrostatic systems."

The basis of the design for the Leak "FREL" was described in a paper published in Wireless World (October 1956), the salient feature being the use of a central perforated plate, with two diaphragms placed either side of it. More conventional designs for electrostatic speakers use a single diaphragm with two perforated plates placed either side of it.

At the 1956 London Audio Fair, Leak, Quad and Goodmans all demonstrated electrostatic loudspeakers. The Leak unit was not for full-range, being for use above 1kHz only, with a 15" bass unit for low frequencies. (Quad showed the new full range unit that would become famous.) A similar 2-way electrostatic/moving-coil speaker system was demonstrated by Leak at the 1957 London Audio Fair. The bass enclosure featured "specially designed damping diaphragms" for smooth middle and bass. Even as late as 1959, just before D. A. Barlow was engaged on the Sandwich project, Leak was still demonstrating this hybrid loudspeaker.

Low output, fragility and un-reliability were the prime reasons why the electrostatic speaker project was stopped. Materials technology was a constraint that hampered Leak's efforts here, although this is often true of leading-edge development work (for example, the Leak Sandwich resulted from a constraint in existing materials technology being relaxed through the use of new materials).

Staff who contributed to the development work on electrostatic and ribbon speakers were Ted Ashley and Phil Biggs. Ananta Sarkar, who was previously from STC, and who had written a thesis on acoustic impedance measurements, was engaged by Leak to assist with this work for a period of about 18 months.

This measurement hut was established on top of the Brunel Road factory for making loudspeaker measurements in a reasonably reflection-free environment. Ted Ashley is shown here making final microphone adjustments just prior to a test run. Taken in 1955, this view faces north east – the top of a factory opposite Leak in Brunel Road can be seen on the left. (Reproduced by permission Leak Electronics Ltd.)

"The microphone was spun-out to about three feet away from the measurement hut. When you got about half-way through the measurement sequence, which was done manually by adjusting controls and jotting results down on a graph, then the 'four-twenty-two' train from Paddington would go through . . . it was great fun!"

Ted Ashley, 1998

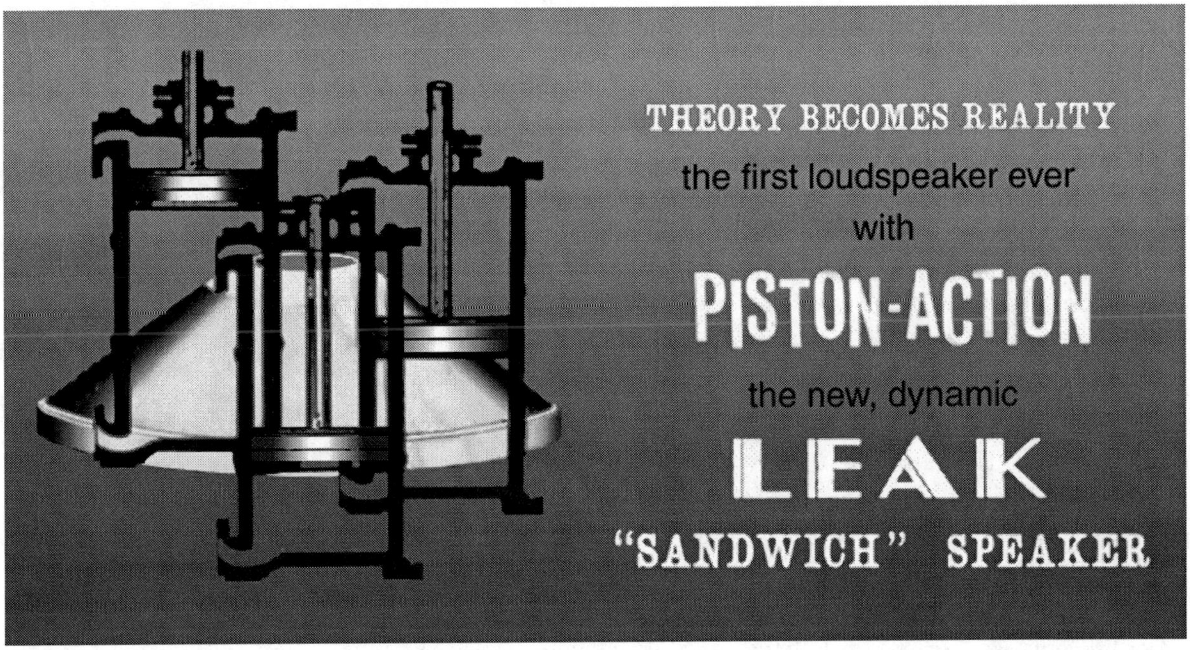

Leak Sandwich

In 1961 Leak released a revolutionary loudspeaker design that gave piston action of the diaphragm over the working frequency range of the drive unit. This was the Leak "Sandwich" loudspeaker, so named because the diaphragm was made from a thick section (approximately 10mm) of expanded polystyrene sandwiched between two outer skins of aluminium. Finally, after many years of transducer research and some false-starts, this product provided the company with the technical and commercial success it had been seeking in the field of loudspeakers. Leak promoted the Sandwich technology with a description prepared by Don Barlow and others as follows:

> "To understand the impact of this invention one must remember that to obtain low distortion the movement of a loudspeaker diaphragm or cone should follow as accurately as possible the waveform of the signals present applied to its speech coil. The ability of the cone to follow these impulses depends on the lightness and stiffness of the cone assembly.
>
> "Conventional cones (paper, impregnated fabric, plastics, aluminium) suffer from low stiffness. This results in significantly large areas of the cone vibrating in motions uncontrolled by the speech coil. The uncontrolled motions produce transient 'hangover', amplitude distortion and intermodulation distortion. These shortcomings are inherent in all direct radiator loudspeaker systems using conventional cones, irrespective of cost.

Above: The Leak Sandwich was well marketed, with graphic images bringing home the piston-action message.
Below: Another graphic representation of the immense stiffness of the Sandwich cone - the weight of 100 halfpennies is easily supported by the Sandwich material, with normal fibre-cone material flexing under the weight of just one halfpenny. (Reproduced by permission Leak Electronics Ltd.)

"All of the above shortcomings of conventional cone loudspeakers are overcome by the Leak Piston-Action Sandwich construction. Using engineering principles as applied to air-frame construction, the 'Sandwich' cone comprises stiff aluminium skins for the outer surface where stresses are greatest, bonded to a thick core of featherweight expanded polystyrene where stresses are lowest, giving far greater stiffness than the same total weight of either material used separately. This use of the most suitable material in optimum proportions and dispositions gives immense stiffness and rigid piston-action over a range of more than 6 octaves.

Opposite: The original Leak Sandwich speaker won recognition for its appearance from the British Council of Industrial Design and the Fashion Foundation of America. It received accolades in the hi-fi press for its neutral character and became a clear market success story. The sandwich construction technique was patented under US patent 3,111,187.

The search for piston-action had been a constant goal through loudspeaker development at Leak – the Ionophone, ribbon, and electrostatic types all have piston-style action. After the research into these types had not resulted in a viable product by the late 1950's, there was quite a degree of excitement at Leak when a paper published in the December 1958 issue of Wireless World discussed a new technique to achieve piston action with a moving coil speaker. This paper was by Don Barlow, and it was titled "The Rigidity of Loudspeaker Diaphragms". In it he discussed the use of a lightweight expanded polystyrene core for the bulk of the diaphragm, bonded to an aluminium foil skin for strength.

Harold Leak went straight over to visit Don Barlow just before Christmas 1958. Don Barlow was a Research Scientist with Aluminium Laboratories Ltd, in Banbury. He initially saw sandwich construction techniques for loudspeakers as a sideline

Don Barlow

Born in 1924, Don studied Metallurgy at Birmingham University and began his career in 1943 with Aluminium Laboratories Ltd, in Banbury, Oxfordshire. Don Barlow exhibited natural talents for research and product design that were evident well before he established his career in metallurgy. He also enjoyed applying these talents to various projects in his domestic life.

In 1955 Don Barlow read a paper by Hunt that sparked his interest in the deformation of record grooves from stylus pressure. This lead him to become more interested in other aspects of hi-fi, and the realisation that a combination of aluminium and polystyrene could make a promising basis for a loudspeaker diaphragm. After construction of a prototype, Don published a paper in Wireless World describing the potential of this new development.

Late in 1958, Harold Leak approached Don Barlow with an offer of employment. It took around six weeks for Don to decide whether he should accept the offer. Don had been working in a pure research environment, whereas the role at Leak would be one of 'production research' requiring refined products, suited to a demanding consumer market. But Don decided to take on the challenge and the Sandwich speaker range that was marketed by Leak or Rank-Leak for over 15 years resulted from his endeavours.

Over the years Don published around 25 papers in technical journals, 10 of which were as a result of activities in his own time. In 1972 he completed his PhD in Engineering which was undertaken as an external activity to his role with Rank Hi-Fi Ltd.

His employment with Rank Hi-Fi came to an abrupt end when his research department was closed in 1974 due to financial rationalisation. For a time he was with Fane Acoustics Ltd, but eventually he retired. Don died unexpectedly in 1988, survived by his wife Sylvia and family.

application. However Harold Leak made an offer that Don Barlow couldn't refuse, even though it meant a significant shift in Don Barlow's career pattern and also moving house. As it turned out, Don Barlow remained with Leak for quite some time, with his major contributions being on the Sandwich and Mini-Sandwich speakers. Don Barlow also undertook quite a lot of work on stereo pickups but a commercial product never eventuated from this research. (Leak did advertise a stereo pickup on the back of amplifier manuals but the product was never launched due to unresolved production problems.)

A special version of the Sandwich was supplied to the Australian Broadcasting Commission by H. J. Leak (Aust.) Pty. Ltd. The extra height added by an extension to the cabinet, put the high frequency unit closer to ear-level when listening, and also provided a space for a power amplifier. In time the ABC adopted the Sandwich as a universal monitor throughout many parts of the production and broadcast chain.

During the Sandwich speaker project, Harold Leak contributed at a high level to the guidance of the project. All the Sandwich development work was done at Brunel Road but the manufacture of Sandwich speakers moved to Downham Market in Norfolk around 1964. The Sandwich project received high priority in the work done at Leak and Don Barlow reported direct to Harold Leak. Improved acoustic measurement facilities were installed, with a full suite of Bruel and Kjaer test equipment.

Don Barlow was very motivated to get the best possible results from the underlying technology of the Sandwich speaker cone and to complement this with a well thought-out cabinet design incorporating many novel features. Given that the Sandwich technology effectively removed the possibility of uncontrolled cone motion, it was imperative that the cabinet be on a par, with well-damped panels. To achieve the precision result envisaged by Don, as much, or possibly more time, was spent on development of the cabinet as was spent on the Sandwich cone technology.

Don Barlow at the Brunel Road factory in the mid 1960's, with Steve Wilkins (far left), Bill Symons (left) and George Hill (right) looking on. (Photo kindly supplied by Sylvia Barlow.)

The Sandwich cabinet size was based on the maximum that was considered acceptable in a domestic environment, especially when two were needed for stereo. This was a working constraint for the project – such commercial decisions ensured that the product would be marketable and were determined mainly by Harold Leak, Frank Hawkins and Don Barlow.

The Sandwich bass unit departs from the norm of 12-inch or 15-inch diameter, being a 13-inch unit, mainly because the parameters Don Barlow wanted to meet meant that this diameter was required. All the castings were newly designed and made for this drive unit, so there was no need to make it compatible with any previous basket castings. The woofer roll-surround was supplied by a company in Dusseldorf, Germany, and was treated by Leak with a Bostik adhesive to make it air-tight. Two different type of surrounds are seen on Sandwich woofers according to age – yellow with an open weave, and black with a finer weave. Experiments were undertaken with red rubber surrounds, but these were not used in production.

The cabinet wall of the Sandwich speaker is deliberately not too thick – around 10mm of plywood with 12mm of bituminous damping material. Don Barlow spent

The Sandwich tweeter is mounted on the front frame to minimise diffraction, and the rear-foam lining protrudes from the cabinet to maximise acoustic absorption. The woofer has a high-compliance suspension and a die-cast basket. The cabinet is damped with bitumen impregnated "canite" and multi-cellular mineral wool. It is braced at the rear by a timber block clamped between the woofer magnet and the cabinet wall. A six-element 900Hz crossover is used.

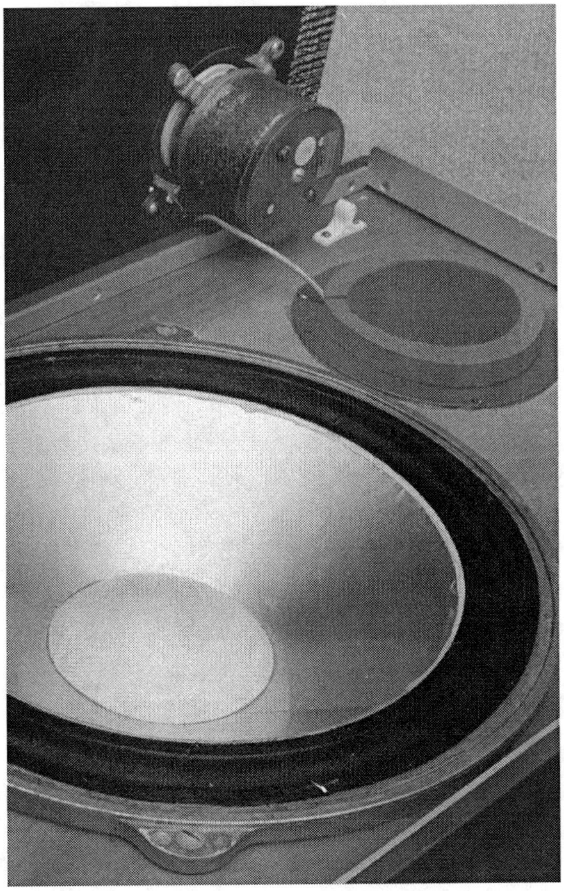

a lot of time measuring the Q (an indication of the lack of damping in a resonant system) of various strips of timber of differing thicknesses. These test samples included composite materials made from timber and tacky-bituminous damping material. These laminates were clamped at the edges to ensure that the test panel was homogeneous, rather than acting as a timber panel with damping material attached at a few locations. From these tests, the construction of the Sandwich cabinet was determined.

One of the biggest problems with cabinet production was get the cabinet manufacturer to place enough "goo" on the walls, and to make sure the entire surface area of the each wall was covered (cabinets were supplied to Leak by an external supplier). In some instances, cabinets were supplied with the damping material only attached in a few places – in which case it didn't do the job it was designed to do.

The Leak Mini-Sandwich claimed to sound identical to its larger brother with the exception of the lowest octave.

Another innovation was the bracing of the speaker magnet to the back panel of the cabinet through a block of timber, which also provided a convenient place to mount the crossover.

To minimise frequency response irregularities from diffraction effects, the inside of the tweeter chassis is chamfered, and the tweeter chassis casting is shaped to allow flush-mounting to the baffle board from behind. All production tweeters had the frequency response measured before assembly into the cabinets. A foam insert that surrounds the rear cavity for the tweeter is designed to contact the front baffle board to minimise acoustic interaction with the woofer.

Initially drive units for the Sandwich were assembled at Brunel Road. When the Downham Market factory opened in 1964, the drive units were assembled into cabinets there – but, after a year or two, the actual drive units were also assembled at Downham Market. After the closure of Brunel Road in 1971, drive units continued to be assembled at Downham Market, but then they were shipped to the Rank-Leak-Wharfedale assembly facility in Yorkshire.

Basket castings, magnets and un-laminated polystyrene cones were supplied to Leak from external companies (e.g. the basket casting came from a company in the Midlands). The design of the cone called for the aluminium skin to be attached during moulding but this never worked. Instead, the expanded polystyrene cones were delivered to Leak, coated with epoxy resin and laminated with aluminium skins at Brunel Road or Downham Market. In this way the crucial Sandwich technology was kept in-house.

The Leak Mini-Sandwich was portrayed as being easy to lift, despite some signs of distress from the model. (Reproduced by permission Leak Electronics Ltd. Source: HFN March 1968.)

The Sandwich woofer-cone is barely noticeable beneath the knee of the model in this promotional photo, taken on the occasion of the 1961 Audio Fair in Toronto, Canada. (Photo courtesy of the Leak family.)

The first Sandwich speakers had a conventional fibre-cone high frequency unit, but around 1965 Sandwich construction was introduced. One of the triggers to introduce this design revision was from commercial pressures. Some larger customers had pointed out that sandwich construction should be equally applicable to the high frequency unit, if the claimed advantages for sandwich construction were valid. The revised drive unit, which also featured an aluminium voice coil, didn't produce a significant change to the technical specification of the Sandwich speaker, although the audible results were a little different.

Harold Leak, in his office and presentation area, standing on the apex of a Sandwich cone with just a plate of perspex between him and the cone. (Reproduced by permission Leak Electronics Ltd.)

The design of the treble sandwich unit was a challenging task, as cone break-up at high frequencies can be utilised by speaker designers to extend the high frequency response of cone speakers* – although careful attention must be paid to the choice of materials and the construction technique to obtain smooth response. With the Leak Sandwich tweeter, the increased piston-action meant there was less opportunity to make use of break-up in the tweeter cone to extend the response. (See also a discussion of modifications in Chapter 12.)

The original Sandwich was produced up until just after the take over by Rank (in Mark 1 and Mark 2 form). The Mini-Sandwich was developed at Brunel Road and released in 1964. It used a 8-inch by 12-inch elliptical drive unit with a pressed steel chassis bought in from an external supplier.

Cabinets for the Sandwich speaker were supplied mainly by Design Furniture of Banbury. On occasions they could not keep up with demand, so GKD of Luton supplied the shortfall – but these cabinets were finished in Formica, not timber veneer. All the Mini-Sandwich cabinets were made by GKD, who were able to work with timber veneer by 1964. In the late 1960's prototype cabinets were made up with orange and lime green finish, but were never marketed. However a version finished in black vinyl and timber was advertised, but never sold because of a disagreement between Frank Hawkins and John Leak.

The Mini-Sandwich tweeter radiates at the rear into an open cavity, but an unusually shaped baffle was specially designed to optimise the frequency response, by making the acoustic path-length from front-to-rear vary across a range of values.

* P. G. Voigt made use of this characteristic and consequently patented a twin-cone wide-range speaker design under British Patent 413,758.

The Mini-Sandwich with its elliptical drive unit, mounted on a slight angle off-vertical, compared here to its big brother. A circular drive unit was not used because it would have been small enough to attract a higher tax rate than the elliptical unit. (Photo courtesy Philip Spicer.)

"Harold Leak was instrumental in much of the company's publicity. He was fair-minded and our advertisements reflect this. For instance, he was keen to emphasise the benefits of sandwich speaker construction, hence the well known image of him standing on a cone. In fact, his first idea was to stand Don Barlow on the cone, however Don was, to put it politely, on the heavy side and quickly dispatched the poor cone to the promised land! Harold was very disappointed, but realising that he was getting on for half the weight of Don, substituted himself with full success."

Ray Whitcombe, 1999

Leak had considerable success with the Sandwich loudspeaker, with it being used extensively in the professional audio industry and in homes both in Britain and overseas. Angus Gilchrist recalls that "There was a big demand for the Sandwich speaker and producing it was quite hectic at times. I used to help out with final testing on occasions. I went to Downham Market on one occasion for a week to help with a backlog of speakers awaiting test".

The marketing of the speaker brought home the benefits of sandwich construction with graphic images. At one of the London Audio Fairs in the 1960's Harold's daughter Hannah and John Leak's wife Janet both wore eye-catching aluminium foil garments that were very successful at marketing the Sandwich speaker, although it was hot work under the heat-reflective aluminium.

The Sandwich technology was used in Leak speakers up until the demise of the brand in 1978.

11. Rank-Leak

The Selling of Leak to Rank

Around 1968, Harold first spoke of the possibility of retiring. Several factors were at play here. As many people know, Harold had been a heavy smoker for some years and also had a collapsed lung (Harold died from emphysema in 1989). Also during the late 1960's, his wife contracted cancer. These health factors, plus the taxation arrangements being imposed by the Government on his business, led to a decision to retire. His wife underwent an operation, which fortunately was a total success. They both enjoyed a long retirement together on Jersey, having moved there in 1969. From time to time, Ted Ashley and his wife Alma visited Harold and Muriel during their retirement. On occasions other colleagues from Leak – John Bolingbroke, Julian Walcuch and Ernie Hord – also visited them.

Leak was bought by the Rank Organisation in January 1969. Rank had a diverse range of business interests and also owned Wharfedale, which it acquired in 1959. But there were other contenders for the purchase of Leak:

- An American company that manufactured automated car-park equipment was the first to register an interest.
- Ferrograph, which was owned by Wilmot Breeden, the car components company indicated interest through Dick Merrick.
- A. C. Farnell (of Leeds), who were the UK distributors of Tandberg.
- Associated Power Transformers (APT): this company was not in the hi-fi business at all and was partially Government funded.

Both Farnell and APT offered around £900,000 for Leak, and a deal was nearly finalised with Farnell when John Leak contacted Tony Swain and Tony Morpeth at the Rank Organisation. Rank had cash available from a foiled takeover of De La Rue (the bank-note printers) and John Davis was on the phone to Harold Leak within hours. Rank offered £1,034,000, so the deal was finalised.

As with any transitional period, staff morale can be affected. Some staff would have preferred the company to have been sold to Farnell or APT, as it was perceived that Leak could retain its culture and identity better that way. The tie-in with Wharfedale at Rank highlighted some differences and similarities between the companies. Leak were more advanced on the electronics front at this time and also had an advanced loudspeaker, although a much smaller market than Wharfedale.

"I joined the Rank Organisation in Bradford as Marketing Manager for Wharfedale and Leak in 1970. I met Frank Hawkins and visited the factory at Brunel Road. He showed me around and I remember asking him what was the average employment duration of factory staff; he told me 7.5 years. This was a good statistic for a small factory. . . . (I knew) that my Lords and Masters were planning to move the whole thing up to Bradford very shortly thereafter. True to form, the deed was done and not surprisingly, some of the technology and manufacturing consistency fell off the removal trucks."

Gordon Provan, 1999.

Gordon Provan, previously of the Rank Organisation, and currently Chairman for the British Federation of Audio. (Photo circa 1998 and courtesy of the BFA.)

Ted Ashley continued with the newly acquired company, leaving the technical role for a managerial role and becoming acquainted with John Collinson (previously of Quad) who was part of the Rank-Wharfedale operation.

LEAK
Ask someone who knows.

The Rank-Leak logo reflected the fashionable geometric shapes of the early 1970's. (Reproduced by permission Leak Electronics Ltd.)

The Rank-Leak Era

Brunel Road was closed in 1971, factory operations were moved to Airedale Works, Apperley Bridge, Bradford, Yorkshire. After the take-over, about 50 staff from Brunel Road left the Leak company, plus other losses from Downham Market. Before the move to Yorkshire, a coach-load of staff and their wives travelled to Yorkshire to assess the possibility of moving but eventually most staff opted for redundancy.

Harold Leak left in January 1969 immediately the company was taken over, and Fred Blackman (Sales Manager) left after eight months. Frank Hawkins agreed to stay until the operations were moved out of London, so he eventually left in 1971. Norman Rowe, the Service Manager, moved to the Rank servicing operation at Chiswick. Ian Pullen and Ted Ashley both moved to Yorkshire – Ian remained with Rank until late 1970, while Ted (who had moved into management at the instigation of Frank Hawkins) was with Rank until 1981. Don Barlow re-commenced his loudspeaker research with Rank after leaving Leak around 1968.

Soon after the move, Ted Ashley had to build a new engineering team, mainly from staff that had not come from Leak and so had not had any exposure to the Leak idiom. For a time this caused some problems but with assistance from Ian Pullen, he had assembled a skilled team of engineering staff.

Ted Ashley recalls that during this time Downham Market was used for assembly of loudspeaker drive units (previously done at Brunel Road), and that these were then shipped to Yorkshire where the Rank cabinet making facility was located. Downham Market was also used for electronic sub-assembly work, both by Rank-Leak and Rank-Bush-Murphy (located at Skegness).

The Japanese electronics industry was really beginning to bite into the market by the 1970's. Many parts for production were also bought from Japan, such as FM-tuner front ends. So the Japanese industry was selling product both to end-customers and manufacturers in the UK. Another area in which the Japanese industry was well

H. J. LEAK & CO., LTD.
BRUNEL ROAD,
WESTWAY FACTORY ESTATE,
LONDON, W.3

Please note as from 14th January 1969 our new telephone number is 01-749 1641 (10 lines).

H. J. LEAK & CO., LTD.

A card advising of a revised phone number from January 14 1969, issued two weeks after the Rank take-over of H. J. Leak & Co Ltd. (Card courtesy of Ray Whitcombe.)

The Delta range of equipment as it was in 1973 allowed a customer to choose from several different models to build a complete system. (Reproduced by permission Leak Electronics Ltd.)

advanced was in manufacture of appearance-parts (knobs, escutcheons, etc). In the UK, these parts were mainly made by companies associated with the automotive industry but Japanese appearance-parts were made by small niche businesses that could provide a wide range of options.

Leak Delta Series

In 1971 Rank face-lifted the appearance of the Stereo 70 and Stereo 30 Plus amplifiers, but left the electronic design untouched, to create the very modern looking Delta 70 amplifier (and its sister the Delta 30).

The commercial focus from Rank-Leak at this time was to provide a complete range of hi-fi products, as well as to allow for some choice in price and performance, so customers could tailor a system to suit their needs. Accordingly, three new loudspeakers were introduced (150, 250 and 600), all using Sandwich diaphragms, with the exception of a new plastic-dome tweeter.

Opposite: The three-way Sandwich 600 released by Rank-Leak included a plastic dome tweeter that was also used in some Rank-Wharfedale loudspeakers.

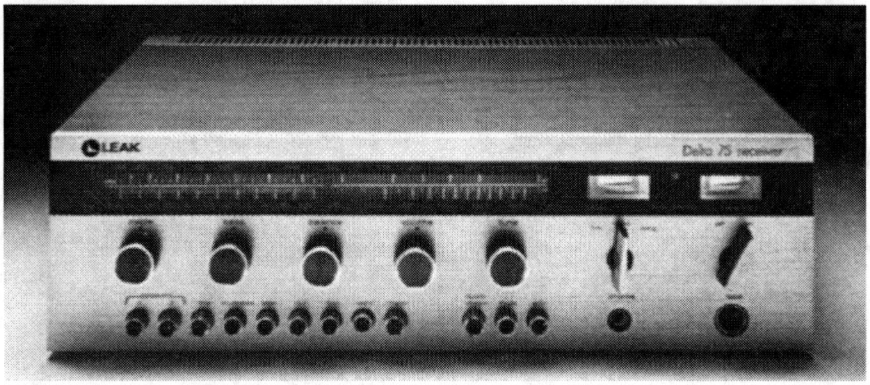

Leak's first Tuner-Amplifier appeared in the Delta range of equipment as the Delta 75 Receiver. (Reproduced by permission Leak Electronics Ltd.)

Two radio tuners were available – the Delta FM and the Delta AM-FM tuner with long and medium wave bands on AM. A novel feature for reducing noise on distant FM stations was the "quasi-stereo" function, which Rank promoted as a Leak invention (it was first introduced in the pre-Rank Stereofetic FM tuner).

For the first time Leak was able to offer a tuner-amplifier as part of its product range – the Delta 75 Receiver. Leak had considered developing a tuner-amplifier as far back as 1960 but the idea did not proceed due to the size and weight of a combined unit based on valves.

A Leak Delta turntable was also included in the range, being supplied by Lenco in Switzerland and fitted with a Shure M75/E cartridge.

The Delta 70 amplifier was a face-lifted version of the Stereo 70. (© Philip Spicer 1999. Reproduced by permission.)

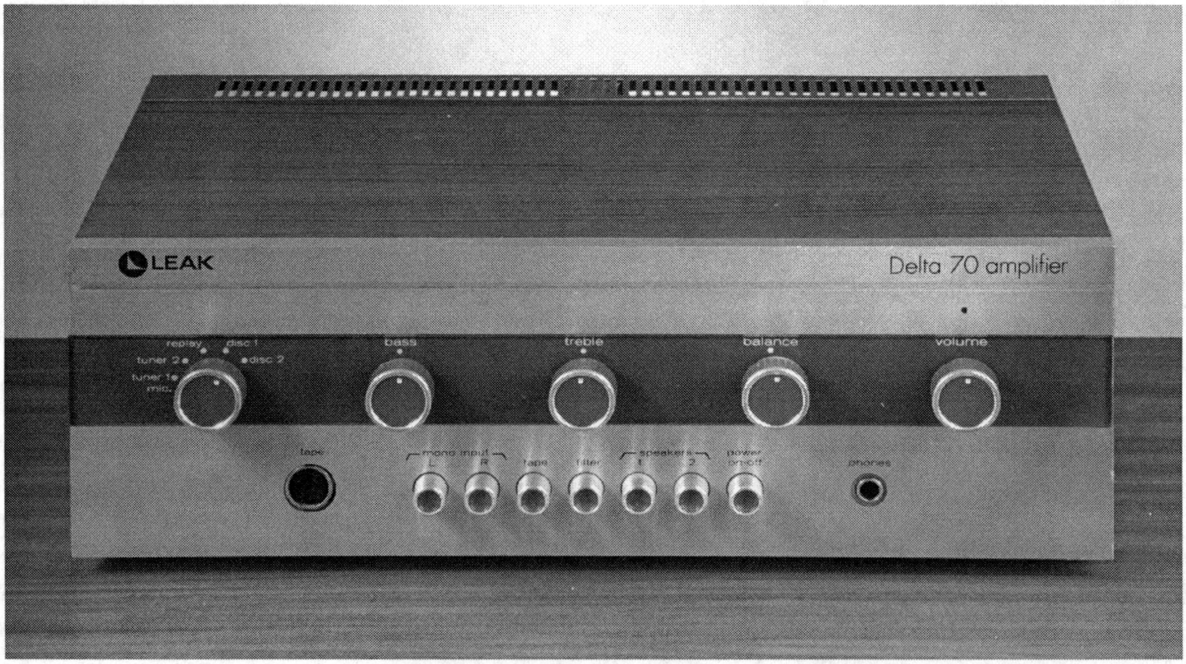

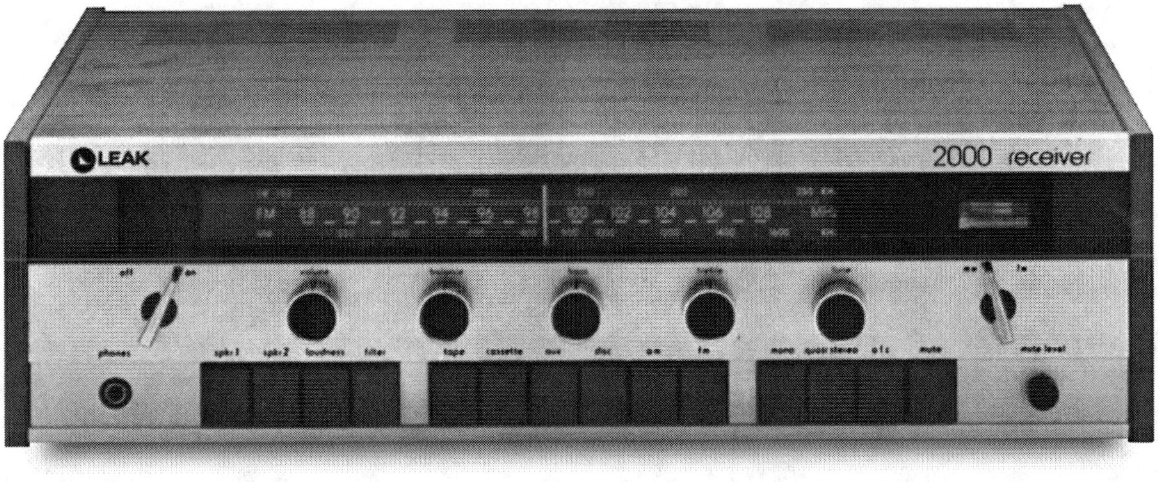

The 2000 series receiver with 30 Watts per channel. Like the 2300 Tuner, the 2000's FM section used an FET front end, while the IF-strip used IC technology, in a configuration based on the original Stereofetic from 1968. (Reproduced by permission Leak Electronics Ltd.)

Ian Pullen worked on the design of electronic products in the Delta range which was primarily manufactured in Leeds. However, all solid state amplifiers were made at Brunel Road up until 1971, after which the factory was closed and operations moved to Leeds.

Around this time Leak (and Wharfedale) were transferred into Rank Radio International, a division focused on the manufacture of Bush and Murphy television sets. Gordon Provan of Rank Hi-Fi recalls: "Leak and Wharfedale became lost and neglected in this huge single-product division, and by 1974 Rank Radio was losing huge amounts of money . . . At this time I was asked by Sir John Davis if I would take Leak and Wharfedale out of Rank Radio, and put them with my other (audio-visual) businesses – and that is how Rank Hi-Fi was born."

Leak 2000 Series

After Ted Ashley and Ian Pullen established the new electronics engineering department in Yorkshire, there was a very significant period of work in which the new Leak 2000 range was designed. The product plans for this range of equipment were extensive, requiring a committed team of designers. On the whole Ted Ashley considers this range represents the finest electronics that was ever produced under the Leak name (although the Leak 2002 top-loading cassette deck was imported from Nackamichi). The summary specification for the electronic components in the range is:

1800 Tuner-Amplifier: 20 Watts per channel
2000 Tuner-Amplifier: 30 Watts per channel
2100 Amplifier: 30 Watts per channel
2200 Amplifier: 45 Watts per channel
2300 Tuner: FM with long and medium AM wavebands

The 2000 series included four different loudspeaker models. The 2075 loudspeaker was the top of the range, being a very large four way system – on a par in size with many refrigerators. The mid-range and high frequency speakers were housed in a separate, rotatable cabinet on top of the bass cabinet. Other speakers in the range were the 2020, 2030 and 2060. All drivers in the 2000 range used Sandwich cones as follows, except the treble units which were generally a dome style unit:

2020 speaker: 2 way, 170mm bass/mid, 30mm plastic cone treble, 15 litre sealed cabinet

2030 speaker: 3 way, 200mm bass, 100mm mid, 25mm dome treble, 25 litre reflex cabinet

The 2000 series was very extensive. Loudspeakers shown are (smallest to largest) 2020, 2030, 2060 and 2075. A 2000 series turntable on the left, and the 2001 transcription turntable on the right. In the middle, left to right are: 2100 amplifier, 2000 Receiver, and 2200 amplifier. In the front is the 2300 tuner and the 2002 Leak cassette deck sourced from Nackamichi. (Reproduced by permission Leak Electronics Ltd.)

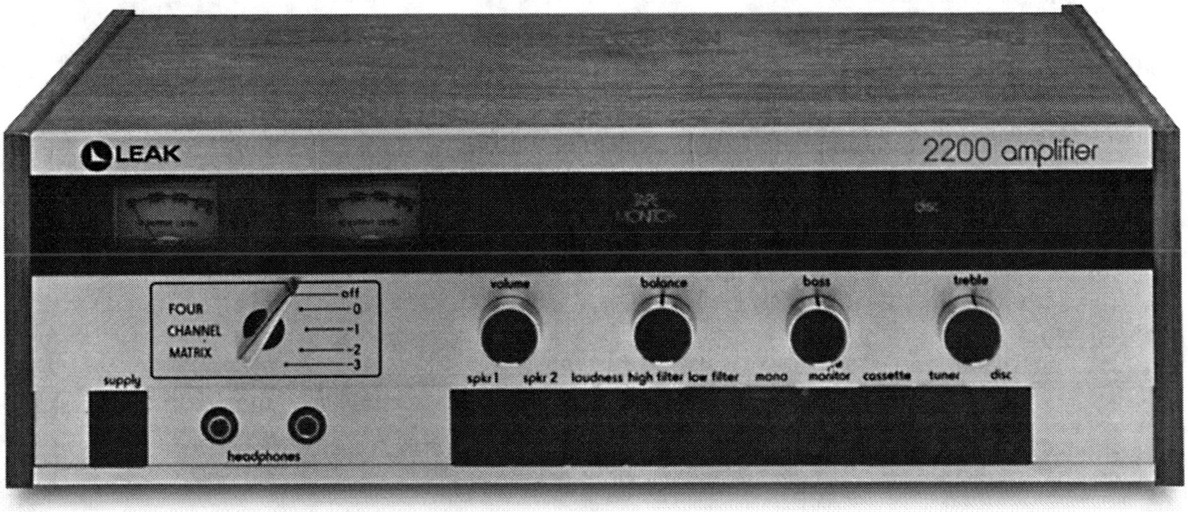

2060 speaker: 3 way, 300mm bass, 100mm mid, 25mm dome treble, 60 litre sealed cabinet

2075 speaker: 4 way, 380mm bass, 170mm mid, 100mm upper mid, 25 mm dome treble, 125 litre transmission line cabinet.

The 2000 range of amplifiers were direct-coupled to the loudspeaker, unlike the Delta 70 and its predecessors which were capacitor-coupled. This is the 2200 amplifier, rated at 45 Watts per channel. (Reproduced by permission Leak Electronics Ltd.)

For a time the Delta turntable appeared in Leak brochures for the 2000 range alongside a new 2001 transcription turntable but eventually only the 2001 was listed as part of the range. The 2001 turntable was belt-driven from a 10-pole motor, was fitted with an Audio Technica 1005 pickup arm, and had a novel suspension system. The suspension was developed by Don Barlow and his team while he was with Rank Hi-Fi (although many samples were taken home by Don to test under practical circumstances). The suspension was marketed under the name "spherodynamic" and it had a very low resonance of around 1Hz, which made it very effective at isolating the turntable from vibration and knocks.

Leak 3000 Series

In September 1976, Rank decided to cease manufacture of all Leak electronic equipment in the United Kingdom, so all 2000-series amplifiers and tuners were discontinued. To replace these products, Rank entered into an arrangement with Roland Electronics (Rotel) of Japan for supply of "factored" electronics that would bear the Leak name and be sold only in the United Kingdom as part of the new 3000-series.

In parallel with this revised approach to marketing of electronic products, Rank applied considerable resources to develop the 3000-series loudspeakers. For the first time since 1961, Leak moved away from Sandwich cone technology, and developed a range of loudspeakers that used more conventional bextrene-coned bass and mid-range drivers, accompanied with plastic soft-dome tweeters. The cabinets for these drivers featured a novel polyurethane front panel that had a "stepped" front face, allowing the drive units to be "time-aligned" by placing the acoustic centres in the same vertical plane. Four new models were introduced as a result of this development work – the 3020, 3030, 3050 and 3080.

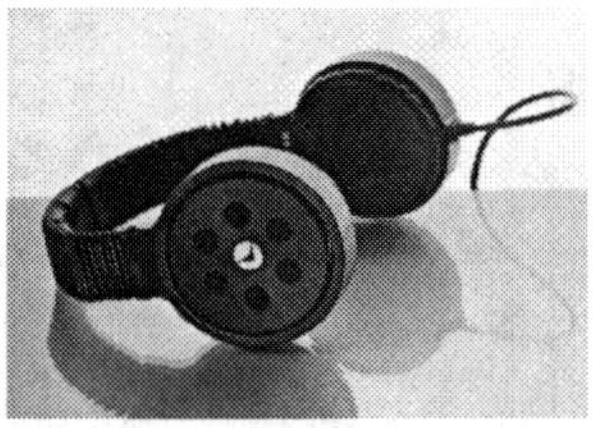

Leak 3000 Iso-dynamic headphones used the same operating principles as the 3090 tweeter. (Reproduced by permission © Leak Electronics Ltd.)

However, Sandwich technology was retained for the top model in the 3000 speaker range. The 3090 was a revised version of the 2075 loudspeaker, and had the same 380mm bass unit with transmission line loading. The lower mid range was a Sandwich unit of 170mm diameter. The upper mid-range unit was re-designed using lasers to examine cone break-up behaviour, which resulted in a cone that featured a series of patterned perforations to reduce cone break-up. Finally the dome tweeter of the 2075 was replaced with an "iso-dynamic" tweeter for the 3090. This tweeter had a flat diaphragm 25 microns thick, with a thin electrical circuit etched on its surface. A special arrangement a rare-earth magnets behind the diaphragm, and a slotted front plate established the necessary magnetic field, allowing the diaphragm to be driven evenly over its surface (in a manner similar to conventional ribbon tweeters).

The 3090 had a much improved appearance compared to the 2075, and due to the resources applied by Rank is probably one of the finest Sandwich speakers ever made. They have been acclaimed by recognised authorities, but by the time they were released the Leak name was less well accepted in the market place and the 3000 series of Leak speakers did not sell as well as anticipated.

From the outside: 3080, 3050, 3030 and 3020 Leak speakers. These models used bextrene cones, with a soft-dome tweeter. The "time-aligned" front baffle was novel, but almost certainly would have also caused a reflected image of the tweeter unit from the large step beneath it. (Reproduced by permission Leak Electronics Ltd.)

The strategy by Rank of obtaining electronic components from Japan to be sold under the Leak name was not well received by potential customers and this would have had an impact on loudspeaker sales, even though these were made in the United Kingdom.

At around this time, both Leak and Wharfedale had not been performing well for Rank, and Gordon Provan was asked to rescue the bigger of the two – Wharfedale. He remarks – "At the end of the day, despite many proposals by my colleagues and myself to salvage it, the (Leak) business was closed down."

The 3090 was a revised version of the 2075 speaker. It was probably the finest Sandwich loudspeaker ever made, although the horizontal layout of the high frequency drivers defies convention and could contribute to an irregular directivity pattern. It was a four way system, with a 380mm sandwich bass unit, 170mm sandwich mid-range, 100mm upper mid-range, and an iso-dynamic tweeter with "ribbon-like" action. (Reproduced by permission Leak Electronics Ltd.)

The Finale

In 1979, Rank-Leak products were discontinued completely. The passing of the Leak name was noted in John Crabbe's "Hi-Fi News" editorial of December 1979, where he reminded readers that "Harold Leak's products were held in great esteem at the apex of British hi-fi". He went on to note that "Leak has faded away despite the backing of a vast business organisation".

The success of the Leak company, particularly from 1945 to 1969, came about from a number of factors. The company was technically very competent, and was always in search of novel approaches to problem solving. Another important factor here was Harold Leak's astute business sense, coupled with an ability to identify suitable collaborators for product development.

The development of the standard-setting TL/12 amplifier surely allowed the company to establish a considerable market lead, with subsequent recognition of other Leak products. However technical competence alone cannot ensure commercial success. Products have to be developed at the right time for acceptance in the market place and be commercially viable. The TL/12 was in the right place at the right time, as was the mass-market TL/10, the Sandwich loudspeaker and the transistor Leak amplifiers which sold so well.

Leak not only demanded high technical performance from its products but also demanded high quality construction, appearance and finish. However, these considerations were tempered always against commercial viability. Leak provided a combination of performance, cost and quality that suited broad market requirements, and so ensured the success of the name.

At the time of writing, International Audio Group Ltd of Huntingdon is the current owner of the Leak name (as well as Quad and Wharfedale). There are plans being developed to re-launch Leak products in the near future, possibly with the release of valve-based amplifiers for high-end audio applications.

There can be no doubt about the extraordinary impact that Harold Leak and his staff had on British hi-fi. His company of just fifty people at Brunel Road, Acton made products that were internationally famous, and recognised as amongst the best available. This reputation will continue for ever.

"Father told me he had a conversation with Frank Hawkins, just before Frank died, about what might have happened had they kept the company. Not surprisingly the discussion centred on the Japanese competition, still negligible when they sold out. Frank said 'we would have seen them off'. The point was that the management team at Leak had worked out for itself just about all the factors that made the Japanese so successful: economical design, efficient production, and not least, treating the staff like human beings and listening to their ideas."

John D. Leak, 1999.

Opposite: A striking example of a Leak Stereo 60 in fully restored condition.

LEAK

.... the first name in High Fidelity since 1934

EL 34
6CA7/
EL34
EL 34
EL 34
STEREO 60
LEAK
MADE IN ENGLAND

12. Reference Data

Circuit Diagrams

Leak circuit diagrams were drawn in accordance with guiding principles laid down by a recommendation for "Inter-Services Graphical Symbols for use in Telecommunication Engineering" (as stated on some Leak circuit diagrams). One of the leading specialists in the area was L. Bainbridge-Bell, who published articles in Wireless World on the subject (during the time he was with the Royal Naval Scientific Service).

At Leak, one of the draughtsmen who drew circuit diagrams was Tom Martin - he was with Leak during the 1950's. Ron Innes was also a draughtsman (along with his role in the production department).

The following Leak circuit diagrams are provided for reference, and have been reproduced by permission of Leak Electronics Ltd. Whilst most audio amplifiers are relatively easy to service using basic equipment, this is not usually the case with radio tuners. If an FM tuner needs re-alignment, best results will be obtained by selecting a servicing company with expertise and appropriate test equipment, rather than attempting to align it yourself.

Tom Martin in 1952. He was a Draughtsman at Leak and a member of the Leak Athletic Football Club. (Photo courtesy E. H. Ashley).

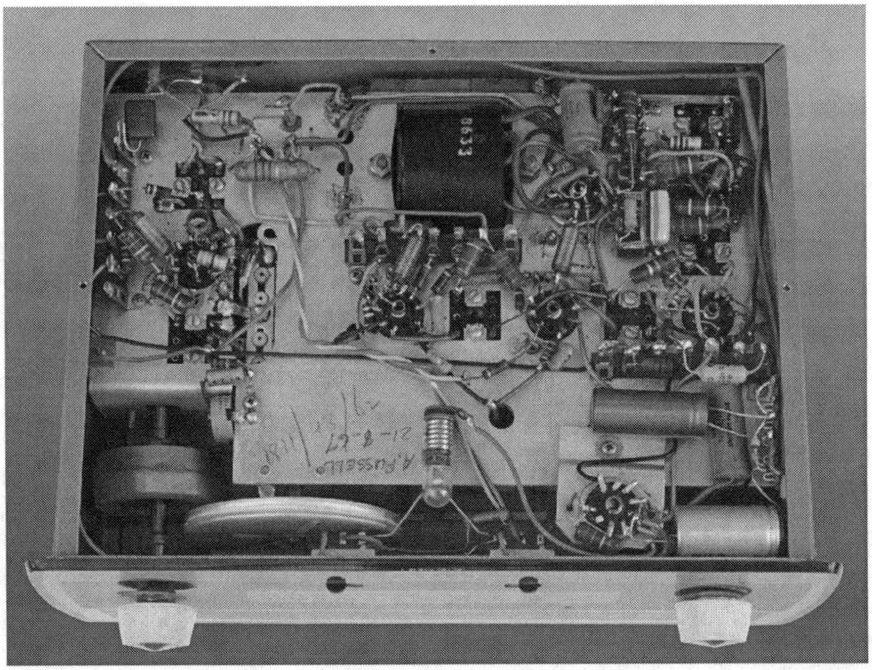

The under side of the Trough-Line II tuner – this unit has been modified to include a panel lamp to illuminate the front panel logo, and has an extra DC supply (derived from the heater winding) for operating an integrated circuit stereo decoder.

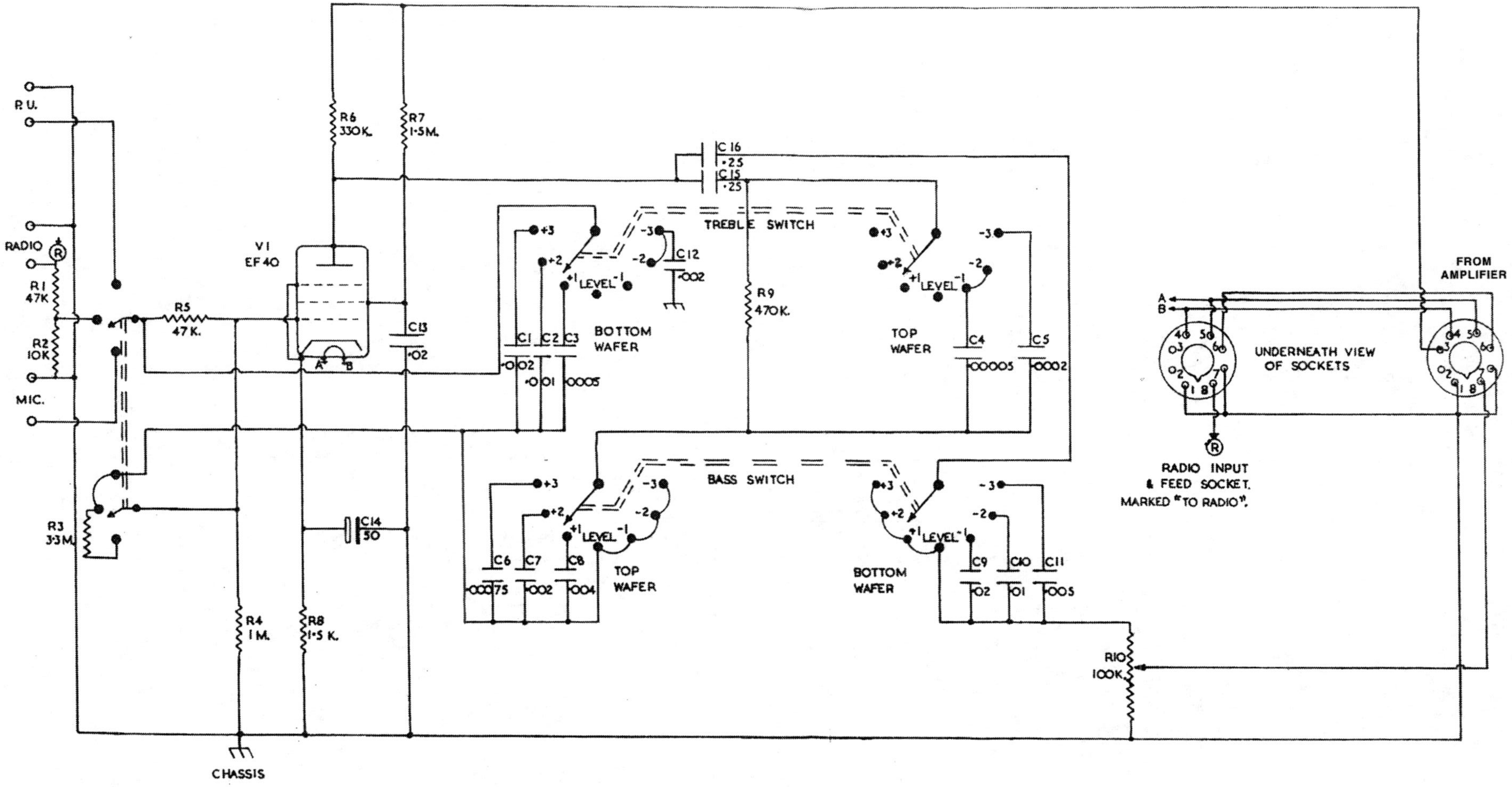

RESISTANCE VALUES SHOWN IN OHMS.

CONDENSER VALUES SHOWN IN MICRO-FARADS.

Reproduced by permission Leak Electronics Ltd.

Leak RC/PA Pre-Amp

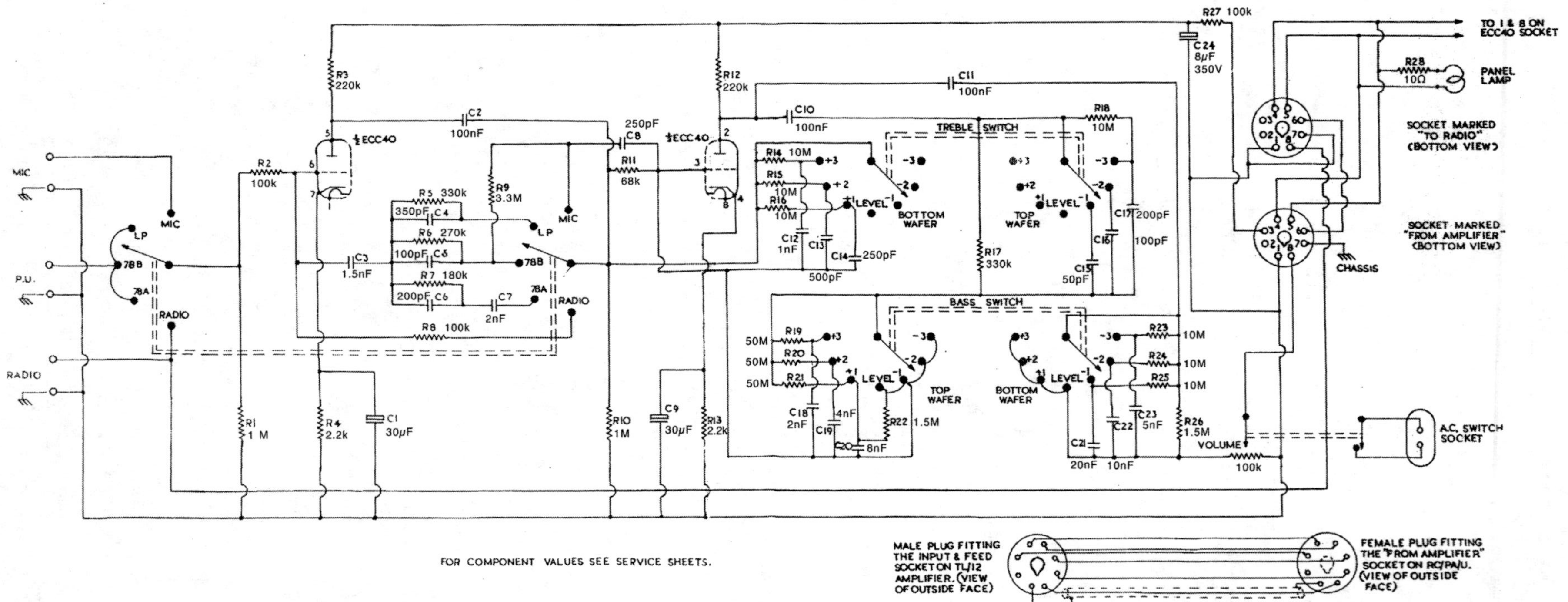

H.J. LEAK & CO., LTD.

CIRCUIT DIAGRAM OF RC/PA/U PRE-AMPLIFIER.

Leak RC/PA/U Pre-Amp

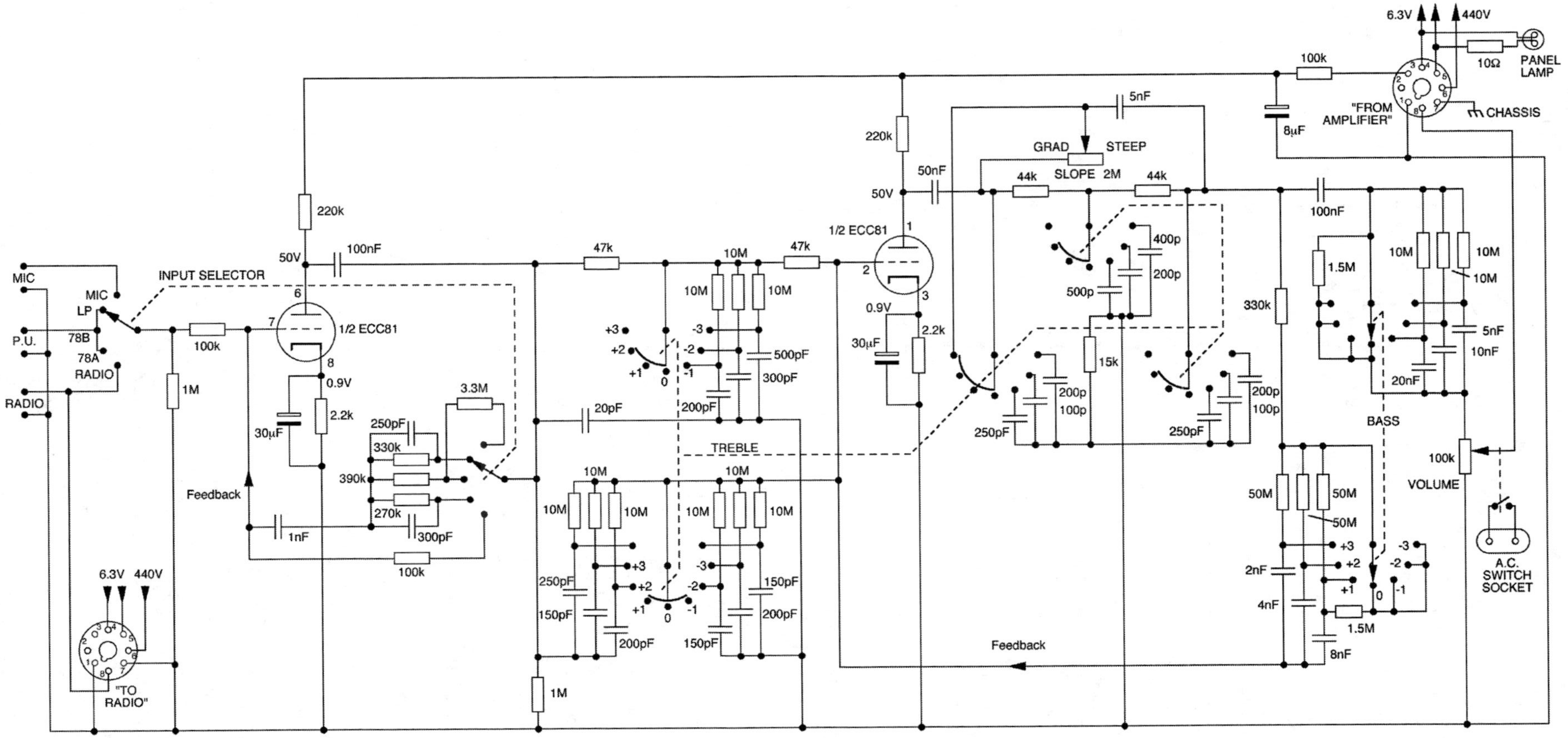

Circuit by S. Spicer, 1999.

Leak Vari-Slope Pre-Amp

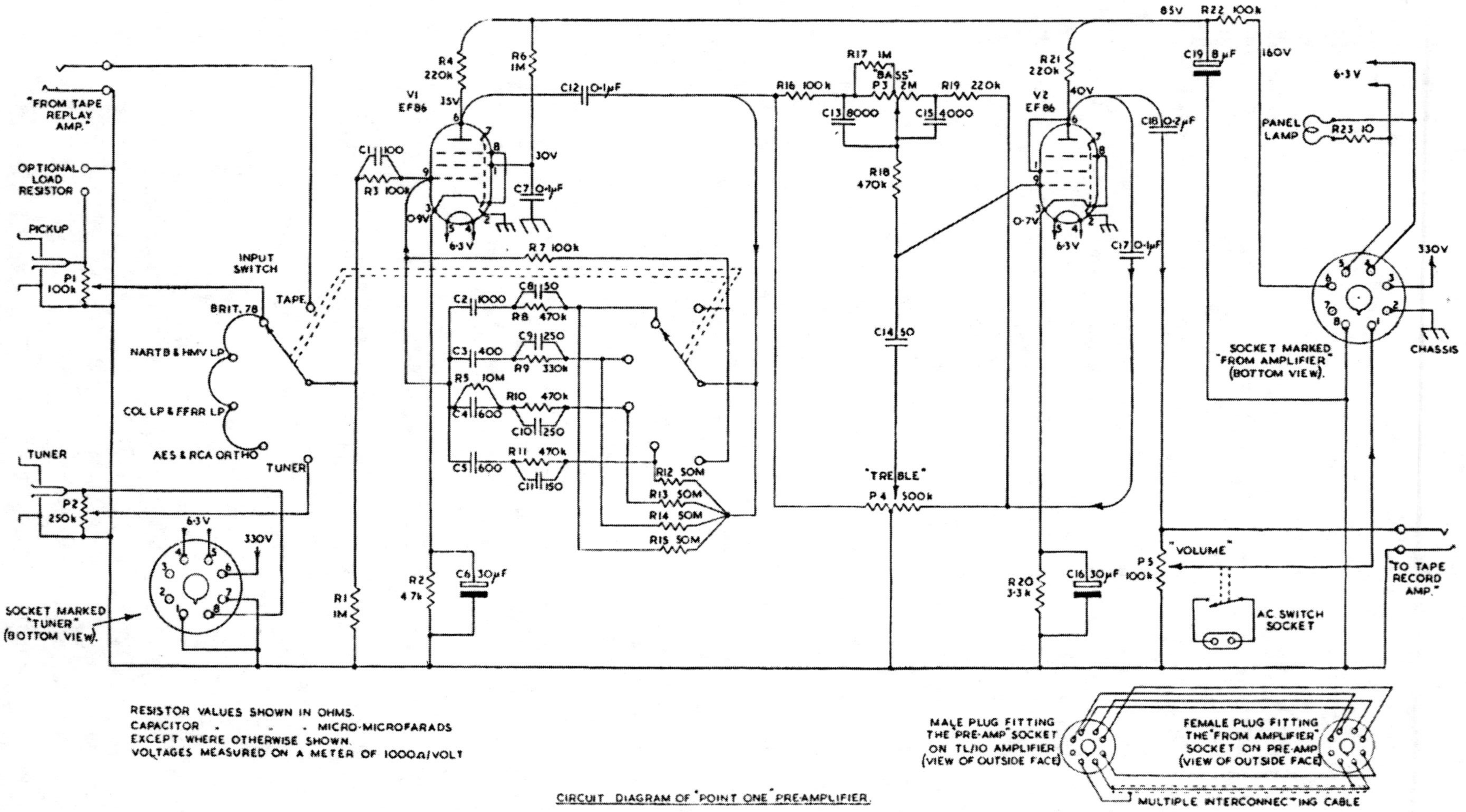

CIRCUIT DIAGRAM OF "POINT ONE" PRE-AMPLIFIER.

Reproduced by permission Leak Electronics Ltd.

Leak Point One Pre-Amp

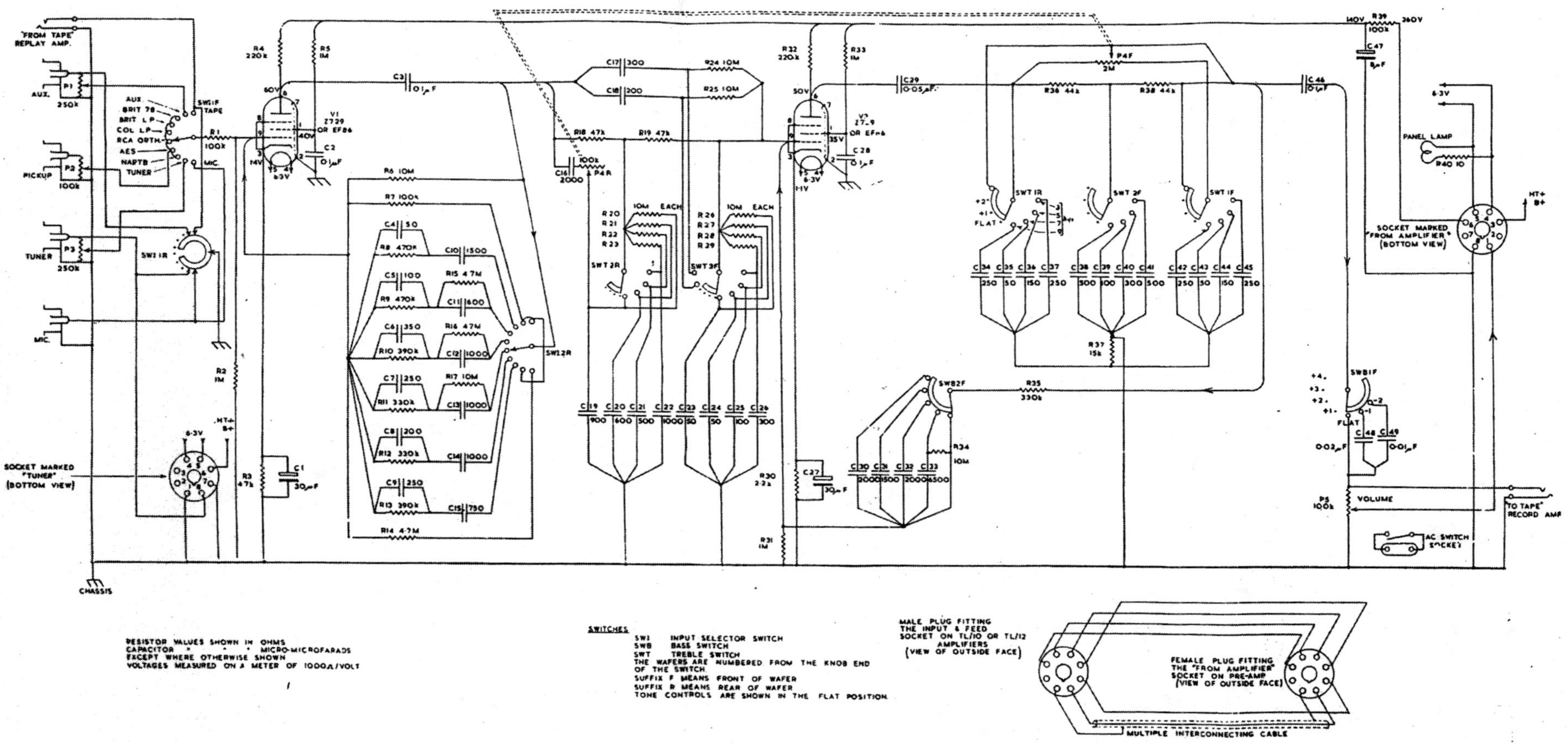

Leak Varislope II Pre-Amp

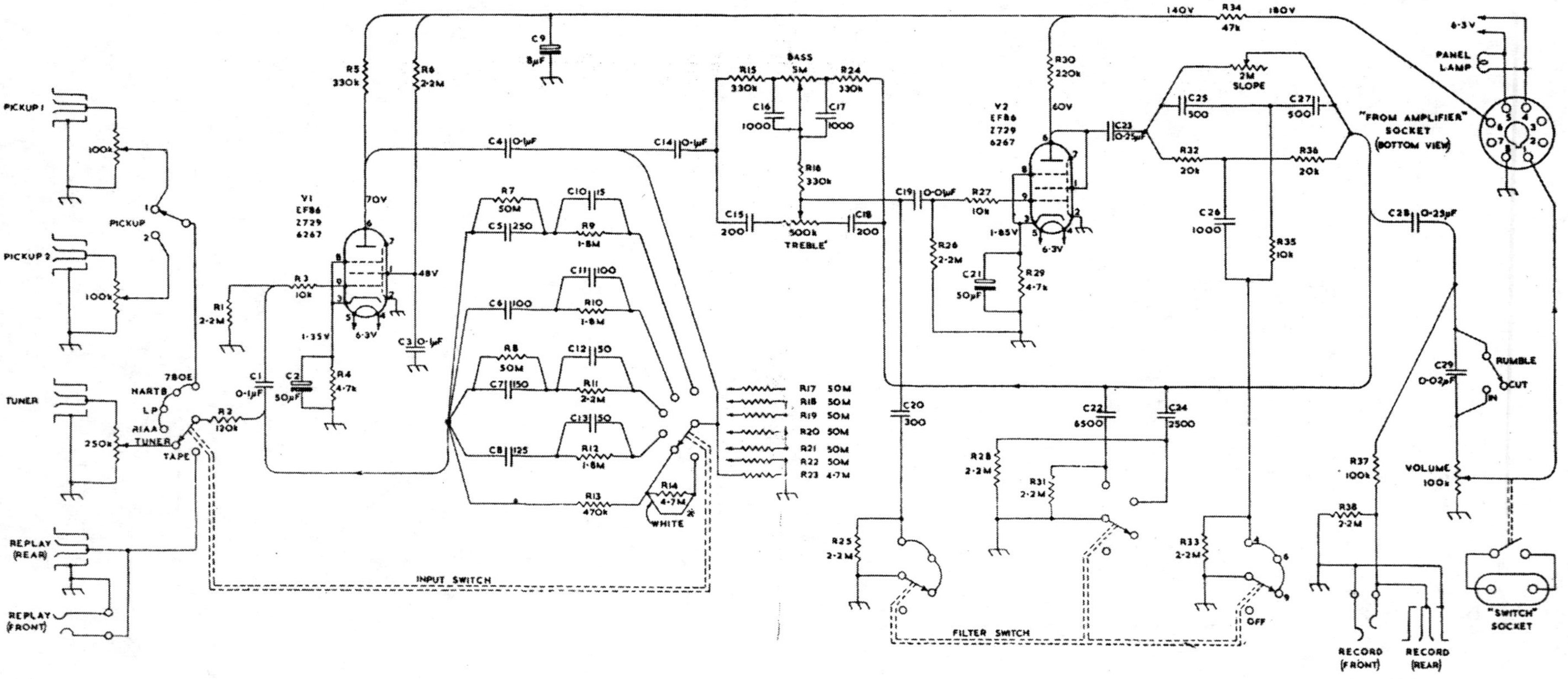

RESISTOR VALUES SHOWN IN OHMS
CAPACITOR VALUES SHOWN IN MICRO-MICROFARADS
EXCEPT WHERE OTHERWISE SHOWN
VOLTAGES MEASURED ON A METER OF 20,000Ω/VOLT

*
TO INCREASE THE SENSITIVITY OF THE TAPE REPLAY INPUT FOR MICROPHONE OPERATION, CUT THE WHITE LEAD ON THE INPUT SWITCH.

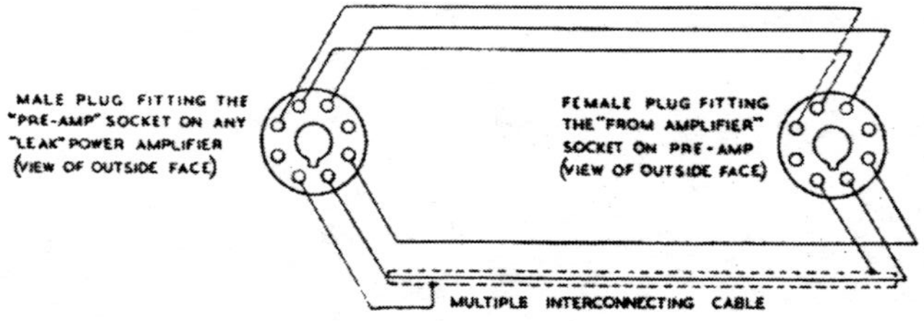

CIRCUIT DIAGRAM

Leak Varislope III Pre-Amp

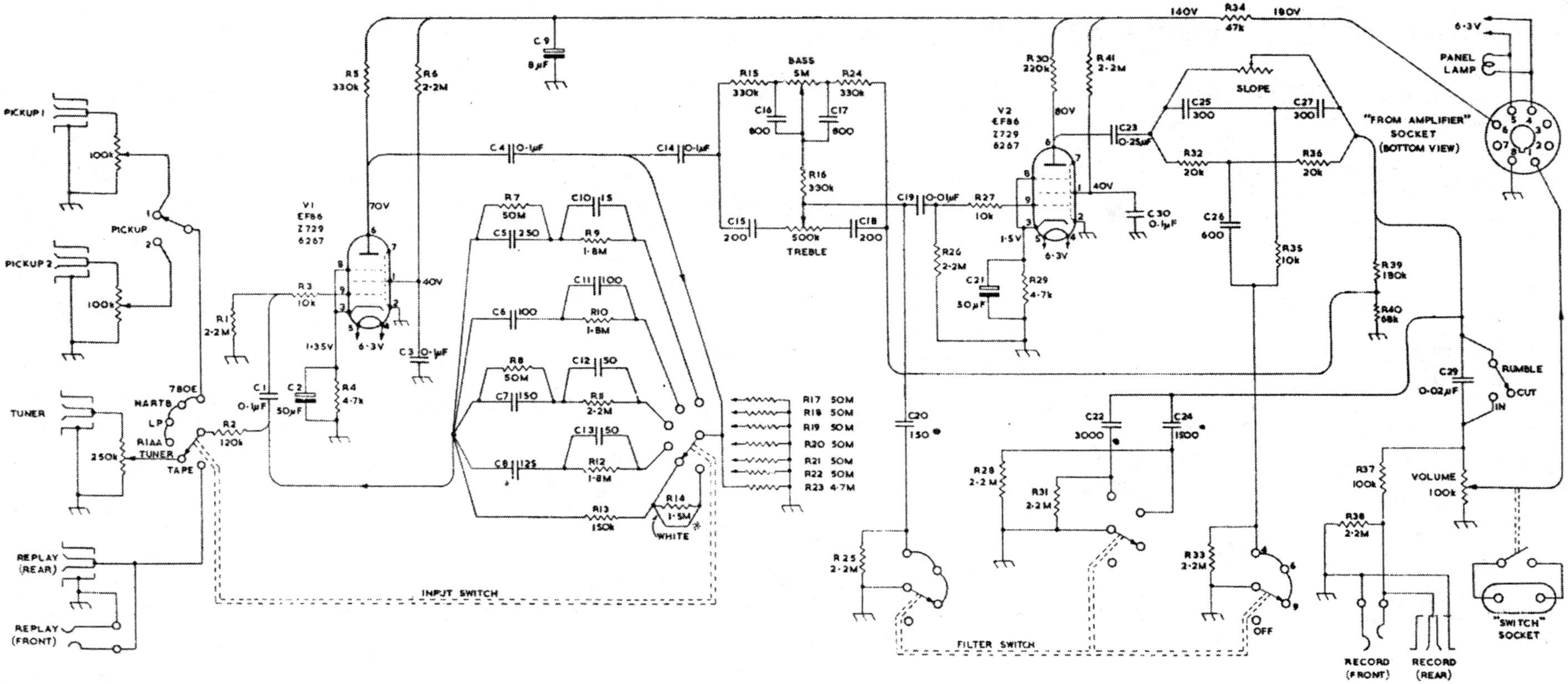

RESISTOR VALUES SHOWN IN OHMS
CAPACITOR VALUES SHOWN IN MICRO-MICROFARADS
EXCEPT WHERE OTHERWISE SHOWN
VOLTAGES MEASURED ON A METER OF 20,000Ω/VOLT

*
TO INCREASE THE SENSITIVITY OF THE TAPE REPLAY
INPUT FOR MICROPHONE OPERATION,
CUT THE WHITE LEAD ON THE INPUT SWITCH.

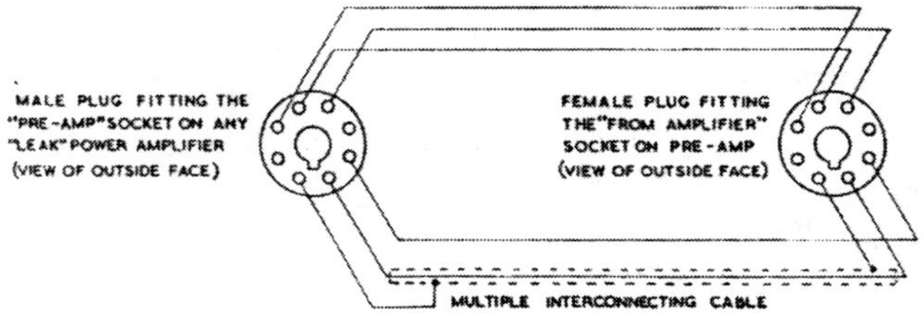

CIRCUIT DIAGRAM

Leak Varislope III (Series N) Pre-Amp

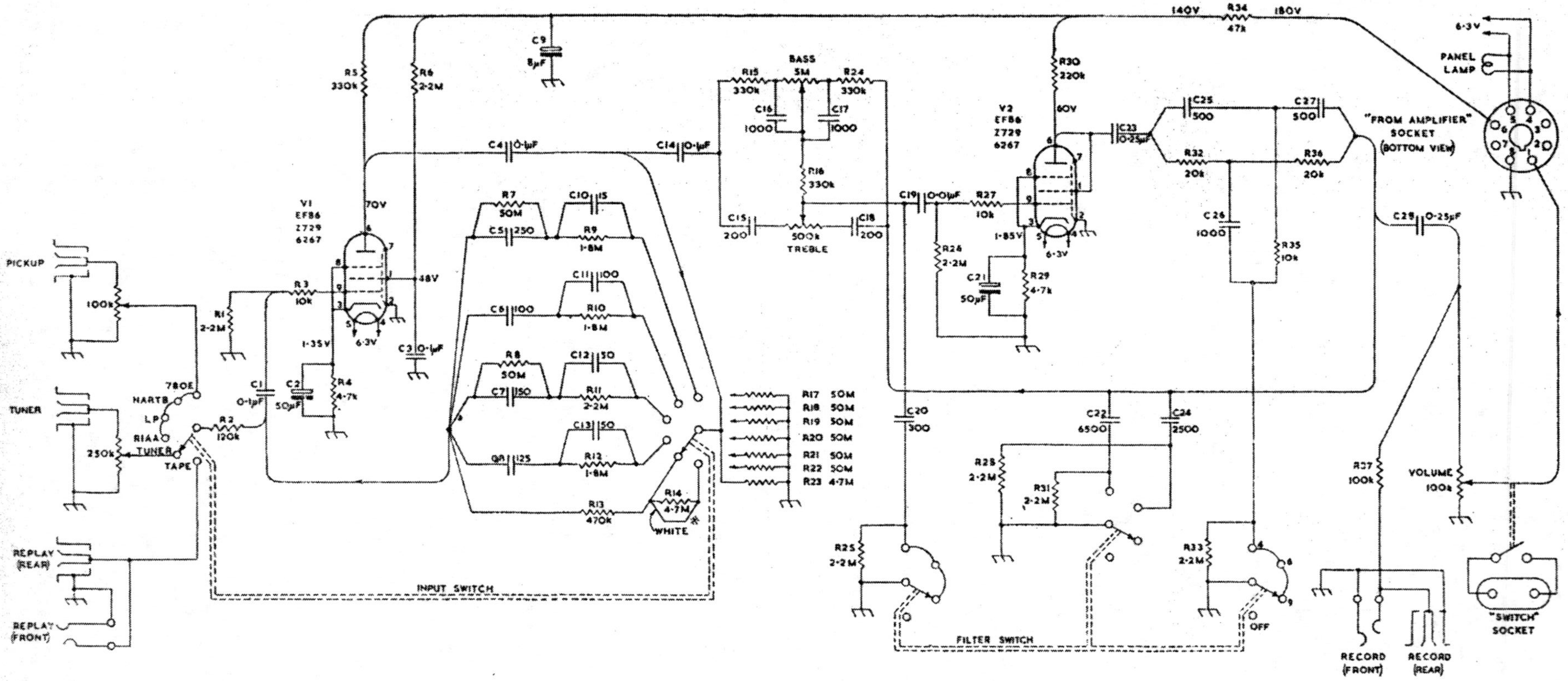

RESISTOR VALUES SHOWN IN OHMS
CAPACITOR VALUES SHOWN IN MICRO-MICROFARADS
EXCEPT WHERE OTHERWISE SHOWN
VOLTAGES MEASURED ON A METER OF 20,000Ω/VOLT

* TO INCREASE THE SENSITIVITY OF THE TAPE REPLAY INPUT FOR MICROPHONE OPERATION, CUT THE WHITE LEAD ON THE INPUT SWITCH.

MALE PLUG FITTING THE "PRE-AMP" SOCKET ON ANY "LEAK" POWER AMPLIFIER (VIEW OF OUTSIDE FACE)

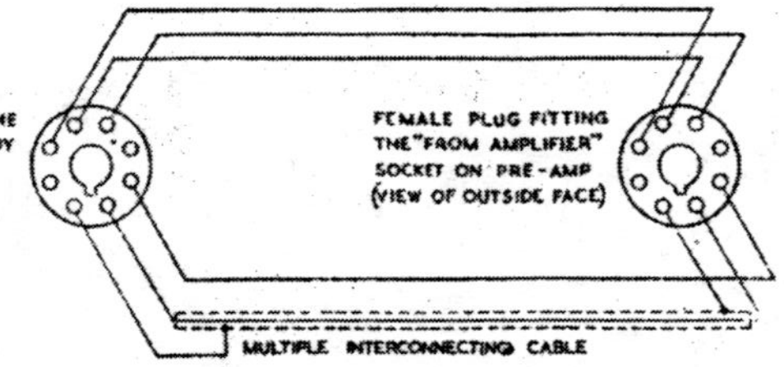

FEMALE PLUG FITTING THE "FROM AMPLIFIER" SOCKET ON PRE-AMP (VIEW OF OUTSIDE FACE)

CIRCUIT DIAGRAM

Leak Point One Plus Pre-Amp

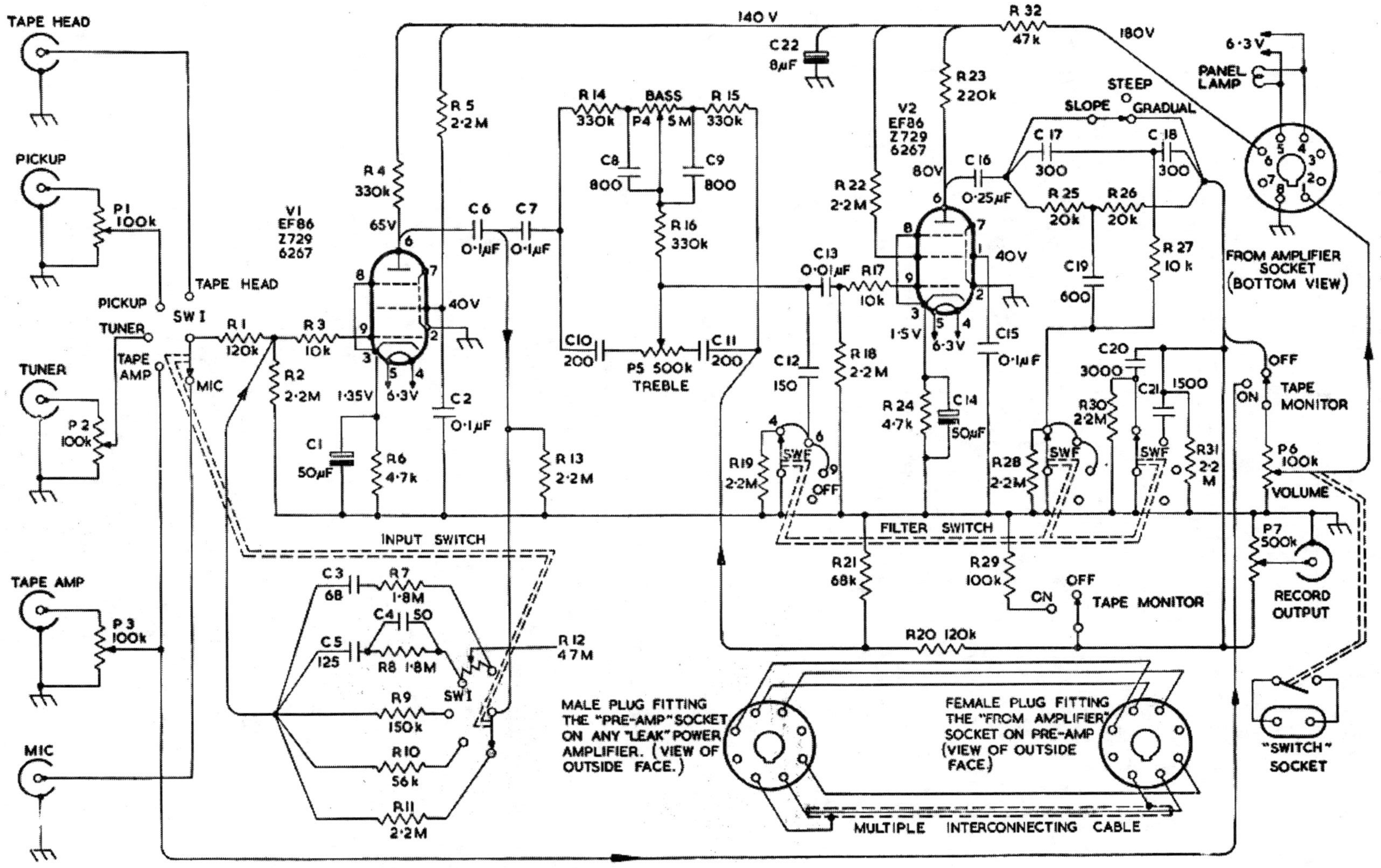

RESISTOR VALUES SHOWN IN OHMS
CAPACITOR VALUES SHOWN IN MICRO-MICROFARADS
EXCEPT WHERE OTHERWISE SHOWN
VOLTAGES MEASURED USING A METER OF 20,000 Ω / VOLT.

Leak Varislope Mono Pre-Amp

Leak Point One Stereo Pre-Amp

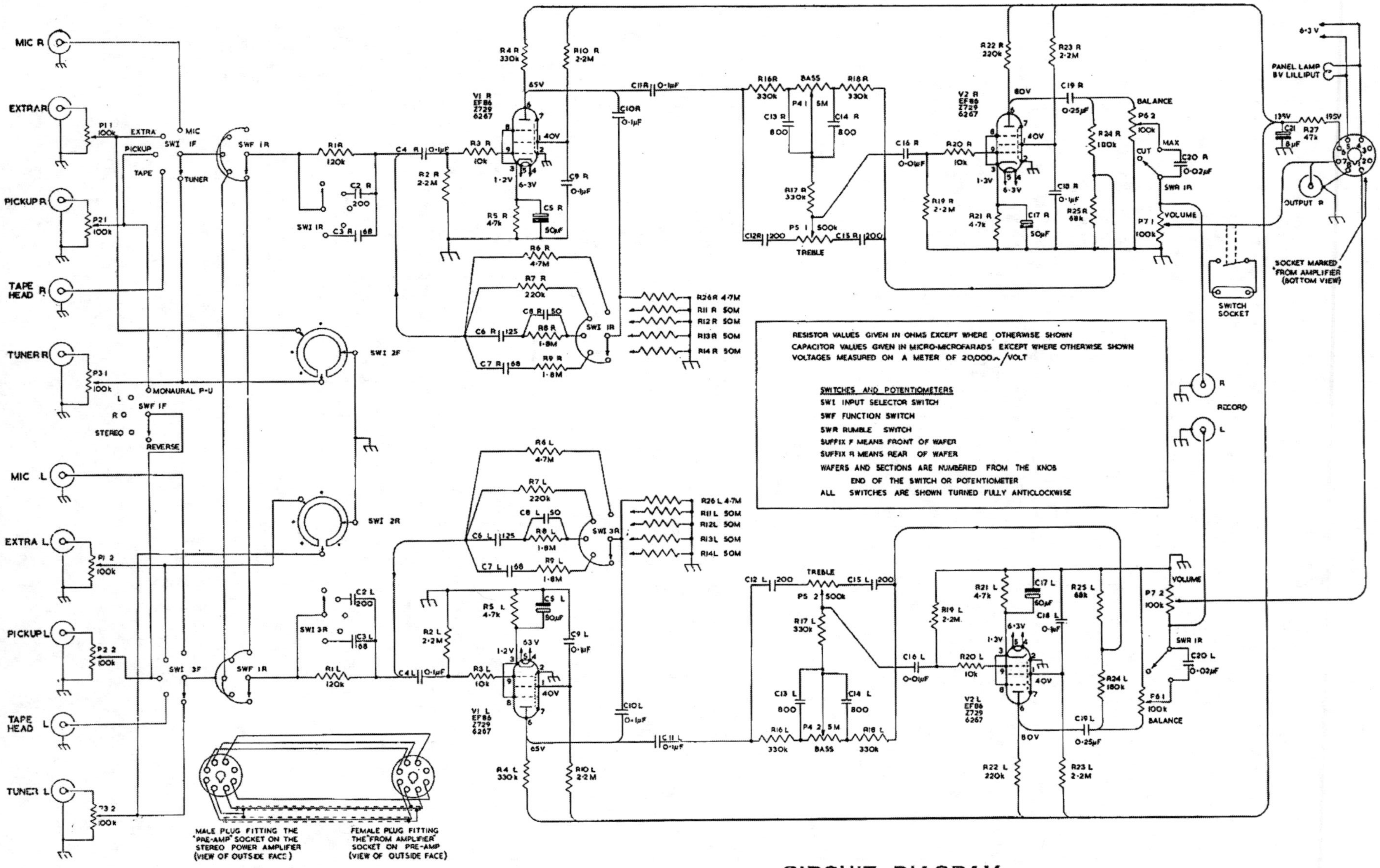

CIRCUIT DIAGRAM

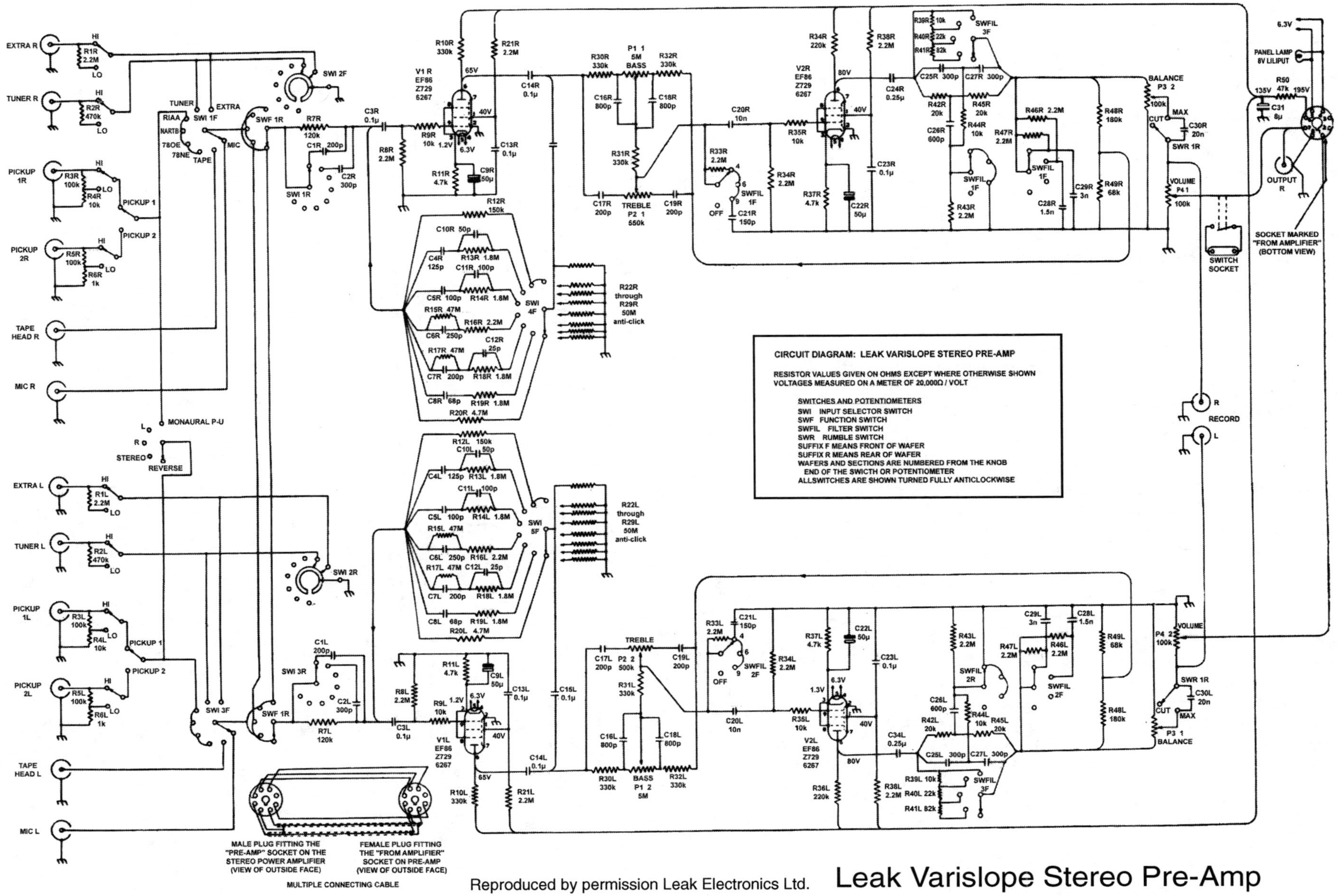

Reproduced by permission Leak Electronics Ltd.

Leak Varislope Stereo Pre-Amp

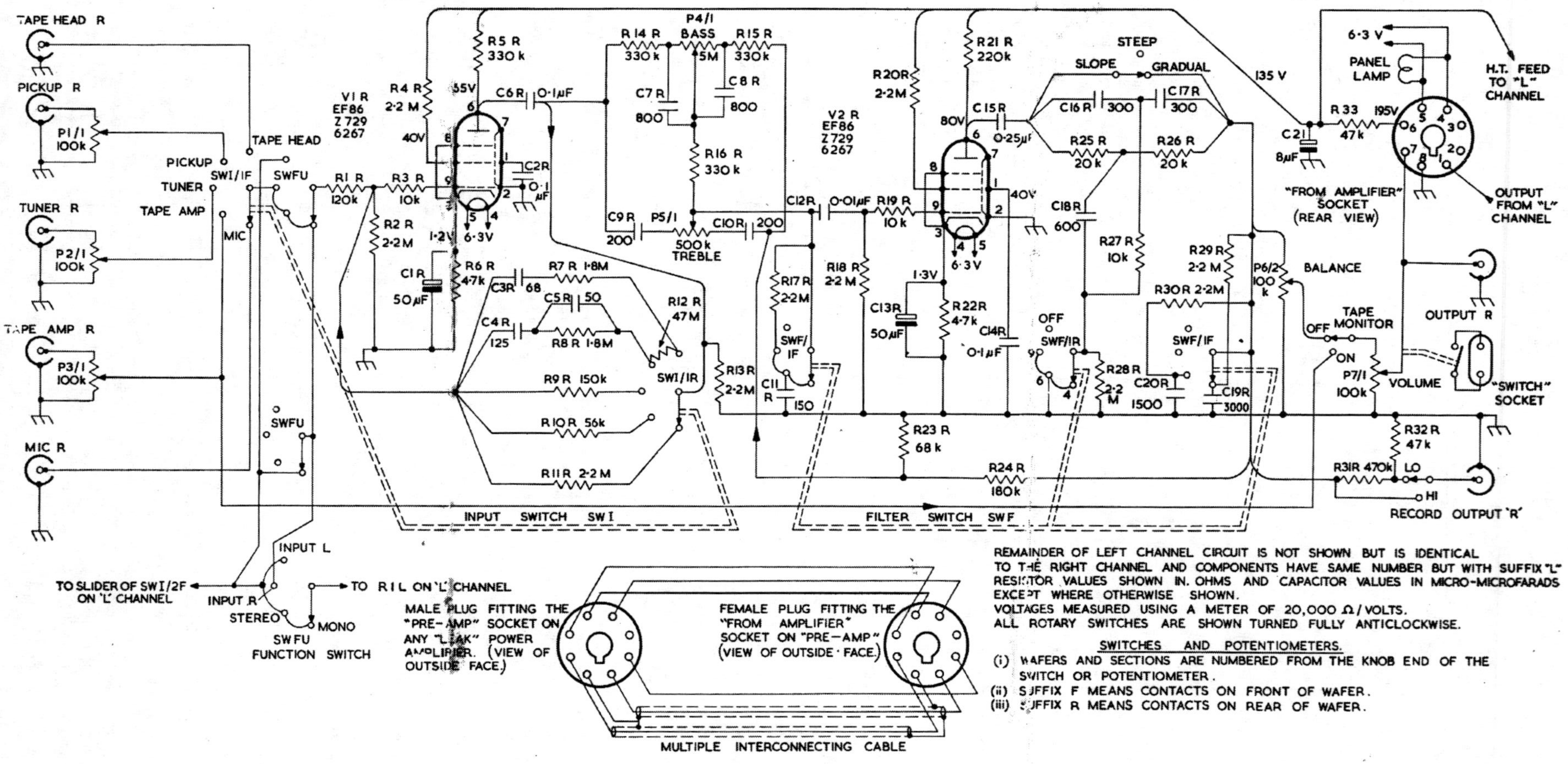

Leak Varislope 2 Stereo Pre-Amp

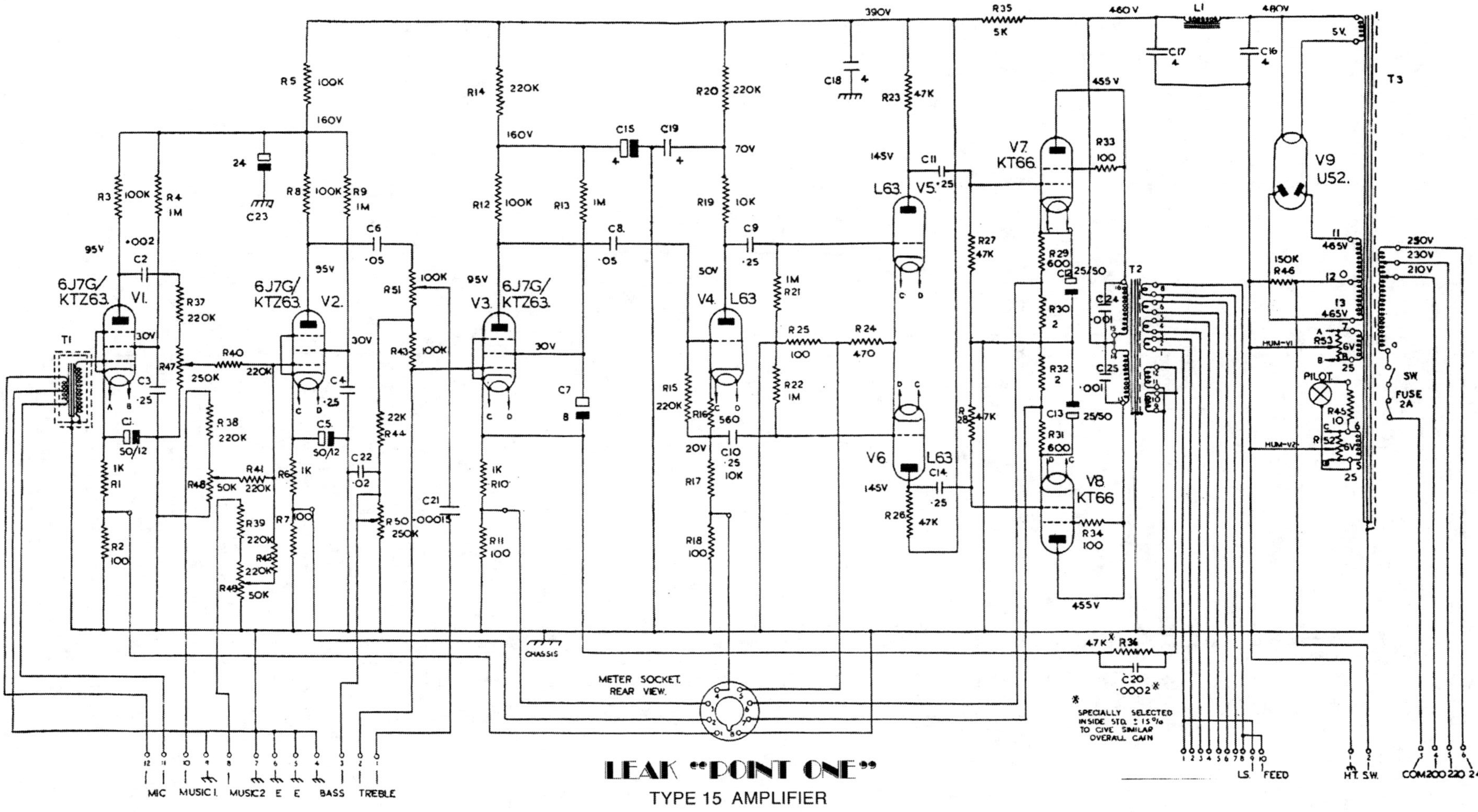

This is the probable circuit for the Type 15, using the 30 Watt circuit as a basis, and incorporating alterations for triode output stage.
Reproduced by permission Leak Electronics Ltd.

Leak "Point One" Type 15 Power Amplifier

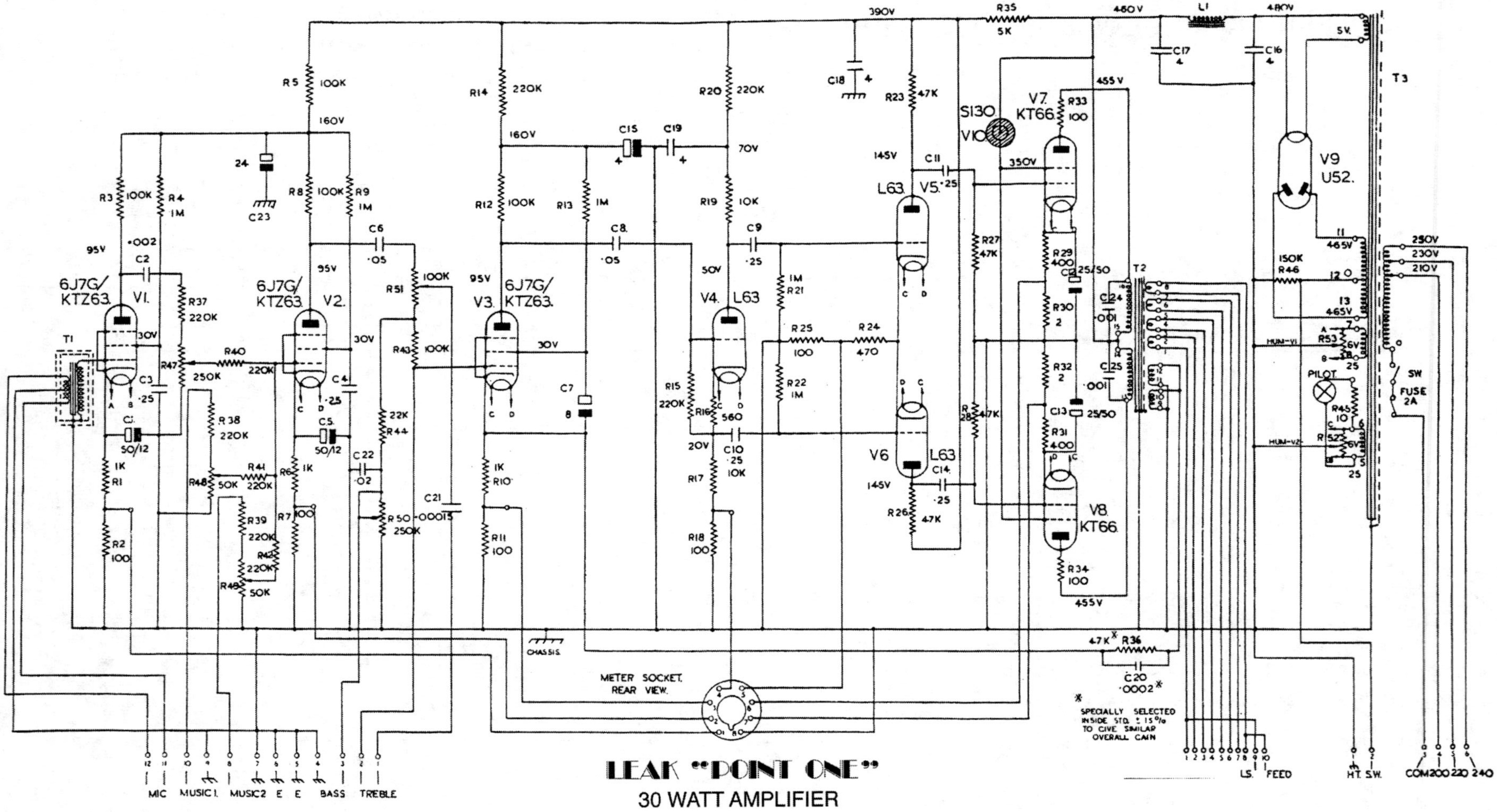

Leak "Point One" 30 Watt Power Amplifier

Note: This model was also supplied to GEC by Leak as Type 2430A.

Reproduced by permission Leak Electronics Ltd.

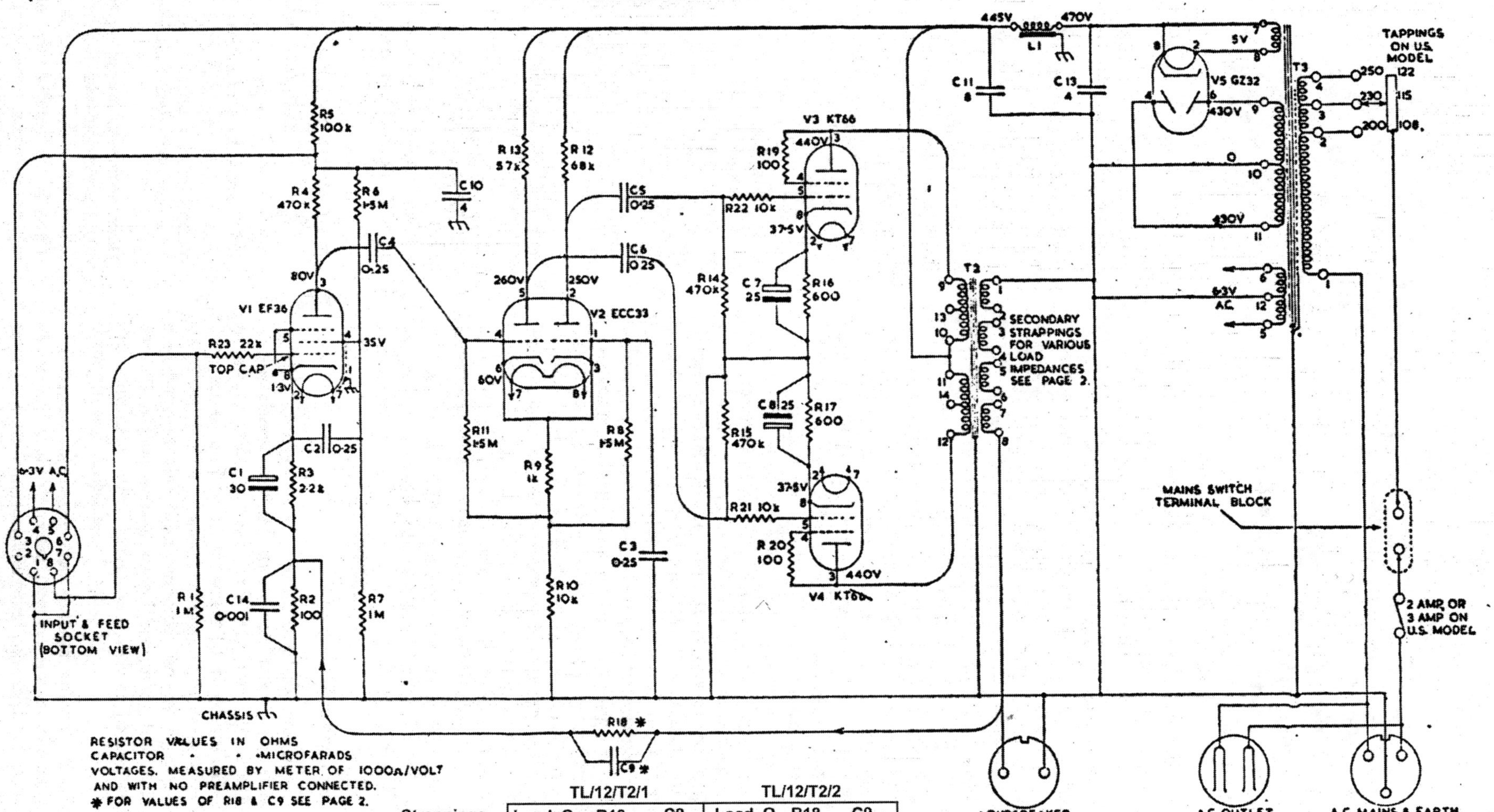

RESISTOR VALUES IN OHMS
CAPACITOR " " MICROFARADS
VOLTAGES. MEASURED BY METER OF 1000Ω/VOLT AND WITH NO PREAMPLIFIER CONNECTED.
* FOR VALUES OF R18 & C9 SEE PAGE 2.

	TL/12/T2/1			TL/12/T2/2		
Strappings	Load, Ω	R18	C9	Load, Ω	R18	C9
1,3,5,7 : 2,4,6,8	1	2.2k	500p	2	3k	200p
1,5 : 2,3,6,7 : 4,8	4	4.7k	500p	8	6k	100p
2,3,5 : 4,6,7	9	6.6k	300p	18	10k	100p
2,3 : 4,5 : 6,7	16	10k	300p	32	12k	100p

Reproduced by permission Leak Electronics Ltd.

Leak TL/12 Power Amplifier

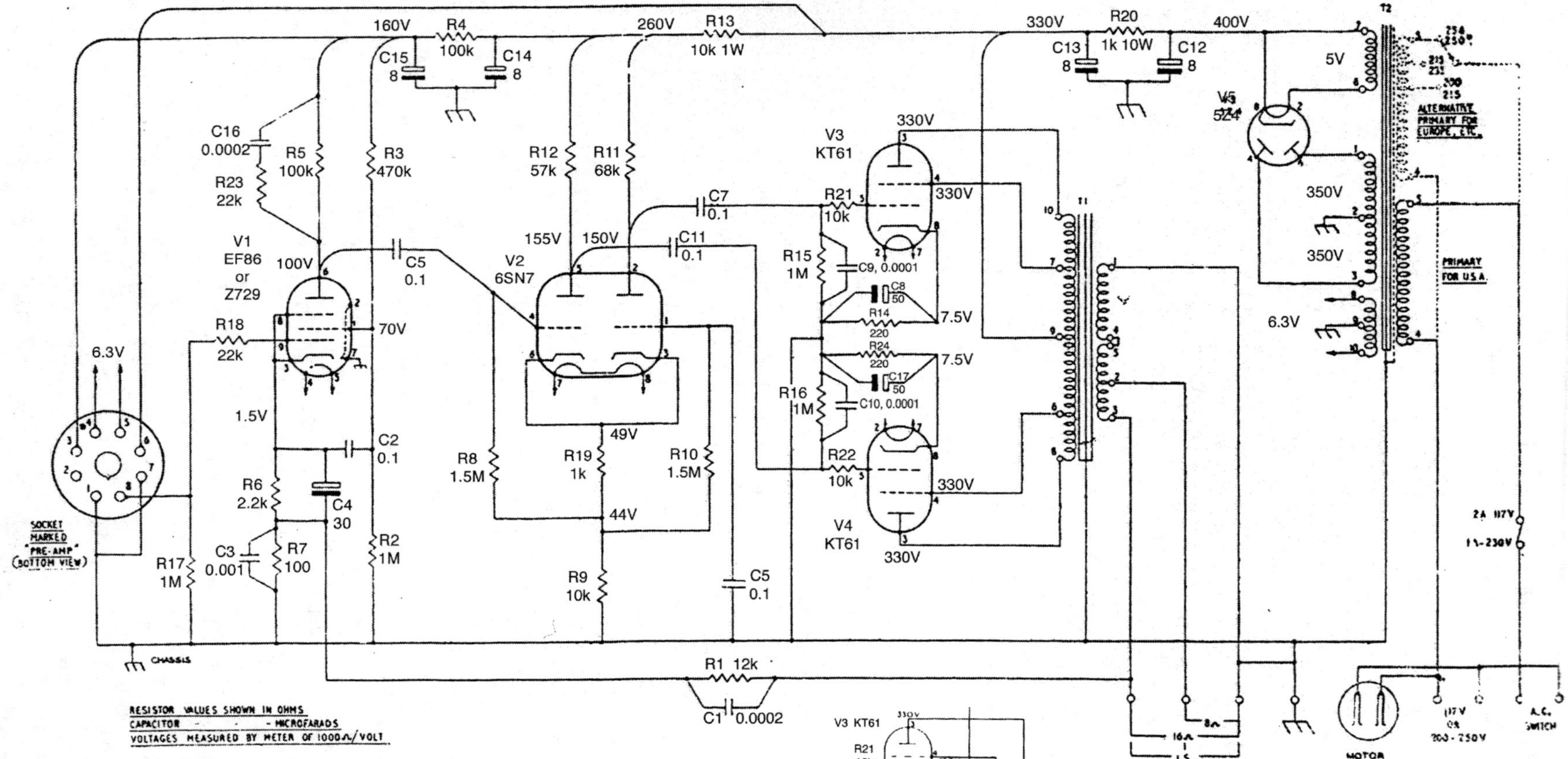

RESISTOR VALUES SHOWN IN OHMS
CAPACITOR - - - MICROFARADS
VOLTAGES MEASURED BY METER OF 1000Ω/VOLT

CIRCUIT DIAGRAM

KT61 is electrically identical to type 6AG6
Type 6L6 may be used instead of KT61 if the following modifications are made:

- EF86 screen resistor, R3, changed from 470k to 1M
- EF86 anode resistor, R5 changed from 100k to 330k
- Output stage cathode resistors changed to 600Ω 5 Watt each, indivually bypassed with 25μF 50Vdc capacitors (or for the single cathode resistor version, R14 is changed to 300Ω 3W, with bypass capacitor C8 being omitted).

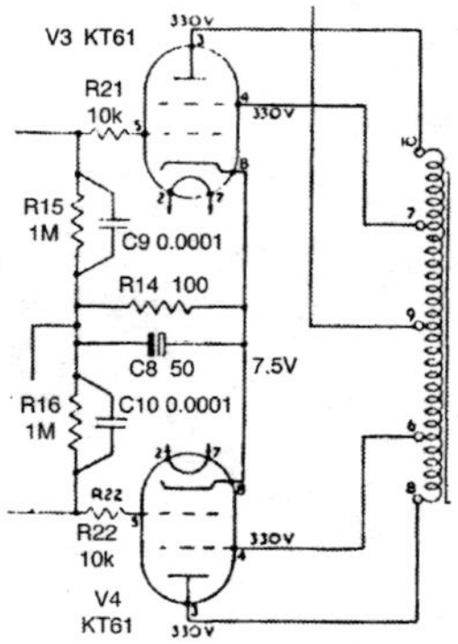

Alternative output stage with single cathode resistor

Reproduced by permission Leak Electronics Ltd.

Leak TL/10 Power Amplifier

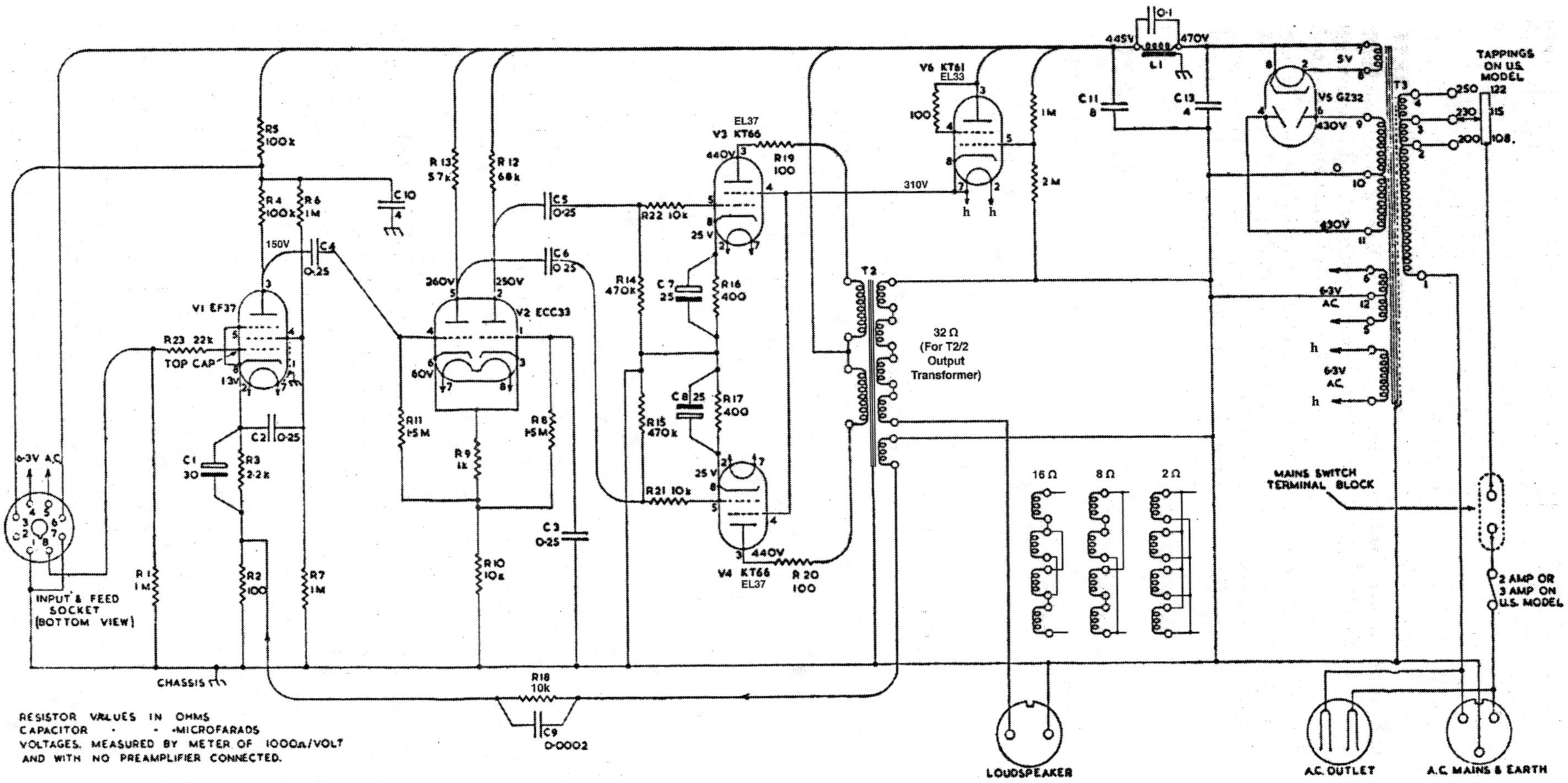

This circuit has been compiled by examining a TL/25A amplifier, and by using image-components from the TL/12 circuit to re-construct a drawing that appears authentic. The original image-components from the TL/12 circuits are reproduced by permission of Leak Electronics Ltd.

Leak TL/25A Power Amplifier

Leak TL/12 Plus

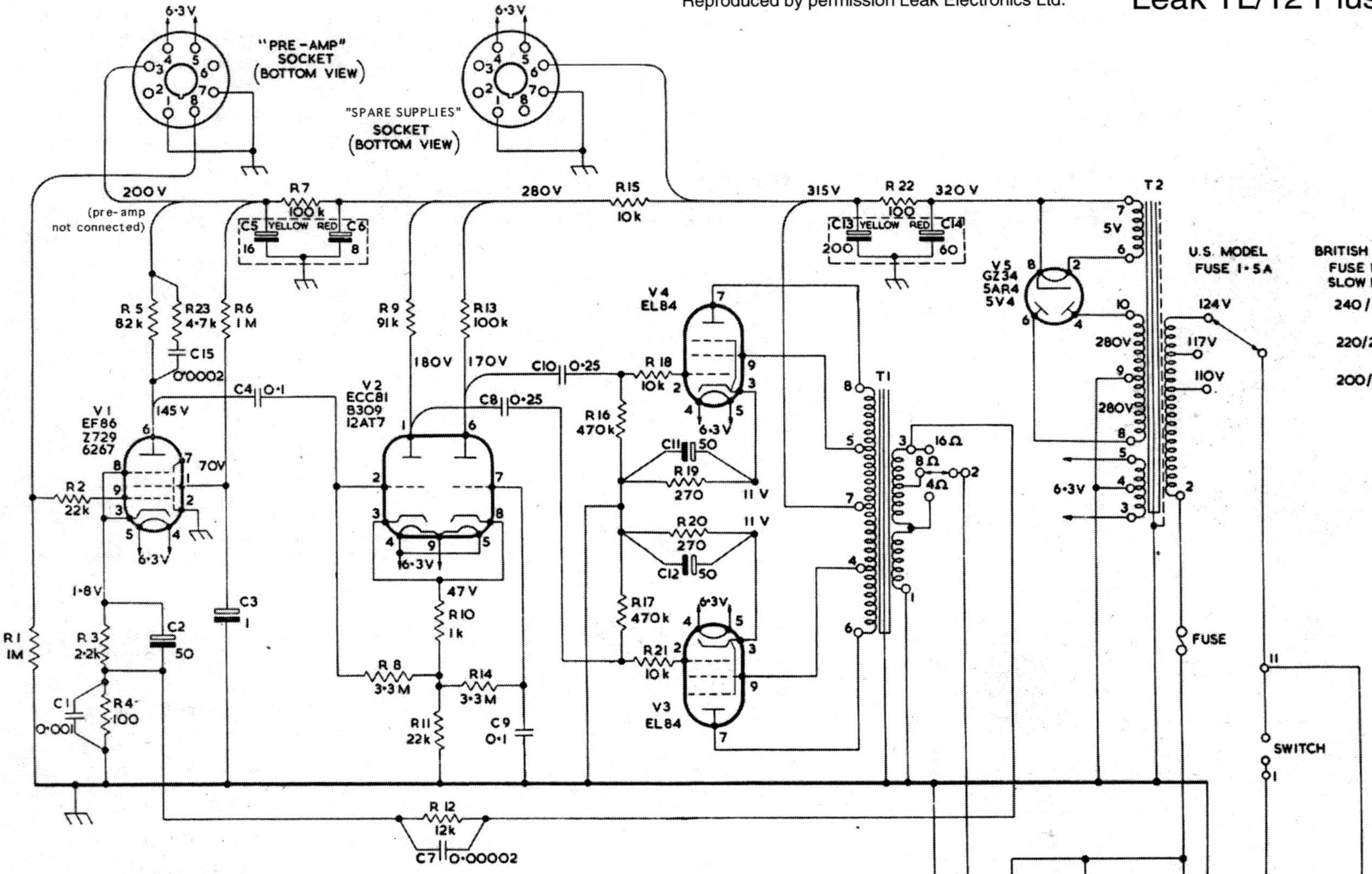

RESISTOR VALUES SHOWN IN OHMS.
CAPACITOR VALUES SHOWN IN MICROFARADS
EXCEPT WHERE OTHERWISE SHOWN.
VOLTAGES MEASURED BY METER OF 20,000 Ω / VOLT
LOWER VOLTAGES TAKEN WITH "LEAK" PRE-AMPLIFIER CONNECTED.
UPPER VOLTAGES TAKEN WITH "LEAK" PRE-AMPLIFIER DISCONNECTED.

Leak TL/25 Plus

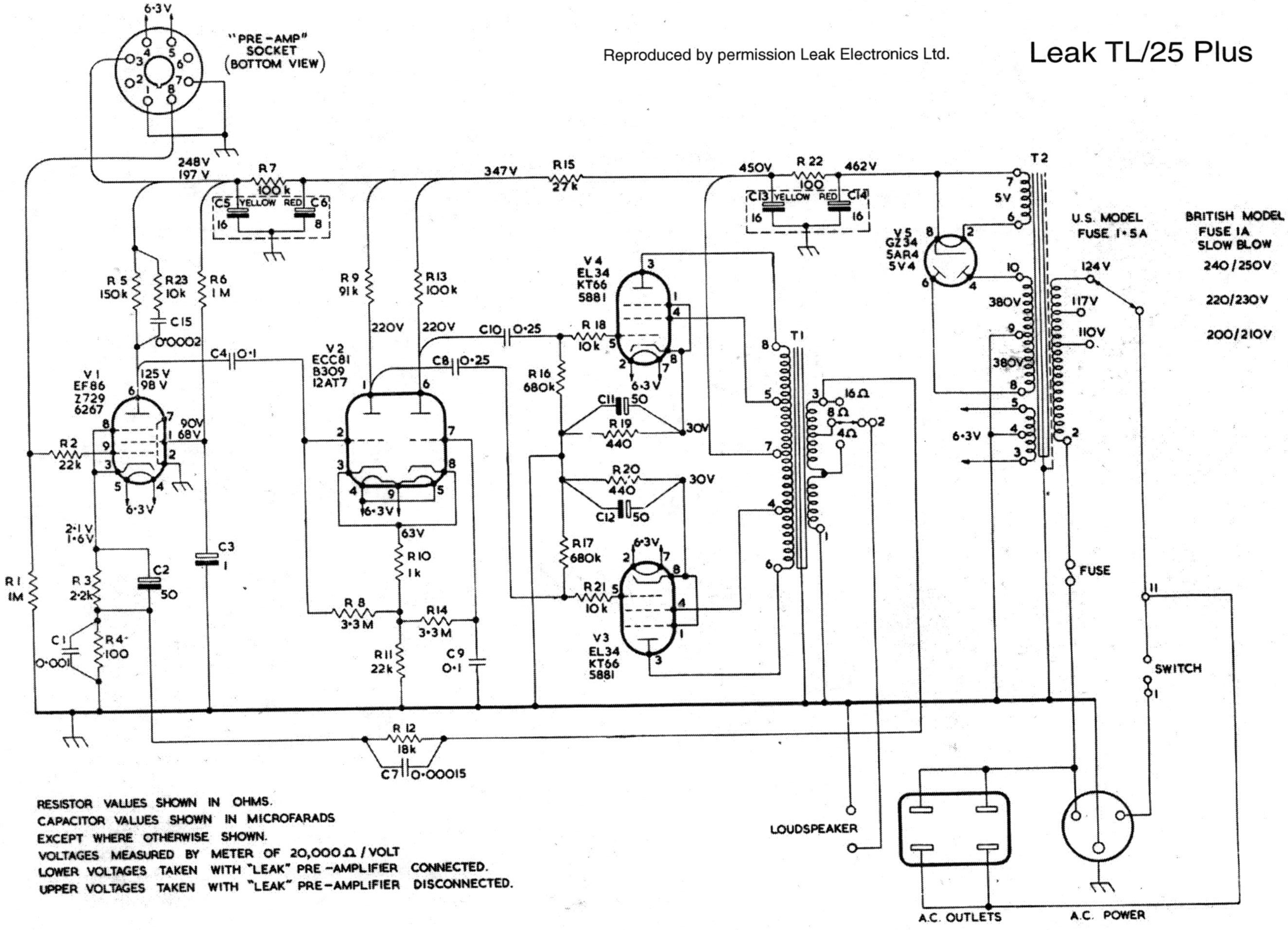

RESISTOR VALUES SHOWN IN OHMS.
CAPACITOR VALUES SHOWN IN MICROFARADS
EXCEPT WHERE OTHERWISE SHOWN.
VOLTAGES MEASURED BY METER OF 20,000 Ω / VOLT
LOWER VOLTAGES TAKEN WITH "LEAK" PRE-AMPLIFIER CONNECTED.
UPPER VOLTAGES TAKEN WITH "LEAK" PRE-AMPLIFIER DISCONNECTED.

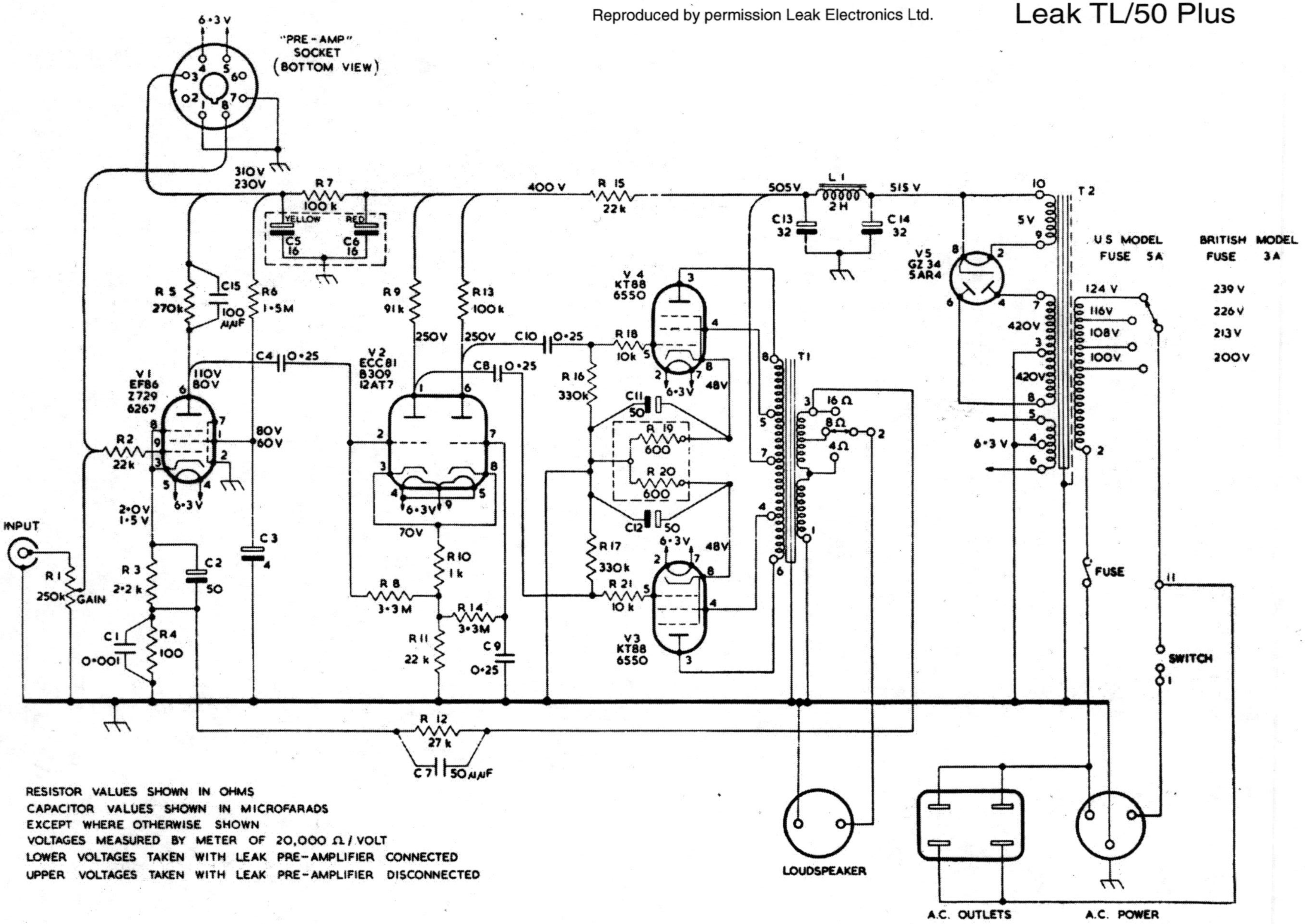

RESISTOR VALUES SHOWN IN OHMS
CAPACITOR VALUES SHOWN IN MICROFARADS
EXCEPT WHERE OTHERWISE SHOWN
VOLTAGES MEASURED BY METER OF 20,000 Ω / VOLT
LOWER VOLTAGES TAKEN WITH LEAK PRE-AMPLIFIER CONNECTED
UPPER VOLTAGES TAKEN WITH LEAK PRE-AMPLIFIER DISCONNECTED

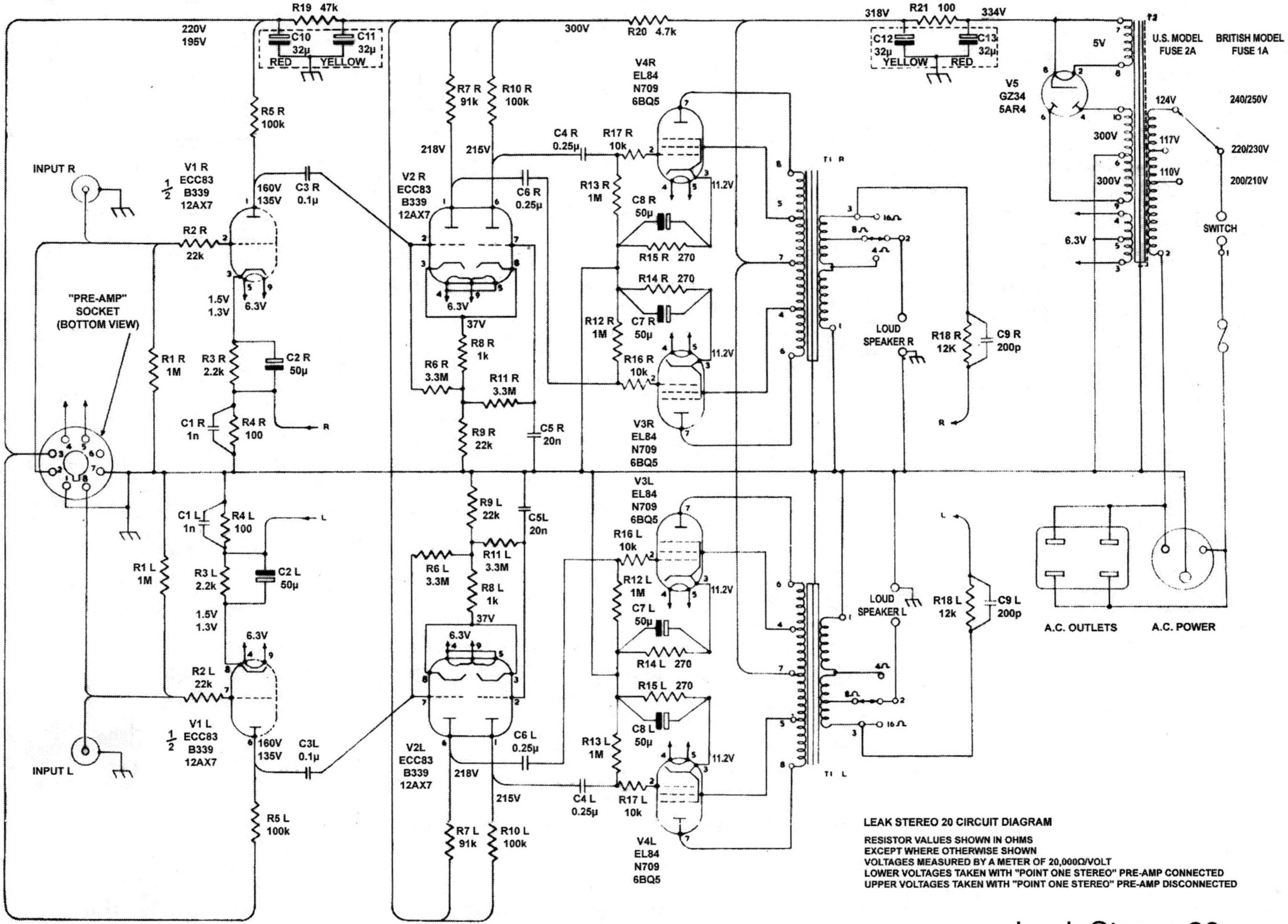

Reproduced by permission Leak Electronics Ltd.

Leak Stereo 20

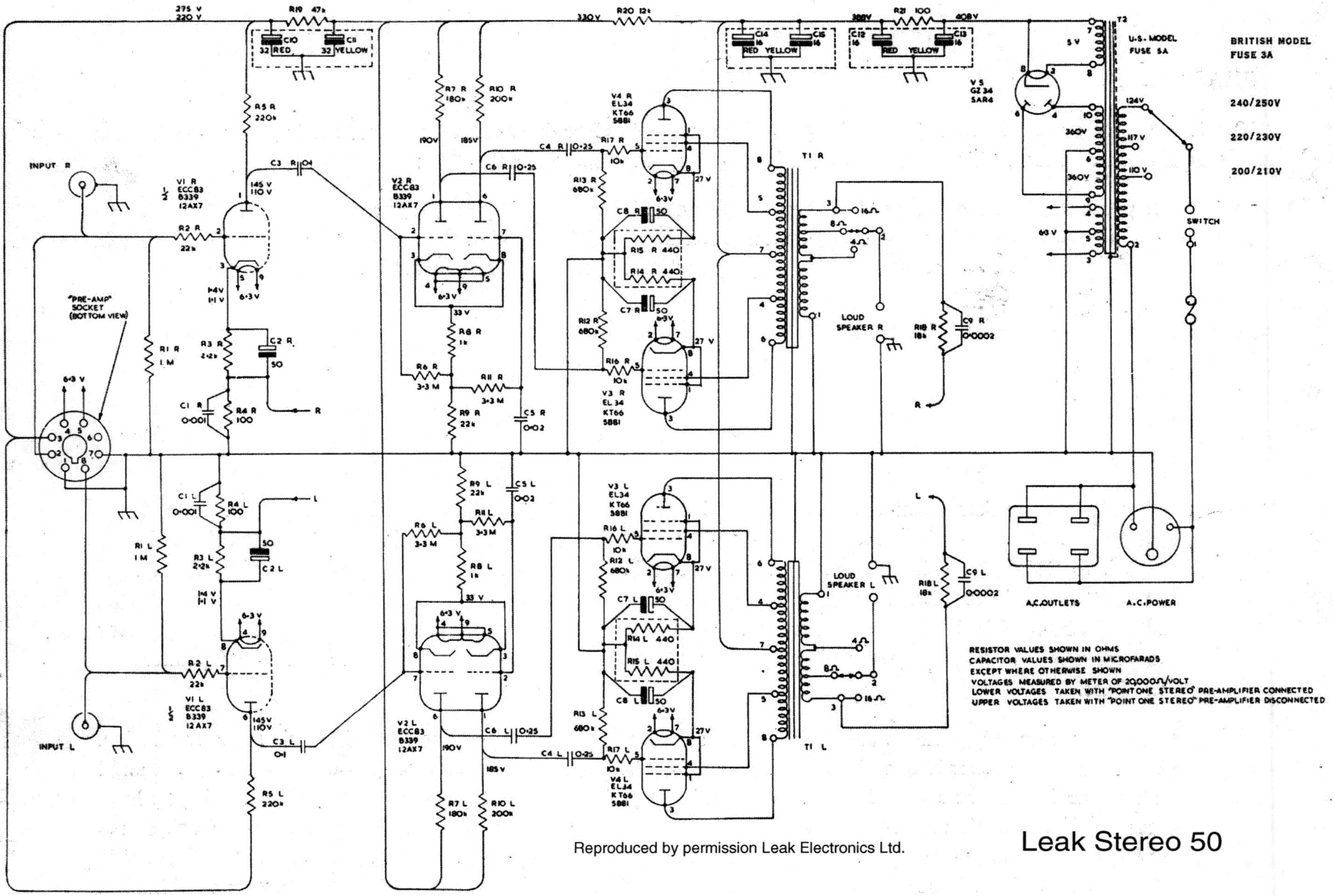

Reproduced by permission Leak Electronics Ltd.

Leak Stereo 50

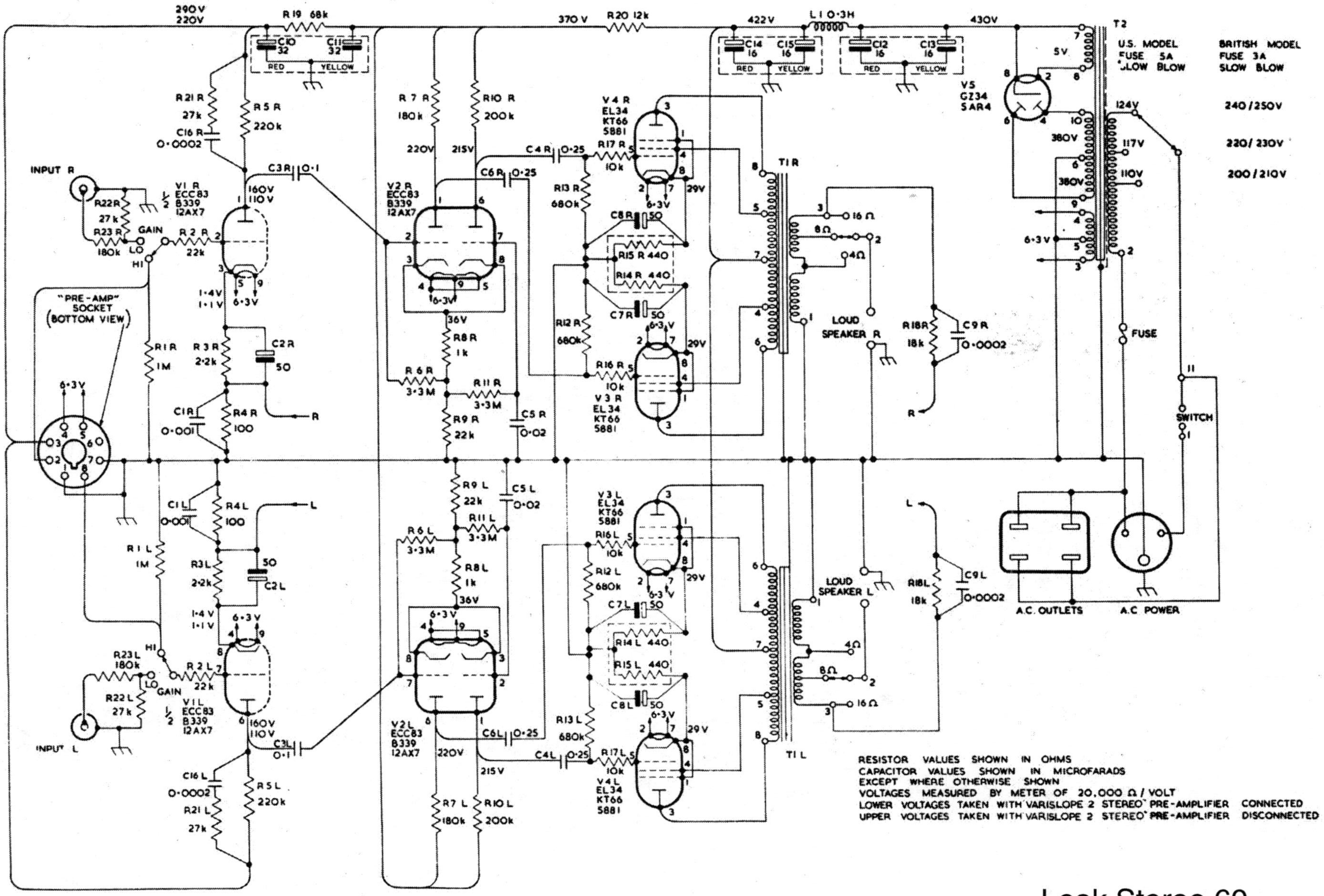

Leak Stereo 60

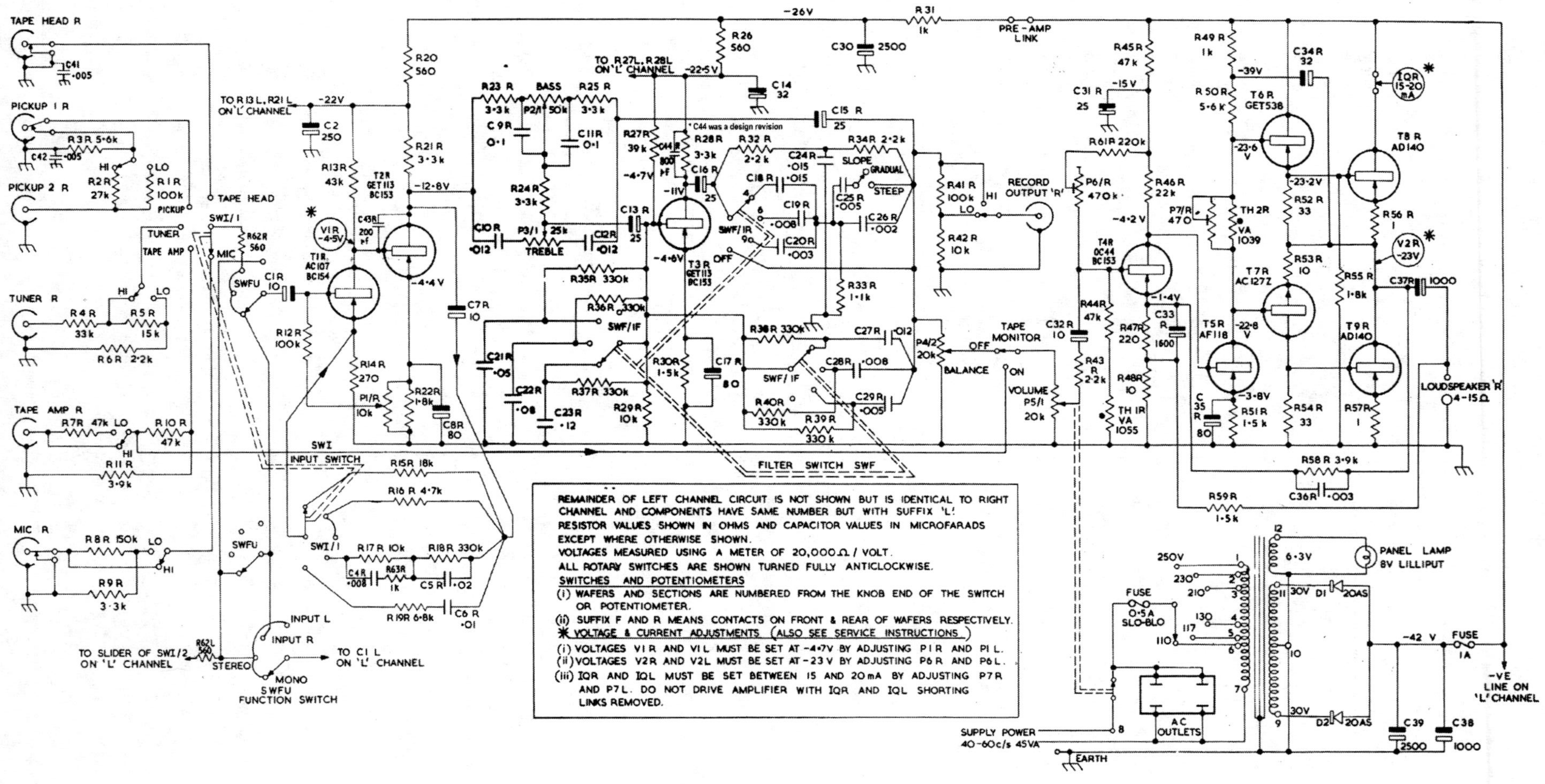

REMAINDER OF LEFT CHANNEL CIRCUIT IS NOT SHOWN BUT IS IDENTICAL TO RIGHT CHANNEL AND COMPONENTS HAVE SAME NUMBER BUT WITH SUFFIX 'L'.
RESISTOR VALUES SHOWN IN OHMS AND CAPACITOR VALUES IN MICROFARADS EXCEPT WHERE OTHERWISE SHOWN.
VOLTAGES MEASURED USING A METER OF 20,000Ω / VOLT.
ALL ROTARY SWITCHES ARE SHOWN TURNED FULLY ANTICLOCKWISE.

SWITCHES AND POTENTIOMETERS

(i) WAFERS AND SECTIONS ARE NUMBERED FROM THE KNOB END OF THE SWITCH OR POTENTIOMETER.

(ii) SUFFIX F AND R MEANS CONTACTS ON FRONT & REAR OF WAFERS RESPECTIVELY.

* VOLTAGE & CURRENT ADJUSTMENTS (ALSO SEE SERVICE INSTRUCTIONS)

(i) VOLTAGES V1R AND V1L MUST BE SET AT -4·7V BY ADJUSTING P1R AND P1L.

(ii) VOLTAGES V2R AND V2L MUST BE SET AT -23V BY ADJUSTING P6R AND P6L.

(iii) IQR AND IQL MUST BE SET BETWEEN 15 AND 20mA BY ADJUSTING P7R AND P7L. DO NOT DRIVE AMPLIFIER WITH IQR AND IQL SHORTING LINKS REMOVED.

CIRCUIT DIAGRAM

Leak Stereo 30

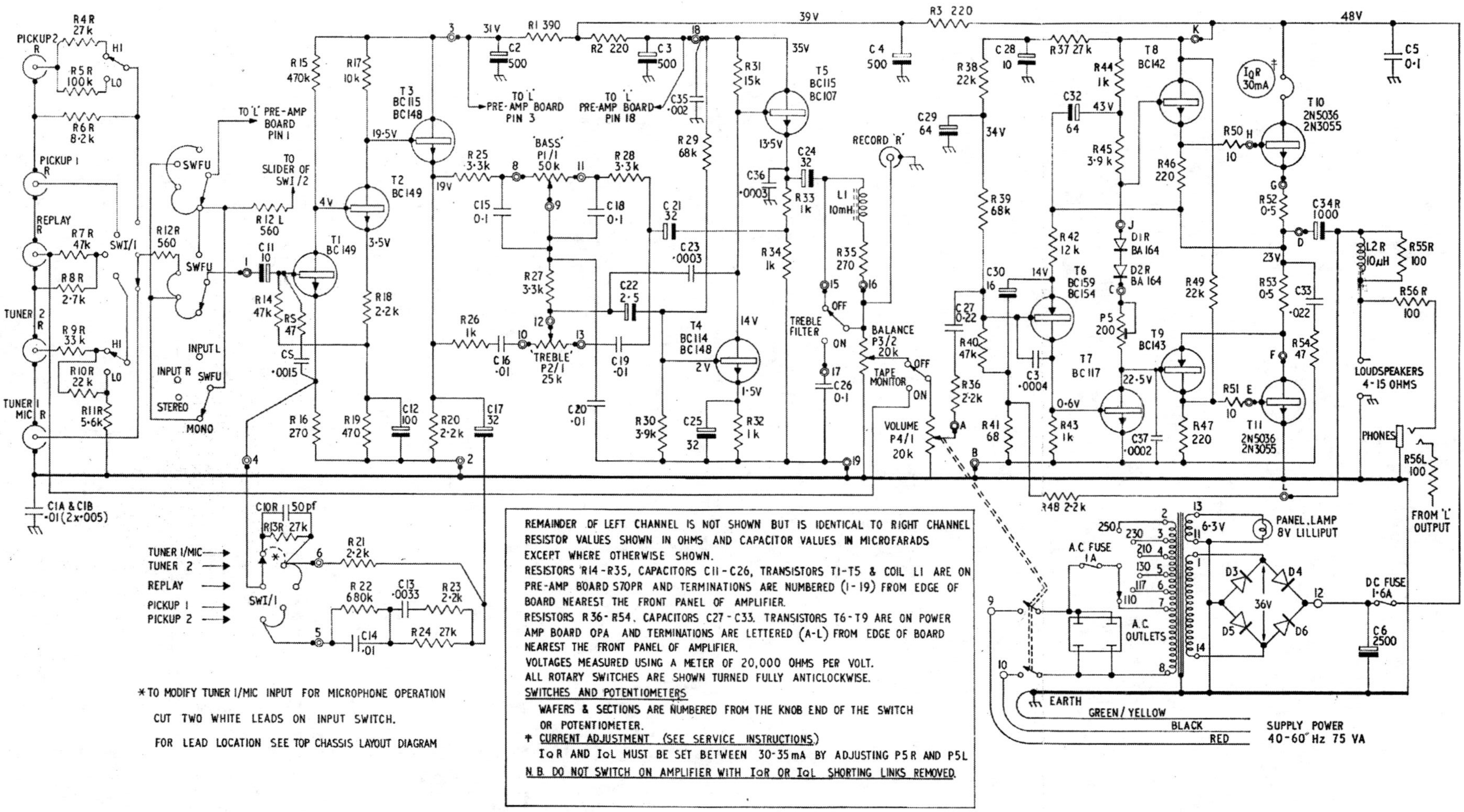

CIRCUIT DIAGRAM

Leak Stereo 30 Plus

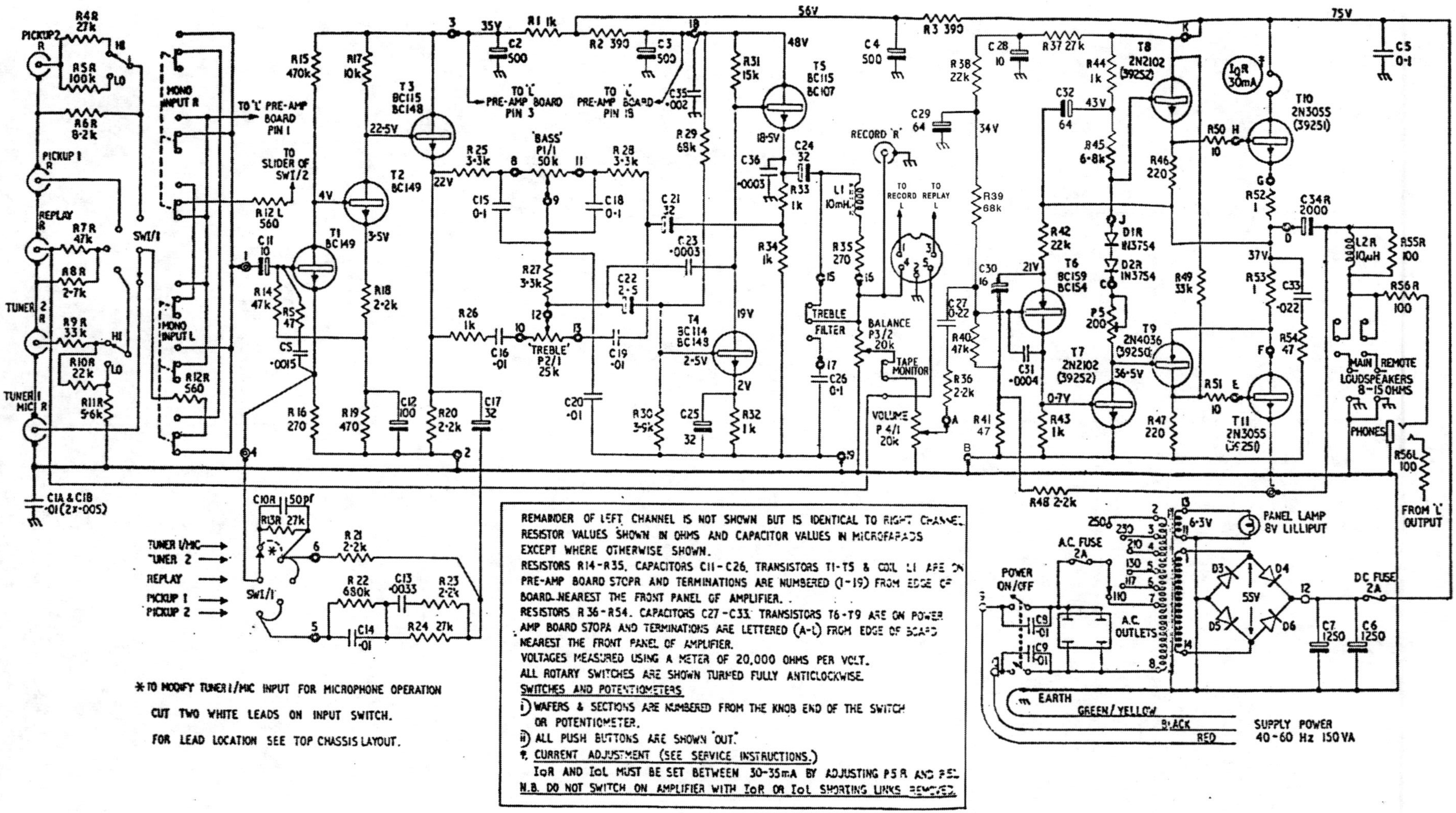

Leak Stereo 70 Amplifier

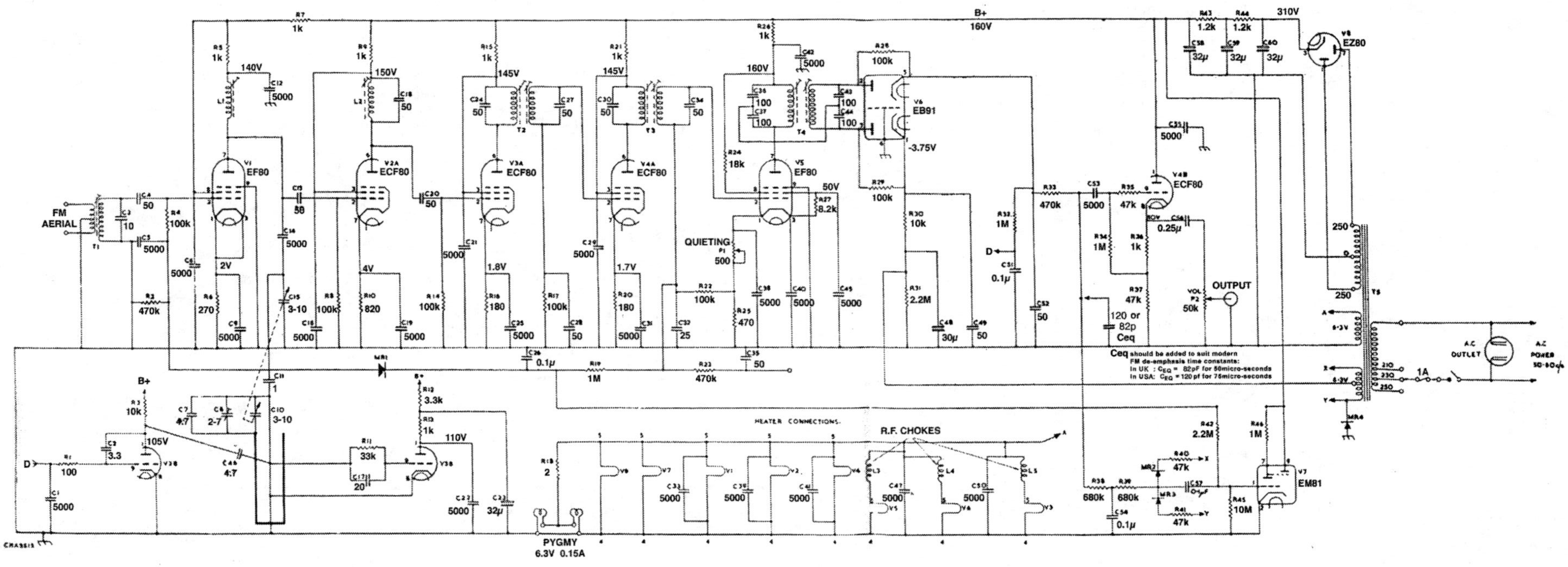

Resistor values shown in ohms except where otherwise shown
Capacitor values shown in micro-microfarads except where otherwise shown
voltages measured on a meter of 20,000 ohm/volt with no aerial connected

CIRCUIT DIAGRAM OF LEAK "TROUGH-LINE" F-M TUNER UNIT

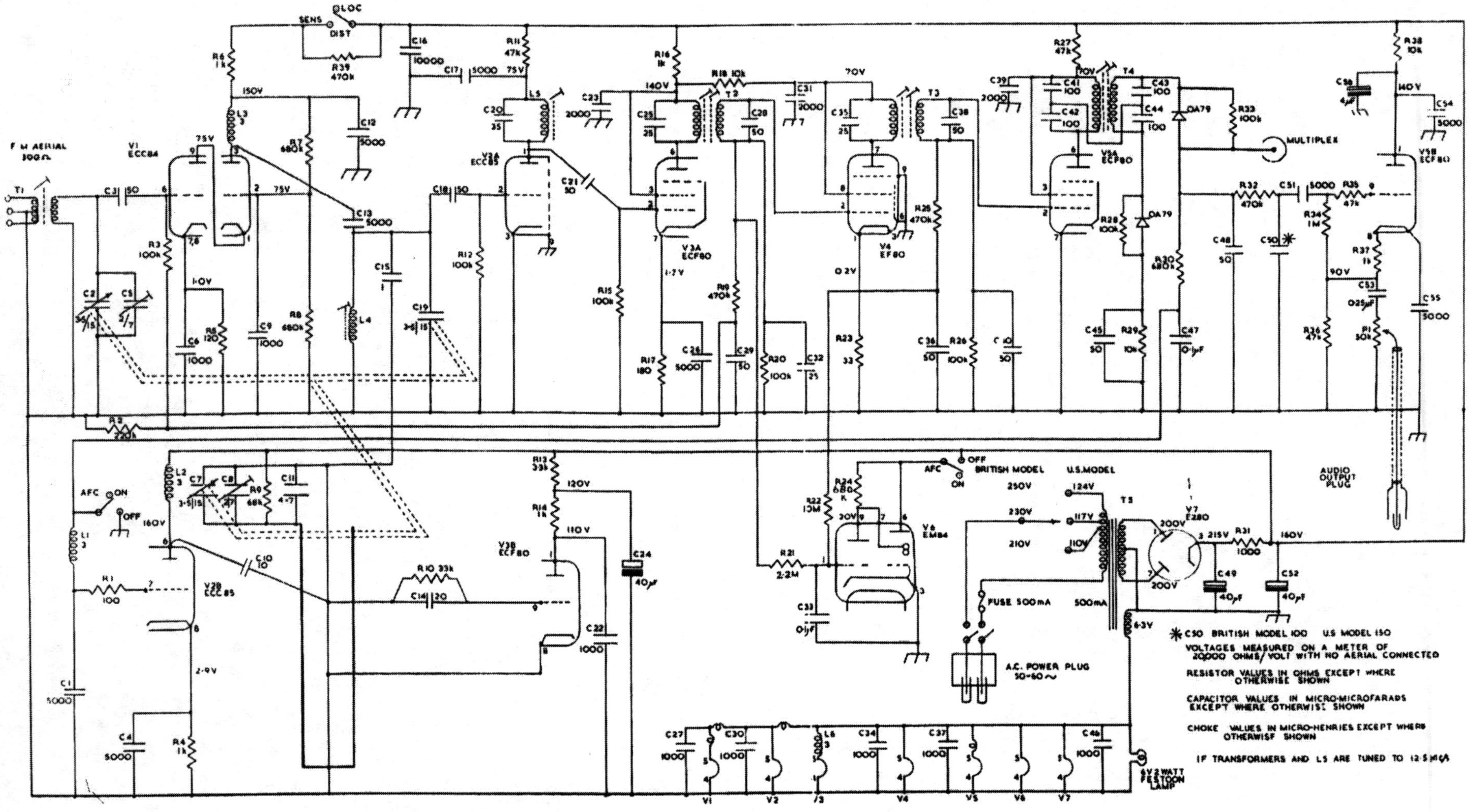

Reproduced by permission Leak Electronics Ltd.

Leak Trough-Line II FM Tuner

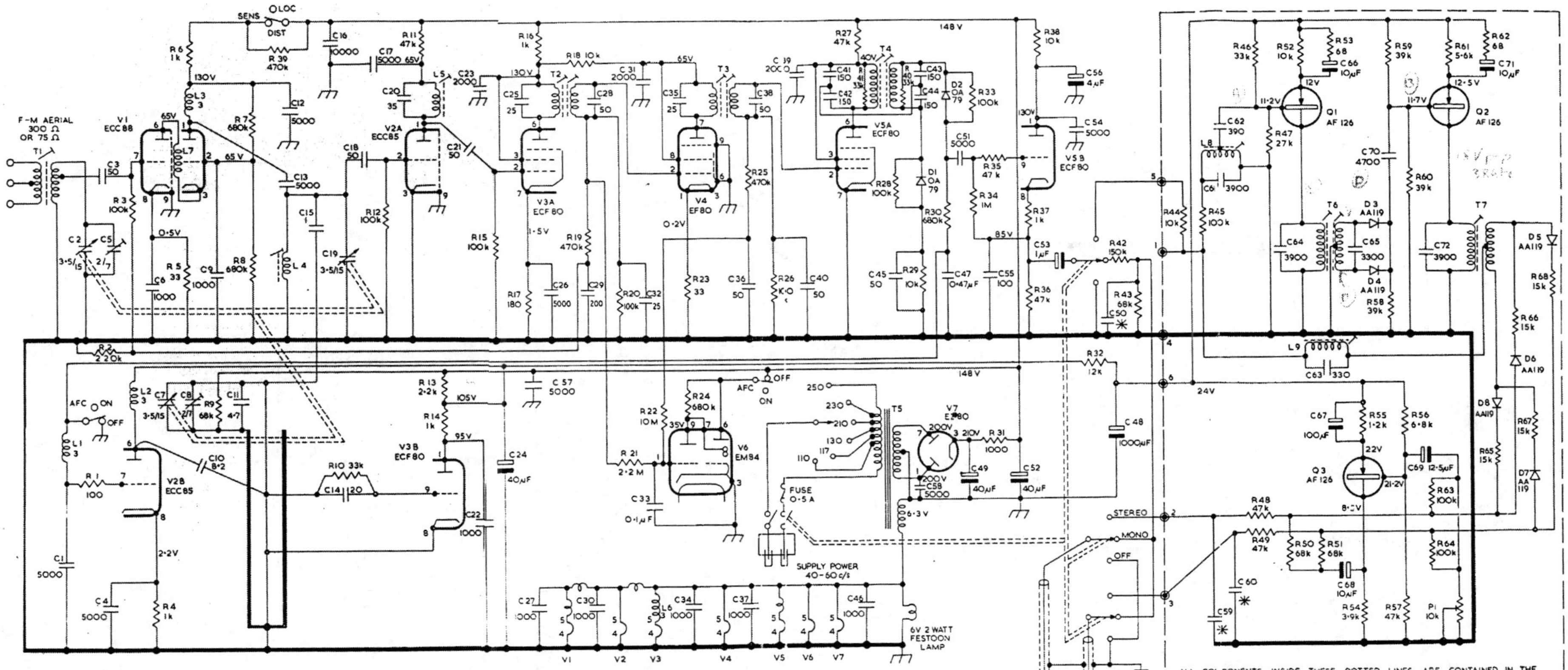

VOLTAGES ARE MEASURED ON A METER OF 20,000 Ω/VOLT, WITH NO AERIAL CONNECTED
RESISTOR VALUES ARE IN OHMS
CAPACITOR VALUES ARE IN MICRO-MICROFARADS } EXCEPT WHERE OTHERWISE SHOWN
CHOKE VALUES ARE IN MICRO-HENRIES
T2, T3, T4 AND L5 ARE TUNED TO 12·5 Mc/s
L8 AND T6 ARE TUNED TO 19 Kc/s, T7 IS TUNED TO 38 Kc/s AND L9 IS TUNED TO 67 Kc/s

* C50 IS 3000pf ON EUROPEAN MODEL
C59 AND C60 ARE 2000 pf ON EUROPEAN MODEL
C50 IS 4000pf ON U.S. MODEL
C59 AND C60 ARE 3000pf ON U.S. MODEL

ALL COMPONENTS INSIDE THESE DOTTED LINES ARE CONTAINED IN THE DECODER UNIT FITTED INSIDE THE TROUGHLINE TUNER
THE CONNECTIONS TO THE DECODER ARE MADE VIA A SIX CONTACT PLUG AND SOCKET AND THE CONTACTS ARE NUMBERED VERTICALLY FROM THE BOTTOM

Reproduced by permission Leak Electronics Ltd.

Leak Trough-Line Stereo FM Tuner

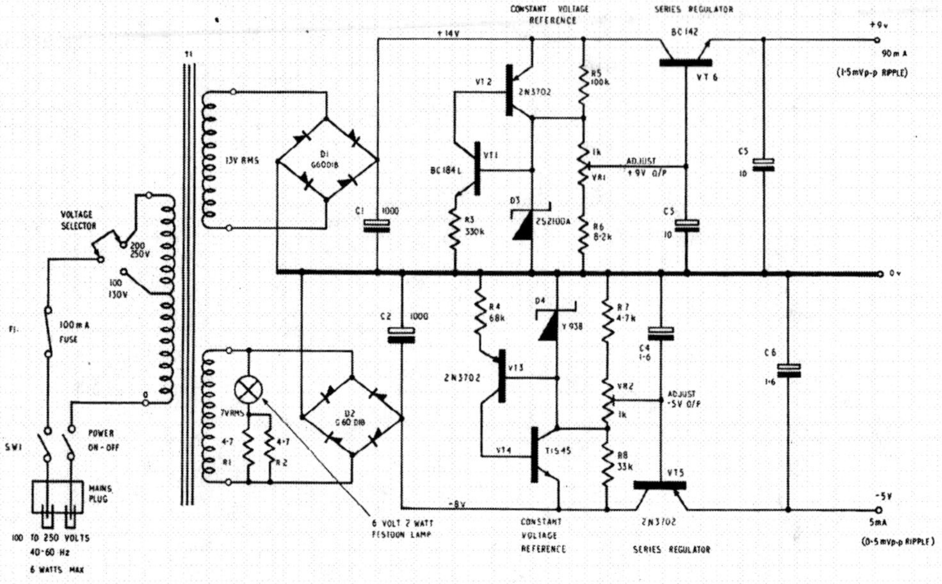

RESISTOR VALUES IN OHMS UNLESS OTHERWISE SHOWN
CAPACITOR VALUES IN MICROFARADS UNLESS OTHERWISE SHOWN
VOLTAGES ARE MEASURED USING A METER OF 20,000Ω/VOLT

POWER SUPPLIES CIRCUIT DIAGRAM

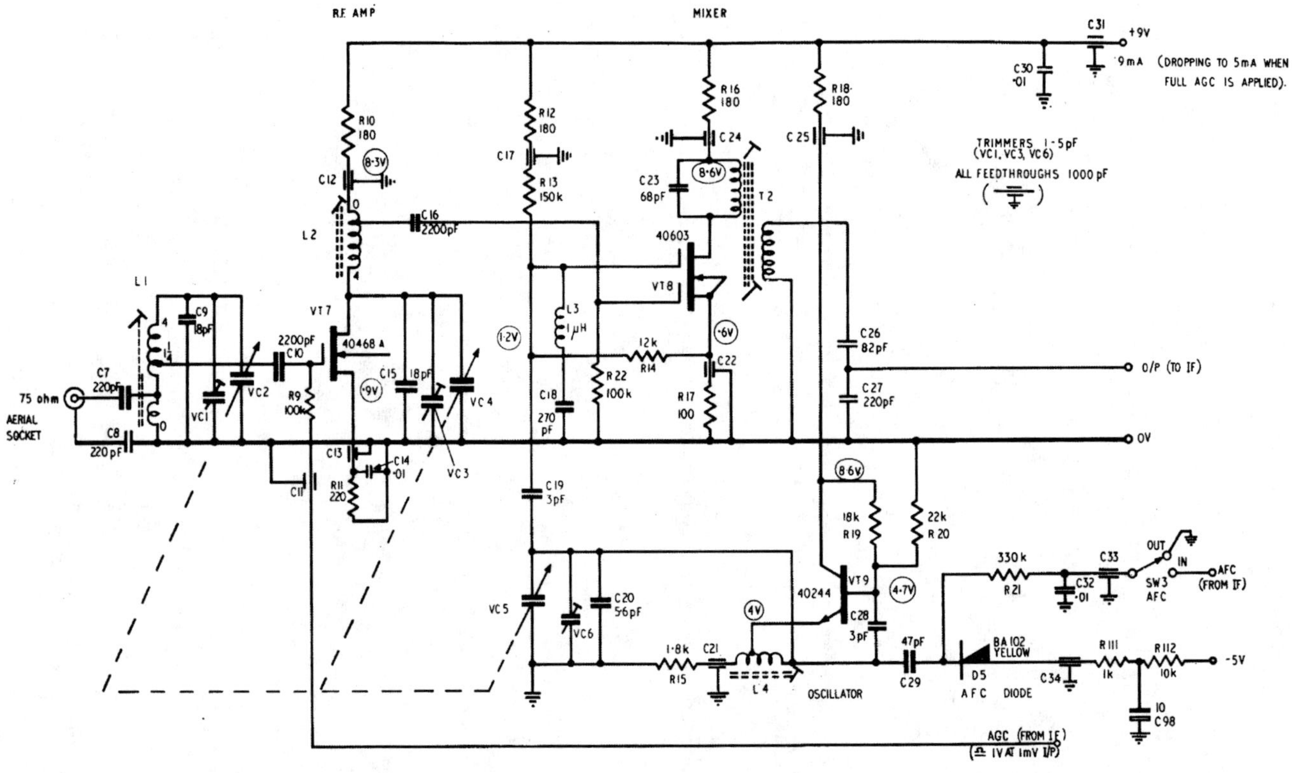

RESISTOR VALUES IN OHMS UNLESS OTHERWISE SHOWN
CAPACITOR VALUES IN MICROFARADS UNLESS OTHERWISE SHOWN
VOLTAGES ARE MEASURED USING A METER OF 20,000Ω/VOLT

FRONT END CIRCUIT DIAGRAM

Leak Stereofetic Tuner

Power Supply and Front End

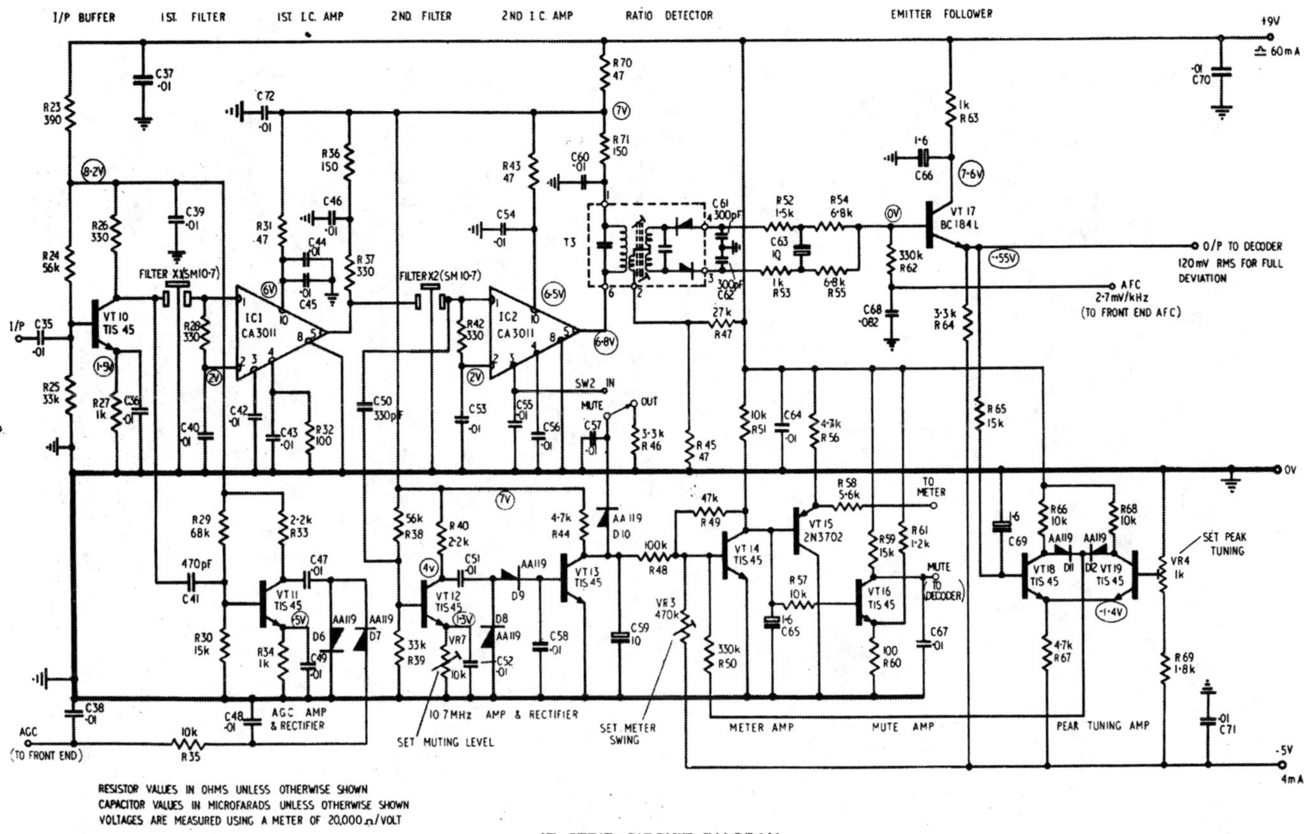

IF STRIP CIRCUIT DIAGRAM

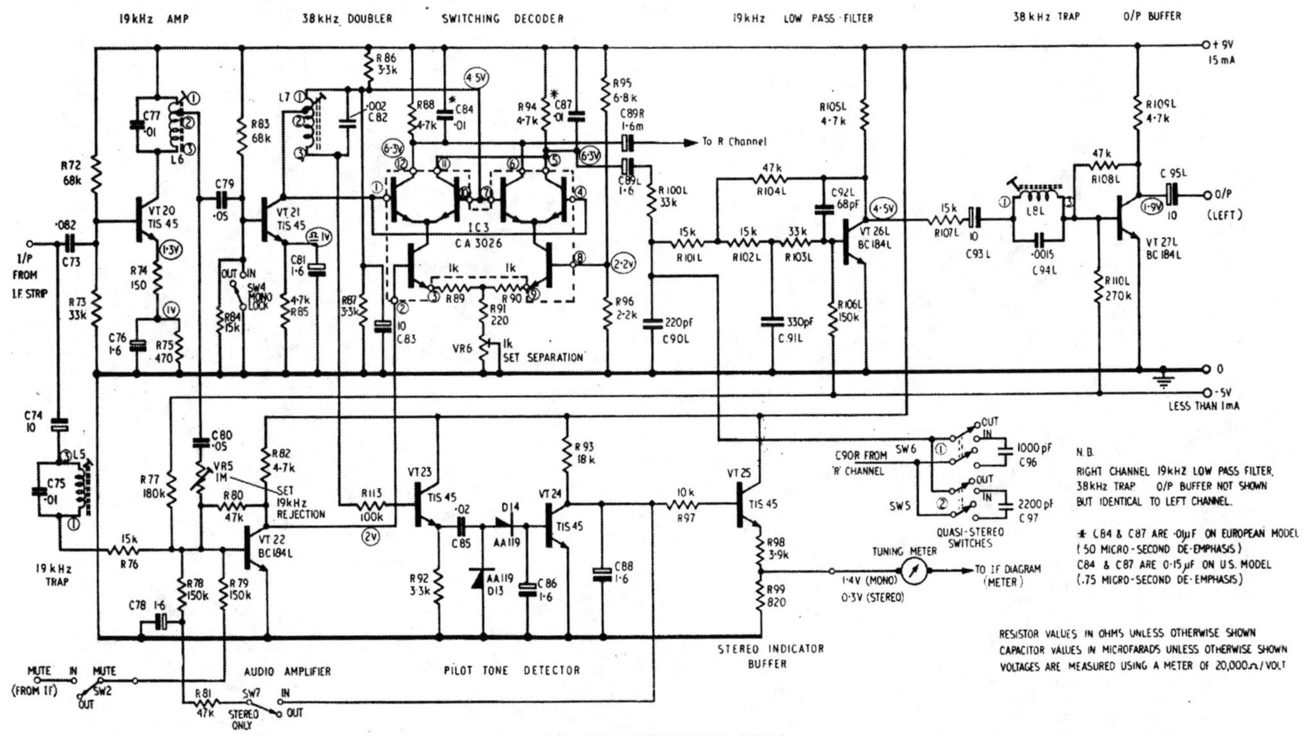

DECODER CIRCUIT DIAGRAM

Reproduced by permission Leak Electronics Ltd.

Leak Stereofetic Tuner

I.F. Strip and Stereo Decoder

L
R
L
R
RECORD
OUTPUT
TAPE
AMP
967
HI
LO
VARISLOPE 2 STEREO
LEAK
MADE IN ENGLAND

Leak Product Listing

1934 to 1945

Various Sub-Contracted Amplifiers for Partridge, GEC, TR, STC etc.	1934-1950	
13 Watt PA Amplifier	1936	£13/13/0
M.2 Microphone	Mid 1930's	£10/0/0
D.3 Loudspeaker	Mid 1930's	£5/15/0

(plus other unknown products, information unavailable)

1945 to 1969

Power Amplifiers

15 Watt Point One – "Type 15"	Sept 1945	£70/0/0 *
30 Watt Point One	Sept 1945	–
TL/12	Dec 1948 to 1957	£25/15/0
TL/10	1954 to 1957	£17/17/0
TL/25	1949	£27/10/0
TL/25A	1951	£27/10/6
TL/12 Plus	1957 to 1967	£18/18/0
TL/25 Plus	1957 to 1967	£25/4/0
TL/50 Plus	1957 to 1968	£33/12/0
Stereo 20	1958 to 1967	£30/9/0
Stereo 50	1958 to 1963	£42/0/0
Stereo 60	1963 to 1967	£45/0/0

Pre-Amplifiers

RC/PA	Dec 1948 to 1950	£6/15/0
RC/PA/U	1950 to 1954	£8/8/0
Varislope	1952 to 1954	£12/12/0
Steep Cutting Filter	1953	£5/10/0
Varislope II	1954 to 1957	£16/16/0
Point One	1954 to 1957	£10/10/0
Varislope III	1957...	£16/16/0
Varislope III, Series N	...to 1963	£16/16/0
Stereo Adaptor	1959	£5/10/0
Point One Plus	1958 to 1963	£12/12/0
Varislope Stereo	1960 to 1963	£25/0/0
Point One Stereo	1958 to 1963	£21/0/0
Varislope 2 Stereo	1963 to 1967	£25/0/0
Varislope Mono	1963 to 1967	–

* Un-confirmed price

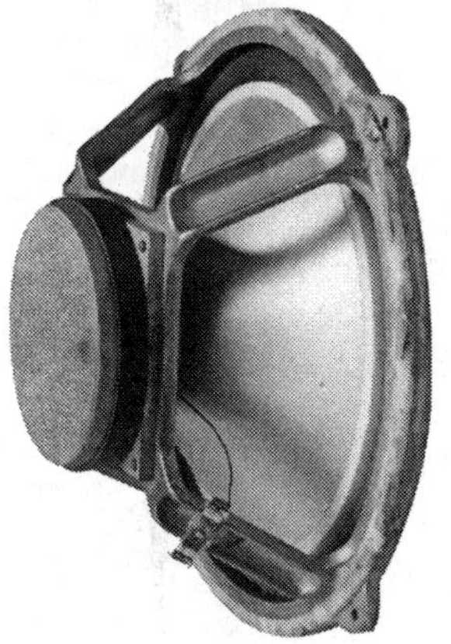

A Sandwich 13-inch woofer fitted with a ceramic magnet. Earlier models had alnico magnets.

Opposite: The rear panel of the Varislope 2 Stereo pre-amplifier.

Radio Tuners

VS-Tuner	1953	£38/5/0
Trough-Line	1955	£25/0/0
Trough-Line II	1959	£29/7/6
Trough-Line 3	1964 to 1969	£29/7/6
Trough-Line Stereo	1966 to 1969	£47/2/1
Stereofetic	1969	£56/11/10

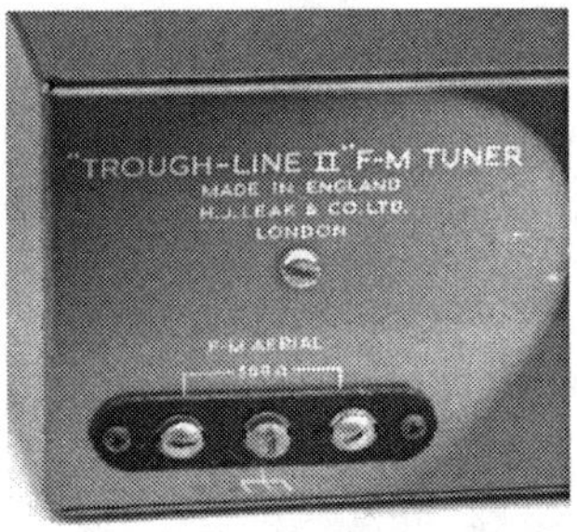

Aerial input on the Trough-Line II FM tuner.

Pickups, Arms, Turntables, Cabinets

Mark 1 Moving Coil Pickup & Arm	1948	£8/8/0
Mark 2 Moving Coil Pickup & Arm	Oct 1954	
LP diamond head		£5/15/0
78 diamond head		£5/15/0
Standard Arm		£2/15/0
Long Arm for 16" records		£3/5/0
Transformer		£1/15/0
Leak Mark 4* (stereo pickup)	1966/67	£26/16/8

(*Listed in 1966/67 Hi-Fi Year Book, although not actually launched.)

Truspeed Turntable, chassis only	1969	£30/3/2
with arm, Goldring 800E cartridge and plinth,		£65/13/3

Console Cabinet	circa 1950	
Southdown Cabinet (manf. D&R)	circa 1957, for TL/10	–
Southdown Cabinet (manf. D&R)	1960	£21/0/0
Southdown Cabinet (manf. GKD)	1962	£26/5/0
Chiltern Cabinet	1960's	
Combined Tuner/Amp case	1968	£9/18/6

Upper: The Stereo 70 pre-amp board. Lower: The Stereo 70 power amp board.

Loudspeakers

550 Loudspeaker	1949 to 1952	55 gns
1961 Sandwich Two-Way		
Mk 1 - fibre hf cone	1961 to 1964	£43/10/0
Mk 2 - sandwich hf cone	1965 to 1969	
Mini-Sandwich	1964	£29/15/0

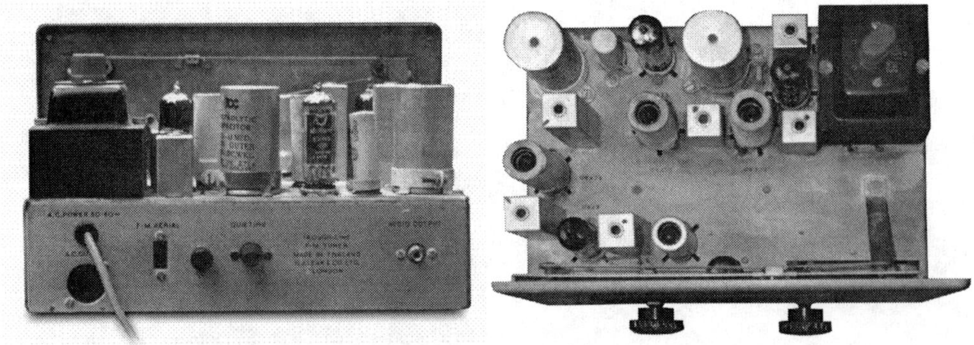

The rear panel and top-chassis view of the original Trough-Line FM tuner showing the quieting control pre-set. (Reproduced by permission © Steve Robinson.)

Integrated Transistor Amplifiers

Stereo 30 with case	1963	£58/10/0
Stereo 30 Plus	1968	£59/10/0
Stereo 70 with case	1967	£69/10/0
chassis only		£63/0/0

An early Leak crossover, with air-cored inductors and oil-filled paper capacitors, as supplied in 1953.

1969 to 1978 (Rank-Leak)

Delta Series 1971 to 1974

Delta 30 amplifier	£65
Delta 70 amplifier	£79.50
Delta 75 tuner-amplifier	£134.74 (plus £25.26 tax)
Delta FM Tuner	£62.13 (plus £10.87 tax)
Delta AM-FM Tuner	£74.04 (plus £12.96 tax)
Delta Turntable	£57.45 (in teak, with cover £9.50 extra)
Sandwich Mk 3 2-way loudspeaker	
Sandwich 150* 2-way loudspeaker	£20.85 (plus £3.65 tax)
Sandwich 250* 2-way loudspeaker	£26.81 (plus £4.96 tax)
Sandwich 600 3-way loudspeaker	£49.50

(*Apparently designated the Sandwich 200 and Sandwich 300 in Australia.)

2000 Series 1974 to 1976

1800 Tuner-Amplifier
2000 Tuner-Amplifier
2100 Amplifier
2200 Amplifier
2300 AM-FM Tuner
2000 Series Leak Turntable
2001 Transcription Turntable
2020 2-way loudspeaker
2030 3-way loudspeaker
2060 3-way loudspeaker
2075 4-way loudspeaker
2002 Cassette Deck

3000 Series 1976 to 1978

3020 2-way loudspeaker
3030 2-way loudspeaker
3050 2-way loudspeaker
3080 3-way loudspeaker
3090 4-way loudspeaker
3000 Iso-Dynamic headphones
• Plus a range of "factored" electronics sourced from Roland (Rotel), Japan

The under chassis view of a restored Stereo 50 amplifier. A previous owner had removed the cable loom and wired the amplifier point-to-point. A new cable loom was made, and the component board re-instated.

De-Emphasis Circuits for the Trough-Line Stereo

FM transmissions apply "pre-emphasis" to the audio signal fed to the transmitter so that an improvement in overall signal to noise ratio can be obtained. This pre-emphasis applies a boost in the high frequency audio signals fed to the transmitter – rather similar to applying treble boost using a tone control on an amplifier.

In Britain, a 6dB per octave boost is applied at frequencies above 3.18kHz. In "time-constant" terminology this equates to 50 microseconds. In the USA, the 6dB per octave boost commences at 2.12kHz (a time-constant 75 microseconds).

As a result all FM tuners incorporate an appropriate "de-emphasis" circuit so that the overall frequency response of the system from end-to-end is flat. The de-emphasis circuit is very simple – usually taking the form of a capacitor to ground from a point in the signal circuit with a well defined impedance (e.g. at the output of a buffer stage that has a series resistor in the output much larger than the output resistance of the buffer, shunted by a capacitor to ground).

In the case of the Leak Trough-Line II the resistor R32 (of 470kOhms), and capacitor C50 (100pF) in the grid circuit of V5B form the de-emphasis circuit. The time-constant of this combination is equal to the product of R and C, or 47 microseconds, which has been deemed by the designer (allowing for stray capacitance and component tolerances) to be accurate enough to apply the required 50 microsecond de-emphasis.

For the Leak "Trough-Line Stereo" the circuit shows R42 (150kOhms) and R43 (68kOhms) forming a voltage divider at the output of V5B. The output resistance of this divider is very close to the parallel combination of R42 and R43, this being 46.8kOhms. C50 shunts the signal output from the divider to ground. So, for a 50 microsecond time-constant that value of C50 should be equal to 0.00005 / 46,800 = 1.07nanoFarads (or 1070pF). The Leak circuit states that C50 is 3,000pF, which appears to be too large in value.

However, unlike the Trough-Line II, the Trough-Line Stereo lacks a cathode follower between the de-emphasis circuit and the input circuit of the pre-amplifier it is

Pre-amp Input Resistance	Cable Capacitance	Correct Values for 50µsec (UK)	Correct Values for 75µsec (USA)
47kΩ	50pF	C50=2.08nF	C50=3.14nF
		C59=C60=1.39nF	C59=C60=2.09nF
100kΩ	50pF	C50=1.51nF	C50=2.30
		C59=C60=1.01nF	C59=C60=1.53nF
470kΩ	50pF	C50=1.12nF	C50=1.71nF
		C59=C60=747pF	C59=C60=1.14nF

Equalisation component values for the Trough-Line Stereo FM tuner, operating with various load resistances, and assuming an approximate cable capacitance of 50pF.

connected to. As a result, the input resistance of the pre-amplifier, and the capacitance of the interconnecting cables affect the time-constant of the de-emphasis circuit, making it difficult to recommend a single correct value as Leak pre-amps varied in input resistance from 47kOhm to 470kOhm.

The Trough-Line Stereo was primarily designed to operate correctly with the Stereo 30, which has a lower input resistance than average. For operation with other load impedances modification of the equalisation circuit would be beneficial. In the tabulation shown here, three different sample cases are given.

Since C50 , C59 and C60 are shunted by the capacitance of the output leads, it may be necessary to allow for this by adjusting on test. Alternatively, it would be better still to install a modern integrated circuit stereo decoder.

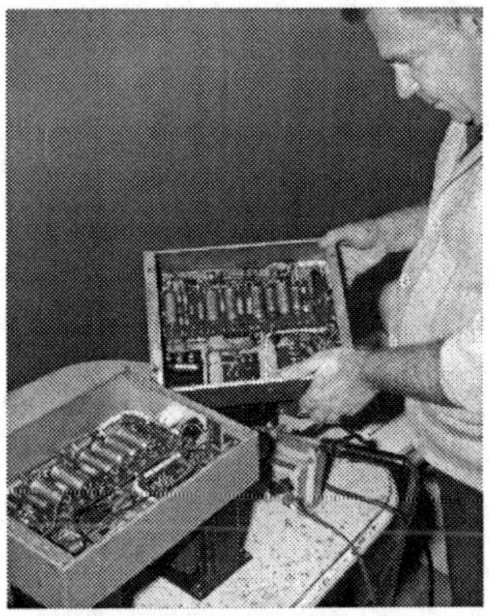

David Harcoan, an Australian audio enthusiast, examines a pair of TL/25A amplifiers with over 30 years service in a theatre sound system, and which are now ready for restoration.

Amplifier Restoration

The underlying value in vintage audio equipment comes from what these products are as they were originally designed and produced. Owners of vintage equipment can decide:

- to leave equipment un-restored (and possibly un-operational)
- to restore it to original condition as closely as possible, as well as to make it fully operational
- to bring the amplifier to operational condition, but modify it in minor ways, so that the amplifier could be re-instated to back to original condition later, if required
- to undertake significant modifications that cannot be un-done, and could be considered destructive to the value and desirability of the amplifier in the long-term

Depending on your requirements, you can choose an approach that suits you best, but in general undertaking significant modifications will result in the equipment losing salient features that make it desirable.

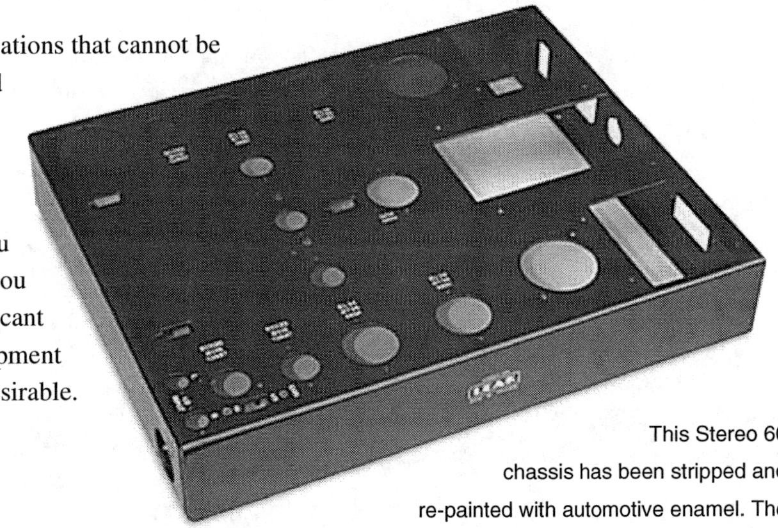

This Stereo 60 chassis has been stripped and re-painted with automotive enamel. The “Leak” logo and other chassis lettering was then applied by a screen printing works. Screen printing artwork can be created with a computer, starting with scanning the original chassis, and then drawing the new artwork using the scanned image as a template.

Resistors

The original Leak amplifiers used carbon resistors, some of which were of the "body-end-dot" variety, and some of which had the more modern coloured bands. The choice of replacement resistor type is to an extent personal, and has been discussed at length in other publications. Leak commonly used carbon-composition resistors, except for some critical resistors (such as those in the first stage of a pre-amplifier, which could be carbon-film). When carbon-film types dropped in price during the 1960's these were adopted for general use by Leak. However, modern metal-film resistors are technically better quality, and will give superior performance for little increase in overall cost. Don't forget to keep your Leak amplifier looking first rate by soldering all resistors in place with the tolerance bands all facing the same direction.

These Standees resistors on Stereo 50 and Stereo 60 amplifiers are often found burnt out and there is no modern replacement. Some enthusiasts have made substitutes by potting wire-wound resistors (using products such as muffler putty) into a steel tube painted with high temperature paint.

Capacitors

In most Leak amplifiers the capacitors that pass the signal between stages are of the "metallised paper in oil" variety (e.g. TCC Metalpack). There is every chance that these capacitors have absorbed moisture and become leaky, or that some capacitors have already been replaced with an inferior type in the past. TCC Metalpack capacitors are very good (provided moisture has not affected them) and should be replaced with the best capacitors available. Unfortunately it is not possible to purchase new types that are identical, but a similar type can be obtained from Angela Instruments, or The Parts Connection in the USA.

A good alternative choice is a capacitor with a polypropylene dielectric, and metal-foil, or metal-film electrodes. Radio Spares and Farnell both carry suitable capacitors with 1000V rating that have the original "tubular" construction, with "pigtail" leads like the originals. The Farnell units are made by LCR in the UK. More exotic varieties can be obtained from Angela Instruments, or The Parts Connection.

A key feature of the Leak TL/12 is the use of paper-in-oil capacitors for filtering and de-coupling the high tension rail. These capacitors have a very long life indeed, and should never need replacing. In

Even "new-old-stock" TCC Metalpack capacitors can be too leaky to use despite the reputed sonic benefits of paper-in-oil dielectric.

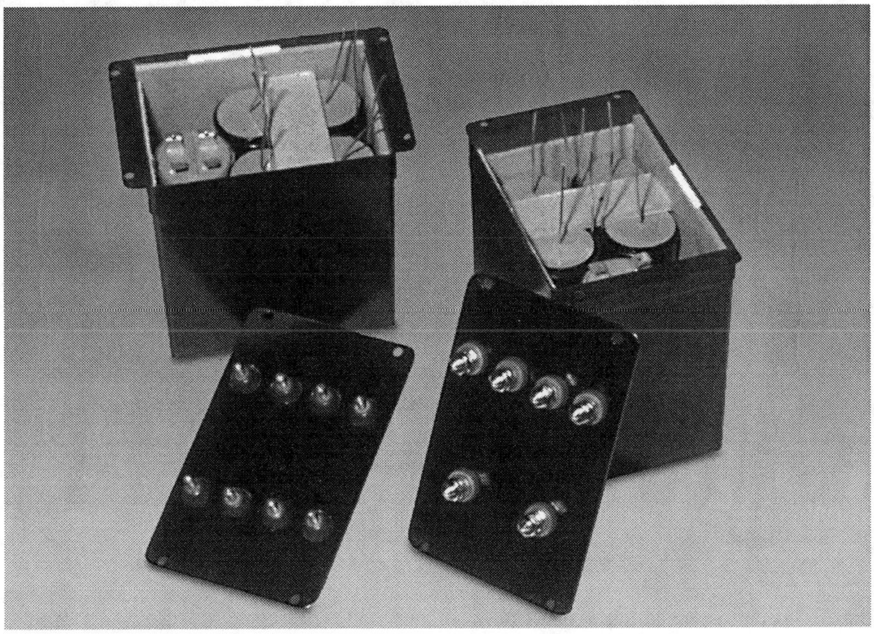

These new capacitor blocks for a TL/12 amplifier have one 4μF motor-start capacitor, and four 10μF metallised polypropylene capacitors ready for connection to the terminal block.

fact it is undesirable to replace them at all, unless it cannot be avoided (e.g. because they have been removed, or have failed from severe leaking of oil).

This capacitor block was custom made for Leak, so finding an identical replacement is not likely. Owners have been known to dismantle this capacitor block, remove the existing capacitors, and re-use the square can to house replacements. Since the oil used in the block is probably a PCB, this is not recommended as PCB's are linked with causing cancer. However, a replacement can could be custom made by a small sheet metal workshop, and new capacitors of your choice housed in the new can.

A few TL/12 amplifiers had a capacitor block that featured a ridged lip around the top, and more rounded corners than the more common square-edged case.

An example restoration is shown here, where a polypropylene motor-run capacitor (4μF/44OVac from Farnell) has been used for the main rectifier filter as it is capable of handling the ripple current in this part of the circuit. It is not desirable to increase the value of this first capacitor in the ripple filter. The other capacitors used were four 10μF/630V polypropylene "Solen Fast Caps". A side effect of this replacement is that the HT ripple in the amplifier is vastly reduced as the capacitance after the filter choke has been increased (from 8μF plus 4μF, to 30μF plus 10μF). The original amplifier had several volts of ripple present on the main high tension supply – which also fed the driver stage. As a result some of this ripple was present on the grids of the output valves. Whilst this signal is cancelled by virtue of the push-pull output stage, it is still beneficial to remove such large amounts of ripple voltage from the signal path.

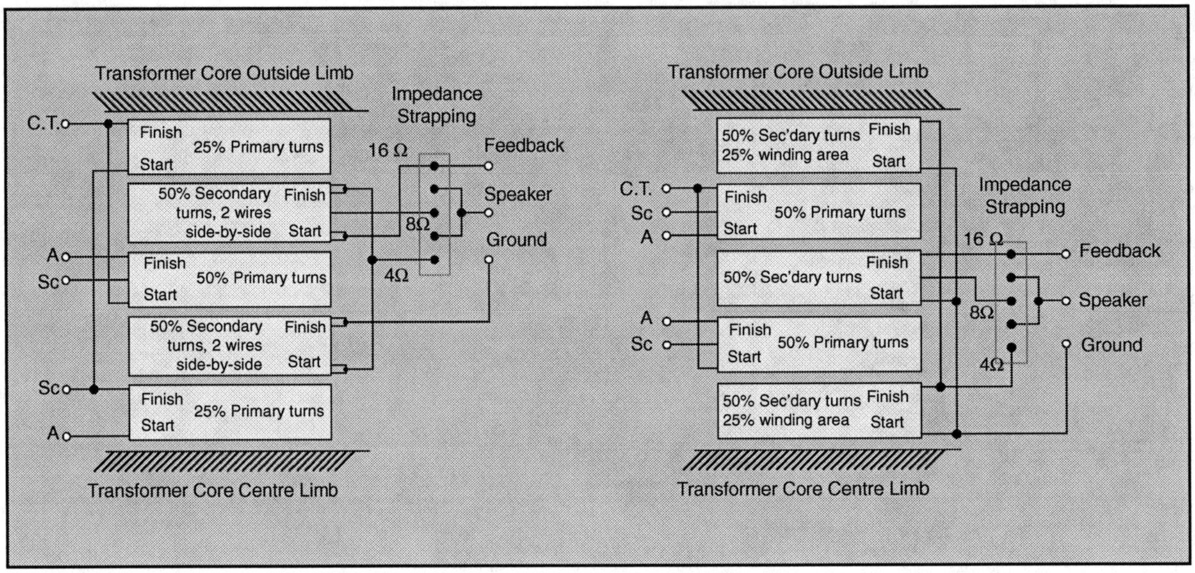

The winding layout for the Leak 3925 (left) and 8615 (right) output transformers. Both types were used in the Stereo 50, with the 8615 appearing in late models. The Stereo 60 was fitted only with the 8615 type. Similar design revisions were made when the 8778 replaced the 3921 transformer used in Stereo 20 and TL/12 Plus power amplifiers.

The replacement capacitors described above were placed in the capacitor can, with medium density fibre board lining the box. This assisted to locate the capacitors physically, and retained the "solid sound" the original box had when tapped with the hand.

Power Transformers and Chokes

Fortunately the iron-cored components are usually fairly reliable, however failures can occur. If the amplifier is operated long term with leaky coupling capacitors this can induce a power transformer failure from the increased quiescent current in the output stage.

If a power transformer or filter choke has failed, these components should be easily re-wound by a small transformer manufacturer in your area. You will need to inform them of special requirements regarding physical exits of winding taps to suit the terminal layout on the baseplate, and make sure they can still fit the original transformer shrouds on the re-wound unit. If necessary, insist on re-using the old core if a new core is incompatible with the original shrouds.

If the output transformer has failed, this is a more serious problem, as selecting a suitable manufacturer for a re-wind will be a difficult task. Some possible manufacturers are listed in the "Suppliers" section.

Transformer I.D. Number	Amplifier Models	Primary Load (anode-anode) +/-250	Resistance (anode-anode) +/-10	Screen Tap +/-2%
TL/12/T2/2 Open-Frame	TL/12 before ~1953	5800	280	No Taps
TL/12/T2/2 Potted/Enclosed	TL/12 after ~1953	5500	280	47%
TL/25A/T2/2	TL/25A	6500 feedback wdg. 1:17.65	110 (55 + 55)	No Taps
1360	TL/10	9100	200	49%
3921	TL/12 Plus Stereo 20 before ~1961	8600	165	49%
3925	TL/25 Plus Stereo 50 before ~1961	7000	102	49%
8013	TL/50 Plus	5200	45	49%
8615	TL/25 Plus & Stereo 50 after ~1961 Stereo 60	5600	102	25%
8778	TL/12 Plus Stereo 20 after ~1961	8950	135 (72 + 63)	25%

In this tabulation of output transformer parameters for Leak valve amplifiers, the data has been determined by measurements on actual transformers, so is subject to experimental error. An indication of the error tolerance is given for each parameter. Leak also produced special amplifiers for Public Address use that had 100-Volt or 70-Volt output connections, a separate feedback winding, and required a load impedance of several hundred Ohms. These transformers are not tabulated here.

Damaged or worn black tape on Leak transformers can be removed and replaced with a self-adhesive plastic cut from a roll as shown on this Stereo 50 power transformer. If required, the end-covers can be easily removed without damaging the internal wiring.

Output Transformers

The information tabulated here is supplied to assist in determining the condition of output transformers. The DC resistance measurements can be useful if it is required to assess a transformer prior to purchase of a second hand amplifier. (In the author's experience, the TL/12 "enclosed case" output transformers have a reputation for going open circuit on the primary).

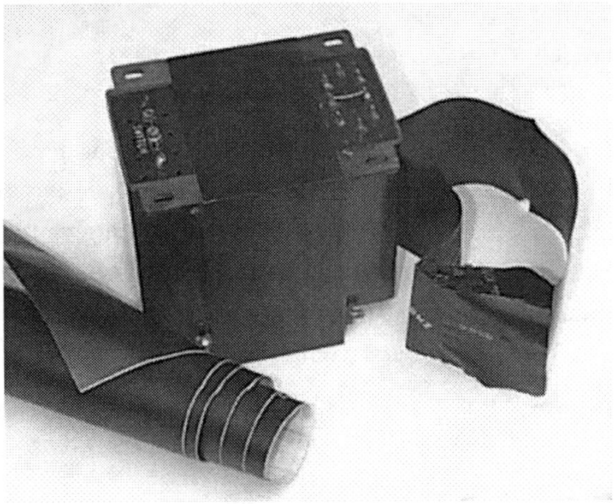

The primary impedance and screen tap percentage can be useful knowledge if you need to have a new component wound, or intend to build/replicate your own Leak amplifier. Note that Leak used two different screen tap percentages – 25% and around 50% (the classic recommendation is for 43% as noted by Hafler and Keroes when they first publicised this output stage, but Mullard noted that other values are also applicable). Later models of the TL/12 apparently included screen-taps, but these were never used in production – all TL/12's were triode connected.

The physical winding layout of an output transformer significantly affects the high frequency performance. Any replacement transformer for your Leak amplifier should have a winding with an adequate number of sections to ensure that the high frequency response (including the phase response) will allow operation within the feedback loop.

As an example, the Leak Stereo 60 output transformer (type 8615) has a small-signal frequency response from around 3Hz to 80kHz with a winding layout comprising five sections as shown here. Note that the 8615 differs in design from the output transformers used in most Leak Stereo 50 amplifiers, even though both use EL34 output valves. The 3925 uses three primary sections and two secondaries, where as the 8615 used two primary sections and three secondary sections. Also note that each secondary section on the 3925 was wound with two smaller wires laid down side-by-side, rather than with one single (and thicker) wire. This has the effect of reducing the winding height, allowing better coupling (and hence better high frequency response).

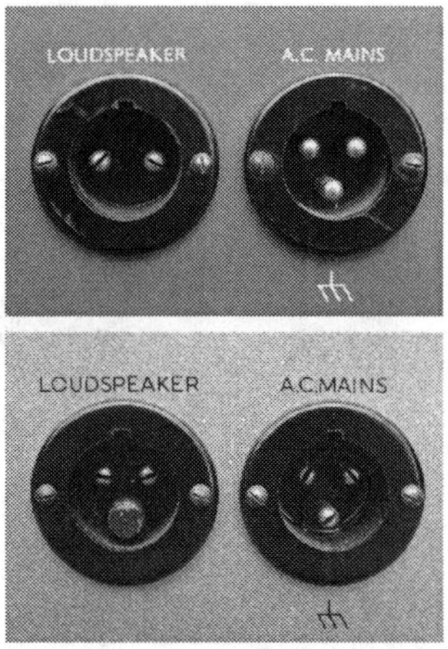

Upper: The original 2-pin and 3-pin Bulgin connectors used on the TL/12. Lower: Some owners convert the speaker outlet to Bulgin 3-pin – but a 6mm bolt shank must replace the third pin to avoid plugging the mains connector into the speaker outlet! The mating speaker connector has the third pin removed and the remaining hole drilled out to accept the 6mm shank.

Connectors

The input connector to the TL/12 and some other models is an octal socket. To avoid drilling holes in the chassis it is possible to make up an adaptor using an octal plug and RCA socket, joined via a short lead. Also it may be possible to mount an RCA socket in the shroud of the octal plug, so the adaptor is a single-piece unit.

The loudspeaker connector on the TL/12 is a 2-pin Bulgin chassis-mount plug. The mating cord-mount 2-pin socket is no longer made and is un-available (except perhaps from second-hand electronic dealers). This is a problem for the enthusiasts requiring to retain the original appearance.

One solution is to replace the Bulgin 2-pin plug with a modified 3-pin type, which is still available. It is very important to make the modification, otherwise there exists a possibility of plugging the mains supply into the loudspeaker output, with catastrophic results! The modification requires that the earth pin on the 3-pin plug is replaced with the shank of a 1/4 inch bolt (the bolt head is sawn-off and the shank mounted in place of the third pin. A mating 3-pin socket is prepared with the earth pin removed, and a 1/4 inch hole drilled out in its place. In this way the

An adaptor can be used for connecting an RCA lead to the octal input connector on the TL/12. This avoids defacing the chassis with a hole for an RCA input connector.

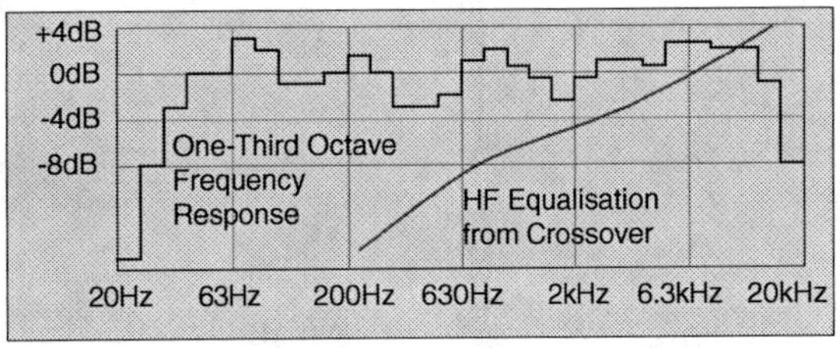

loudspeaker connector will look original, make use of a currently available component, and will not allow the mains input to be plugged into the loudspeaker output.

Another approach is to leave the existing 2-pin connector on the chassis, and dismantle a Bulgin 3-pin connector to recover just the metal sockets. The individual sockets can be wired to the end of the speaker lead, and then individually pushed onto the pins of the 2-pin socket. This won't look very attractive when wired up, but at least it preserves the appearance of the amplifier.

If appearance is not a consideration (but it should be), a metal plate can be made up to replace the existing socket. Connections of your choice are then mounted on this plate.

The frequency response for a modified Leak Sandwich Mark II, measured in a listening room, at 1m distance, with the microphone on-axis horizontally, but 10 degrees off-axis vertically (microphone slightly above the height of the tweeter). The response remains within a +/-4dB envelope, which is a creditable performance. The equalisation applied by the modified crossover is also shown. Be mindful that these speakers are not very sensitive and can be over-driven if high levels are demanded. (As the test above is not taken under anechoic conditions, irregularities in the response can be expected, particularly at low-mid and low frequencies.)

A Modified Sandwich Mark II Loudspeaker

This loudspeaker incorporates many novel design features, making it a serious contender for any modern loudspeaker. While the original design achieved much success, on comparison with more recent loudspeakers the Sandwich has a stronger upper-midrange and less treble extension. The use of a third driver acting as a "super-tweeter" is one modification adopted by some enthusiasts. (In Don Barlow's 1958 Wireless World paper on sandwich construction, he suggested a small 1-inch diameter sandwich driver for frequencies above 5kHz.) Alternatively, the use of a different crossover circuit can yield improvements in performance too. The modified crossover design shown here has a slightly lower crossover frequency, and the rates of cut-off are not as steep. The upper midrange (around 800Hz to 2kHz) is re-balanced, and the output at very high frequencies is increased through the use of an auto-transformer. It is quite practical to use the tapped secondary of an audio output transformer for this function, although a custom-made component should not be very expensive as good performance is more easily obtained at low turns ratios (as used here).

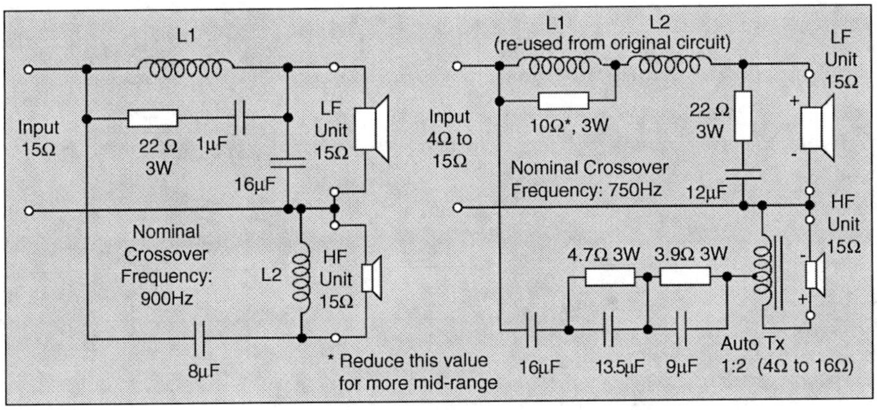

Far Left: The original Leak Sandwich Mark II cross-over circuit (15Ω).

Near Left: A modified cross-over that extends the extreme treble and re-balances the upper mid-range. It also results in a lower (4Ω) impedance at high frequencies.

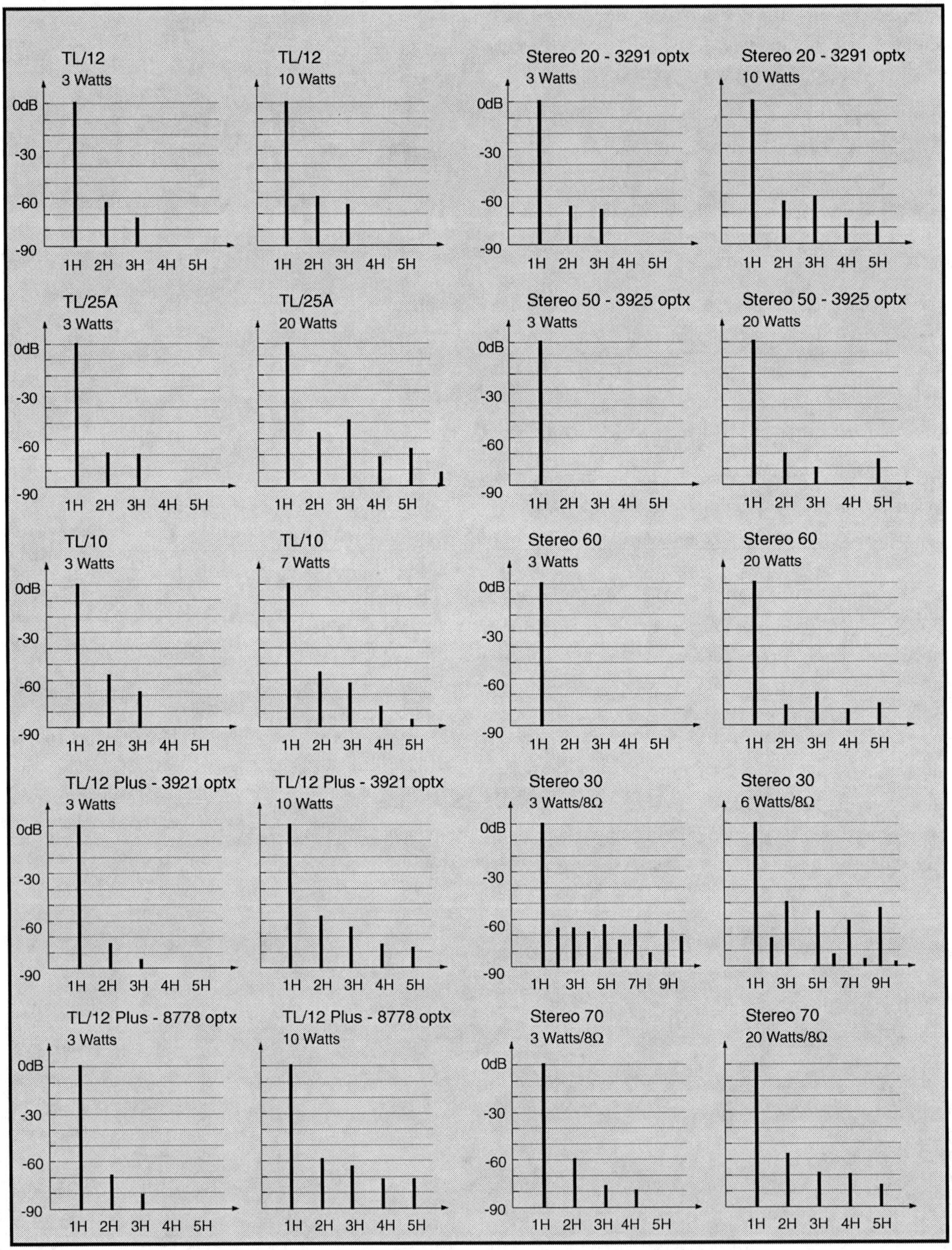
TL/12
3 Watts
0dB
-30
-60
-90
1H 2H 3H 4H 5H
TL/12
10 Watts
1H 2H 3H 4H 5H
Stereo 20 - 3291 optx
3 Watts
0dB
-30
-60
-90
1H 2H 3H 4H 5H
Stereo 20 - 3291 optx
10 Watts
1H 2H 3H 4H 5H
TL/25A
3 Watts
0dB
-30
-60
-90
1H 2H 3H 4H 5H
TL/25A
20 Watts
1H 2H 3H 4H 5H
Stereo 50 - 3925 optx
3 Watts
0dB
-30
-60
-90
1H 2H 3H 4H 5H
Stereo 50 - 3925 optx
20 Watts
1H 2H 3H 4H 5H
TL/10
3 Watts
0dB
-30
-60
-90
1H 2H 3H 4H 5H
TL/10
7 Watts
1H 2H 3H 4H 5H
Stereo 60
3 Watts
0dB
-30
-60
-90
1H 2H 3H 4H 5H
Stereo 60
20 Watts
1H 2H 3H 4H 5H
TL/12 Plus - 3921 optx
3 Watts
0dB
-30
-60
-90
1H 2H 3H 4H 5H
TL/12 Plus - 3921 optx
10 Watts
1H 2H 3H 4H 5H
Stereo 30
3 Watts/8Ω
0dB
-30
-60
-90
1H 3H 5H 7H 9H
Stereo 30
6 Watts/8Ω
1H 3H 5H 7H 9H
TL/12 Plus - 8778 optx
3 Watts
0dB
-30
-60
-90
1H 2H 3H 4H 5H
TL/12 Plus - 8778 optx
10 Watts
1H 2H 3H 4H 5H
Stereo 70
3 Watts/8Ω
0dB
-30
-60
-90
1H 2H 3H 4H 5H
Stereo 70
20 Watts/8Ω
1H 2H 3H 4H 5H

Harmonic Distortion

Musical instruments generate harmonics that are integral multiples of the fundamental note being played. The number of harmonics and the magnitude of each one is what gives each instrument its characteristic sound. An imperfect amplifier will also generate harmonics, but in this case the harmonics are deemed to be distortion.

Good amplifiers generally have very low harmonic distortion (approaching or less than 0.1%), although in the past decade many valve amplifier designs have placed less emphasis on achieving low harmonic distortion. This has seen a return to harmonic distortion figures between 1% and 3%, but with the touted proviso that other, more significant, sonic benefits can be gained by not attempting to minimise harmonic distortion through the use of negative feedback.

Harmonics that are odd-multiples (3, 5, ..) of a fundamental tone are less pleasing to the ear than harmonics that are an even multiple (2, 4, ..) and this is even more applicable if the odd harmonics are high-order ones (7, 9, ..). Traditional harmonic distortion measurements are made by feeding an amplifier with a pure tone (a sine wave) and then measuring the sum of all the harmonics generated by the amplifier. More useful information is gained by examining the spectrum of the harmonic distortion so that the amounts of odd-order distortion can be seen.

Shown opposite are the results of distortion spectrum measurements on a number of Leak amplifiers. In every case a pure sine wave of 400Hz was applied, and the harmonic distortion measured at 3 Watts. An additional measurement was done at a higher power level that was close to full-output for each amplifier. In some instances, the same model amplifier is listed twice, but the output transformer fitted is different

The TL/12 has the notable feature of consistent distortion character at the two power levels, with no harmonics beyond the third. The TL/25A reveals more odd-order distortion that can be attributed to its tetrode output stage. The TL/10 shows its limited power output, but the nature of the distortion is very acceptable. The TL/12 Plus and Stereo 20 designs are similar in performance, although the structure of the TL/12 Plus is marginally better due to its lower odd-order distortion. At low power the Stereo 50 and Stereo 60 both have distortion below measurement detection (ie: more than 90dB below the fundamental), with very low and generally well-structured distortion at 20 Watts. A clear difference can be seen in the nature of distortion from the solid state Stereo 30, with many high-order odd harmonics. But the Stereo 70 is remarkably good in harmonic structure, belying its solid-state heritage.

Opposite: Harmonic distortion spectra for a number of Leak amplifiers measured at 400Hz and at two power levels. (Note the change of scale for the Stereo 30 so that the high-order harmonics can be shown.)

Suppliers

The Parts Connection, USA Phone +1 905 829 5858, Fax: +1 905 829 5388.
Web page: http://www.sonicfrontiers.com/TPC/
(Supplies high end components, valves, sockets, resistors, capacitors, etc).

Angela Instruments, Suite 309, 10830 Guildford Road, Annapolis Junction, MD 20701, USA. Phone: +1 301 725 0451. Web page: http://www.angela.com

P. M. Components, Springfield Enterprise Park, Springfield Rd, Gravesend, Kent, DA11 8HD, UK. Phone: +44 474 560521
Email: 101650.2424@compuserve.com (Supplies electron valves/tubes).

Billington Export, 1E Gillimans Industrial Estate, Billingshurst, West Sussex, RW14 9EZ, UK. Phone: +44 403 784961 (Supplies electron valves/tubes).

Sussex Surplus 13 Station Rd, Horsham, West Sussex, RH13 5EZ, UK. Phone: +44 403 251302 (Supplies electron valves/tubes).

Antique Electronic Supply, 6221 S. Maple Avenue, Tempe AZ, 85283, USA (Excellent valve/tube listings). Web page: http://www.tubesandmore.com

Farnell Electronic Components (outlets in most countries).

Radiospares (outlets in most countries).

Wilson Valves, 28 Bank Avenue, Golcar, Huddersfield West, Yorks, HD7 4LZ, Phone:+44 1484 654650, Fax: +44 1484 655699, Email: wilsonvalves@surflink.co.uk

Langrex Supplies, 1 Mayo Rd, Croydon, Surrey CR0 2QP, UK.
(Supplies electron valves/tubes).

The Emporium, 28 St. Nicholas St., Norfolk IP22 3LB, UK. Phone: 01379 650744

Classique Sound, Vintage Hi-Fi Specialist, 61 Aylestone Drive, Aylestone, Leicester LE2 8QE, UK. Phone: 0116 2835821.

A Stereo 20 power amplifier with its original packing box.

Output Transformer Sources

- Majestic Transformer Co., 245 Rossmore Road, Poole, Dorset BH12 2HQ, United Kingdom, Phone: +44 1202 734 463
- Sowter Transformers specialise in audio transformers and have a range of replacement types for vintage amplifiers: Sowter Transformers, The Boatyard, Cullingham Road, Ipswich IP1 2EG, United Kingdom. Web Page: http://www.sowter.co.uk/
- Harbuch Electronics of Sydney, Australia make excellent wide-bandwidth output transformers: Harbuch Electronics, Unit 9, 40 Leighton Place, Hornsby, New South Wales, Australia, 2077, Phone: +61 2 9476 5854, Fax: +61 2 9476 3231, (will wind to specification). Web Page: http://ourworld.compuserve.com/homepages/harbuch

Later packing cases for Leak power amplifiers used a compacted straw-like material as shown here.

Leak Internet Resources

• H. J. Leak Email Group:	http://www.egroups.com/groups/hjleak
• kiewa Valley Stereo:	http://home.mira.net/~kiewavly/audio1.html
• Patrick's Tube Audio Page:	http://www.users.globalnet.co.uk/~jpchef
• Klaus Schiffer's Web Page:	http://members.aol.com/ikschiffer/index.html

Leak References

Personal Recollections

"Interview with Ted Ashley", ex-Leak Chief Engineer recorded by telephone by S. Spicer, July/August 1998. Related documents and photos supplied by E. H. Ashley.

Information and photos supplied by John Leak and Simon Leak, 1999.

"Interview with Angus Gilchrist", ex-Leak employee, by S. Spicer, Dec 1998. Related documents and photos supplied by Angus Gilchrist.

"Interview with Ian Pullen", ex-Leak Senior Design Engineer, by S. Spicer, Dec 1998. Related documents and photo supplied by Ian Pullen, Feb 1999.

"Interview with Norman Rowe", ex-Leak employee, by S. Spicer, Dec 1998. Related documents and photos supplied by N. Rowe.

"Interview with Ray Whitcombe", ex-Leak Engineer, by S. Spicer, Dec 1998. Related documents and photo supplied by R. Whitcombe.

"Interview with Steve Wilkins", ex-Leak employee, by S. Spicer, Nov/Dec 1998. Related documents and photo supplied by S. Wilkins.

"Interview with Phil Biggs", ex-Leak employee, by S. Spicer, Feb 2000. Related documents and photo supplied by P. Biggs.

"Interview with Sylvia Barlow", by S. Spicer, 1999. Related documents and photos supplied by S. Barlow.

"Interview with George Wise" by S. Spicer, Dec 1998, with related documents supplied by G. Wise and S. Mann.

"Interview with Sydney McClory", ex-Managing Director of H. J. Leak (Aust) Pty Ltd, (Australian manufacturers of the Leak products) recorded by telephone by S. Spicer, August 27 1998. Related documents and photo supplied by S. McClory.

"Interview with J. Cadby", by S. Spicer, March 1999.

"Interview with Joe Middleton", ex-Chief Engineer for Simon Gray Pty Ltd (Australian Distributors for Leak) recorded by telephone by S. Spicer, July/August 1998.

"Interview with Gordon Arbib", by S. Spicer, 1999.

"Interview with Tony Coleman (Tamura Hinchley Ltd.)" by S. Spicer, Feb 1999. Photos supplied by Sue Alexander and Tony Coleman, of Tamura Hinchley Ltd, UK.

General

"The Beginnings of Hi-Fi – A History of H. J. Leak & Co.", Noel Keywood, Hi-Fi World, October 1996. Photos supplied by Richard White, Hi-Fi World.

"Leak Point One Amplifiers", by H. J. Leak (Company Publication), 1951.

"H. J. Leak", Wireless World, Oct 1954, Page 475.

"Audio News and Views – Leak to Rank-Leak", Hi-Fi News, Page 289, March 1969.

"Kaleidoscope", (Notes on Discovering the Leak History), Noel Keywood, Hi-Fi World, October 1996.

"Audio! Audio!", Johnathan Hill, 1995, Sunrise Press, Devon, United Kingdom.

Obituary for "Harold Joseph Leak", by Raymond Cooke, OBE, (of KEF fame), Journal of the Audio Engineering Society, Vol 37, No 10, October 1989.

"High Quality Reproduction", H. J. Leak presentation at the BSRA, Wireless World, April 1950.

"Our Authors", Wireless World, October 1956.

"Sidelines", John Crabbe, Hi-Fi News and Record Review, September 1997.

"After 69 Years – Still Trying", George Wise, Ultra High Fidelity, Issue 1, Page 5.

"Hi Fi News Awards for Achievements in Audio", Donald Aldous, Hi Fi News and Record Review, January 1987.

Pickups

"Pickup Design" – Letter by H. J. Leak to the Editor of Wireless World, Wireless World, July 1950, Page 269.

"Pickup Design" – Letter by H. J. Leak to the Editor of Wireless World, Wireless World, Sept 1950.

"Modern Phonograph Cartridges", Julian D. Hirsch, Radio-Electronics, June 1957, Page 35.

"Hi-Fi Pickup Arms", Julian D. Hirsch, Radio-Electronics, January 1958, Page 41.

Loudspeakers

"Why Electrostatic Loudspeakers", by H. J. Leak, "The Gramophone" May 1955, and also re-printed in "Radiotronics" (AWA Australia) September 1955.

"High Fidelity Loudspeakers: The Performance of Moving Coil and Electrostatic Transducers" by H. J. Leak. British Institute of Radio Engineers, December 1955.

"Full Range Electrostatic Loudspeaker", H. J. Leak and A. B. Sarkar, Wireless World, October 1956, Page 486.

"The Rigidity of Loudspeaker Diaphragms", by D. A. Barlow (later engaged by Leak to develop the "Sandwich Loudspeaker"), Wireless World, December 1958.

Amplifiers

"Laboratory Report on Leak TL/12 Main Amplifiers" by F. Langford-Smith and A. R. Chesterman, "Radiotronics" (AWA Australia) September 1955.

"Leak TL/12 Amplifier with Vari-Slope Control", Equipment Report, Audio Engineering, June 1953, Page 42.

"Restoring the Leak TL/12", Graham Tricker, Hi Fi World, August & Sept 1994.

"Details of New Products – Leak Stereo Amplifier", Hi-Fi News, Dec 1958, P503.

"Leak Varislope III and TL/12 Plus Amplifier", G.W.T., Hi-Fi News, Feb 1959, P661.

"Leak Stereo 20 and Point One Stereo" reviewed by J.C.G. Gilbert, Music Trades Review, reproduced in Hi-Fi News, Sept 1959, P212.

"Leak Stereo 20 and Point One Stereo" reviewed by G.W.T, Hi-Fi News Sept 1959, P249.

"Leak Stereo 60" Equipment Review, "Audio" (USA) January 1960 Page 42.

Editorial (High Quality Transistor Amplifiers), Wireless World, May 1963.

"Details of New Products – Leak Varislope Stereo", Hi-Fi News, June 1960, P37.

"Les amplificateurs Leak", la nouvelle REVUE DU SON, No 139 Juillet/Aout 1996.

"The TL/10 Power Amplifier", Richard Kelly, Hi Fi World, June 1996.

"Fifteen-bit Valve Performance", R. Brice, (Leak TL/25 modifications), Hi-Fi World, Oct 1991.

"Rank's With the Best", D. Church, Audio (IPC Magazines), Oct 1974, Page 49.

"Leak Stereo 30 Plus", John Gilbert, The Gramophone, November 1970, Page 37.

"Audio News – Stereo Adaptor", Hi-Fi News, May 1959, P901.

" Leak TL/12 Plus Rebuild", Hi-Fi World DIY Supplement, Nov 1999, P21.

Tuners

"Equipment Reviewed – Leak Trough-Line II FM Tuner", A. W. Wayne, Hi-Fi News, Feb 1963, P633.

"Equipment Reviewed – Leak FM Tuner", R. S. Roberts, Hi-Fi News, Nov 1956, P231.

"Details of New Products – Leak Trough-Line II", Hi-Fi News, March 1960, P735.

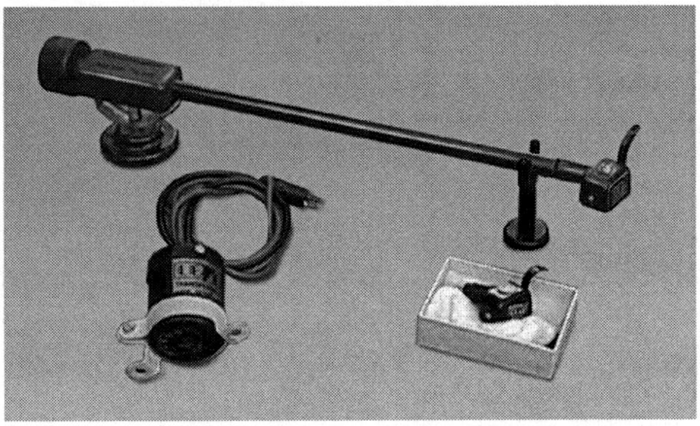

A Leak Mark 2 moving coil pickup and arm, complete with LP and 78 heads, transformer and rest. (Photo courtesy Craig Smith, USA.)

A description of a stereo decoder (Paravicini) for the Leak Trough-Line FM Tuner, Hi-Fi World, July 1992.

"Gateway to the Ether", N. Keywood and K. Bharadia, (discussion and supply of a stereo decoder for the Trough-Line tuner), Hi-Fi World, Sept 1997.

"Trough-Line Tuner" – letter by Ted Ashley (ex Leak Chief Engineer), Hi-Fi World, June 1996.

"Leak Stereofetic", John Gilbert, The Gramophone, December 1970, Page 50.

Loudspeakers and Cabinets

"The Leak Sandwich and the Triple Tone Major", R.L.W., Hi-Fi News, Aug 1961, P168.

US Patent 3,111,187 – Sandwich Construction of Loudspeaker Cones.

"Vintage Virtues", Hayden Boardman, (discussion of the Leak Sandwich), Hi-Fi World, July 1997.

"Southdown Cabinet", Hi-Fi News, May 1960, P909.

"Details of New Products – New Southdown Leak Cabinet", Hi-Fi News, January 1963, P557.

Audio Fairs and Demonstrations

"A Report on the London Audio Fair", Winget and Waitling, Hi-Fi News, May 1959, P901.

"Third Northern Audio Fair", Stanley Kelly, Hi-Fi News, Nov 1959, P377.

"1963 Audio Festival Report", R.N.B, Hi-Fi News, June 1963, P19.

"H. J. Leak & Co. Stereo Competition", Hi-Fi News, July 1958, P89.

"APAE Exhibition", Wireless World, July 1951, Page 259.

"The Audio Fair – 1950", Audio Engineering, October 1950, Page 32.

"The Audio Fair – 1952", Audio Engineering, November 1952, Page 54.

"The Audio Fair – 1953", Audio Engineering, October 1953, Page 84.

"Equipment for Sound Reproduction" – Leak at the BSRA/APAE 1952, Wireless World, Jan 1952, Page 256.

"Audio Engineering Society Convention Program", Audio Engineering, November 1949.

"Developments in Sound Reproduction", Wireless World, July 1955, Page 312.

"Developments in Sound Reproduction", Wireless World, July 1956, Page 335.

"Hi-Fi and Audio Shows in America" (Briggs & HJL), Hi-Fi News, 1956, P91 and P158.

"The Audio Fair", (refers to electrostatic speaker), R. West, Hi-Fi News, May 1957, P487.

"The Audio Fair" (refers to TL/12 Plus) , H. L. York, Hi-Fi News, May 1957, P489.

"The Fair in Pictures" (HJL with Peter Walker), Hi-Fi News, May 1957, P494.

"Audio Fair Photos", Hi-Fi News, May 1958, P708.

"All the Fun of the Fair", Hi-Fi News, April 1965, Page 984.

"Audio Fair, a Preview", Hi-Fi News, April 1968, Page 414.

"Audio Fair 1967", Hi-Fi News, June 1967, Page 38.

"High Fidelity Reproduction", H. J. Leak, Talk presented to Brit. IRE, North-Eastern Section, Neville Hall, Westgate Rd, Newcastle-on-Tyne, 6:00pm, Oct 11, 1950 (see Wireless World).

"High Fidelity Reproduction", H. J. Leak, Talk presented to Brit. IRE, North-Western Section, College of Technology, Manchester, 6:45pm, Dec 7, 1950 (see Wireless World).

"TL/12 amplifier and Radio Feeder Type VS (Variable Selectivity)" described at the APAE and BSRA 1953 Exhibitions, Wireless World, July 1953.

"1968 Audio Fair", Hi-Fi News, Page 694, June 1968.

Front cover photo, Hi-Fi News, July 1968.

"International Audio and Photo-Cine Fair 1969, Hi-Fi News, Page 1200, Oct 1969.

"BBC Stereo" (refers to Stereofetic), Angus McKenzie, Hi-Fi News, Page 1211, Oct 1969.

"Olympic Audio", Page 1552, Hi-Fi News, Dec 1969.

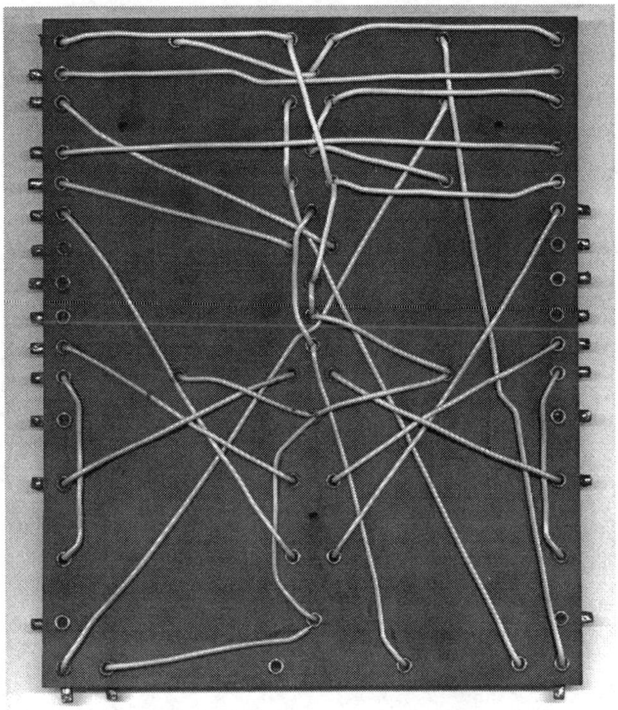

The underneath of a component board from a Stereo 60 amplifier shows how interconnections were made by hard-wiring from terminal to terminal. This practice continued into the late 1960's despite the adoption of printed circuit techniques in Leak solid state amplifiers.

Advertisements

<u>Leak Advertisements</u>

"Leak Hi-Fi", Product line summary and prices, advertisement, Hi-Fi News, Oct 1968, Page 1120.

"Technical Assistant Position" Recruitment Advertisement for H. J. Leak & Co. Ltd., Hi-Fi News, March 1968, Page 316.

"Fan Fair", British Industries Corporation, Audio Engineering, September 1953, Page 3.

"Britain's Best Audio Amplifier", H. J. Leak & Co. Ltd., Audio Engineering, October 1949, Page 7.

"The First 21 Years", H. J. Leak & Co. Ltd., The Gramophone, September 1955.

First "Point One" Advertisement by Leak, Journal of the British Institute of Radio Engineers, September 1945.

"Britain's Best Hi-Fi Equipment", H. J. Leak & Co. Ltd., Hi-Fi News, January 1967, Pages 808-809.

"2000 Series Sandwich Loudspeakers", Rank Radio International Ltd, Audio (IPC Magazines), Oct 1974.

"Leak '73", Delta Series sales pamphlet, Rank Radio International, 1973.

Leak 2000 Series sales pamphlets, Rank Audio Visual 1974.

"Leak Hi-Fi", 3000 Series sales pamphlet, Rank Hi Fi, 1976.

"TL/25 Plus", Audio (USA), September 1957.

"New for 1964", Hi-Fi News, Feb 1964, P674.

"Stereo 30" Hi-Fi News, April 1964, P864.

"Togetherness", Hi-Fi News, Page 1568, December 1969.

Other Advertisements

"Osram KT61", Wireless World, Feb 16, 1939, Page 136.

"Osram KT66", Wireless World, Aug 27 and Dec 30, 1937.

Haddon Transformers, Wireless World, Sept 1956, Page 122.

Hinchley Engineering, Wireless World: Feb 1956 Page 22; Apr 1956 Page 23; Sept 1958 Page 52.

Partridge Transformers, Wireless World, June 1945, Page 24.

Telegraph Condenser Company, Wireless World, Inside Rear Cover, June 1945.

Associated Electronic Engineers Ltd, Wireless World, 1945.

General Audio Bibliography

Radio Tuners

"Design of a Tunable Line Oscillator for 280 - 450 Mc/s", B. J. Salley, Wireless World, June 1945, Page 170, (relates to the oscillator design in the Leak Trough-Line).

"Design for an F.M. Tuner", S. W. Amos and G. C. Jonhstone" Wireless World, April 1955, Page 159, and May 1955, Page 216.

"Stereo by FM Multiplex", A. Csicsatka and R. Linz, Hi-Fi News, Nov 1961, P367.

"Transistorised Decoders for FM Stereo", G. D. Browne, Hi-Fi News, June 1963, P38.

"World of Wireless: Stereo Broadcasting", Wireless World, June 1961.

Loudspeakers

"Ribbon Loudspeaker", P. L. Taylor, Wireless World, January 1951, Page 7.

"How the Ionophone Works", A. E. Falkus, Fane Acoustics Ltd, Hi-Fi News, May 1965, Page 1111.

"Loudspeaker Without Diaphragm", F.L.D, (discusses the Ionophone), Wireless World, Jan 1952, Page 2.

"Wharfedale's Way Ahead", John Crabbe, Hi-Fi News, May 1967, Page 1314.

"Metal Cone Loudspeaker", F. H. Brittain, Wireless World, Nov and Dec 1952.

"Wide Range Electrostatic Loudspeaker", P. J. Walker, Wireless World, May, June and August 1955.

"Frankly Speaking" (refers to the Ionophone), J. Berridge, Hi-Fi News, Nov 1961, P397.

"A Flat Speaker", S. Kelly, Hi-Fi News, Dec 1959, P477.

"Audio News" (refers to Wharfedale use of polystyrene in loudspeaker cabinet wall construction), Hi-Fi News, Dec 1959, P469.

"Loudspeaker Cabinet Design", D. E. L. Shorter, Wireless World, December 1950, P436, (describes the BBC monitor loudspeaker with TL/12 amplifier).

Amplifiers

"An Ultra-Linear Amplifier", David Hafler and Herbert Keroes, Audio Engineering, November 1951, Page 15.

"Ultra-Linear Operation of the Williamson Amplifier", David Hafler and Herbert Keroes, Audio Engineering, June 1952, Page 26.

"Push-Pull Input Circuits: Part 5 – The Cathode Coupled Stage", by W. T. Cocking, Wireless World, May 1948. (Not a direct reference to Leak, but an excellent explanation and analysis of the phase splitter circuit used in Leak valve amps.)

"The Class Problem", Hi-Fi News, Nov 1967, Page 591, and continued in March 1968 issue, Page 317.

"Electric Gramophone", M. G. Scroggie, Wireless World, May 11, 1939, Page 432.

"Design for a 20 Watt High Quality Amplifier", W. A. Ferguson, Wireless World, May and June 1955. Also published in "High Quality Sound Reproduction", a handbook from Mullard Ltd., Century House, London.

"Design For a High Quality Amplifier", D. T. N. Williamson, Wireless World, April and May 1947. Revised circuit with improved stability: Wireless World, Aug 1949, Page 282 and Oct, Nov 1949.

"The Williamson Amplifier of 1947", P. R. Stinson, Sydney, Australia, February 1991.

"Partridge P. A. Manual", N. Partridge, King's Buildings, Dean Stanley Street, London SW1.

"Amplifiers and Superlatives", D. T. N. Williamson and P. J. Walker, Wireless World, Sept 1952, Page 357.

"Quasi-Complementary Transistor Amplifier", H. C. Lin, Electronics, September 1956, Page 173.

"A New Amplifier", A. R. Bailey and A. H. Radford, Hi-Fi News, June, July, Aug and Sept 1962.

"Design for a 50 Watt Amplifier", W. I. Heath and G. R. Woodville, Wireless World, April 1957, P158.

"King Size Quarter Horse Power Stereo Amplifier, R. M. Voss and R. Ellis, Audio, October 1966, P23.

"High Stability Feedback Amplifier", Arnold Kauder, Audio, October 1962, P18.

"Another Power Amplifier", R. A. Greiner, Audio, September, 1961, P19.

"Transistor Audio Power Amplifier", R. Tobey and J. Dinsdale, Wireless World, November 1961, Page 565.

"Views Too: The Tobey-Dinsdale Amplifier", Prof. Jack Dinsdale, Hi-Fi News and Record Review, October 1996, P129.

"Transistorised Equipment Survey" Hi-Fi News, Oct 1963, P322.

Circuit Design

"Output Stage and Loudspeaker", F. Langford-Smith, Wireless World, Feb 9, 1939.

"Negative Feedback Tone Control", P. J. Baxandall, Wireless World, October 1952, Page 402.

"Negative Feedback", E. Griffiths, Wireless World, March 1950, Page 111.

"Pickup Input Circuits", R. L. West and S. Kelly, Wireless World, Page 38.

"When Negative Feedback Isn't Negative", 'Cathode Ray', Wireless World, May 1949, Page 189.

"The Nyquist Diagram at Work", 'Cathode Ray', Wireless World, Jan 1956, Page 43.

"Cascode A. F. Amplifier", Wireless World, June 1956, Page 283.

"Single Ended Push Pull Amplifier", Wireless World, May 1952.

"Output Transformer Design", R. F. Gilson, Wireless World, Apr 1955, Page 195.

"The See-Saw Circuit – A Self Balancing Phase Splitter", M. G. Scoggie, Wireless World, July 1945, Page 194.

"New Phase Splitter", A. R. Bailey, Wireless World, September 1962, Page 411 (a description of the phase splitter used in Radford amplifiers).

"Low Distortion Class B Output", (A description of the Quad 303 output stage), Wireless World, April 1968, Page 67.

"Valve Amplifiers", Morgan Jones, Published by Newnes, 1995.

Other

"H. A. Hartley, Inventor of 'High Fidelity' ", "Audio News", Hi-Fi News, December 1965, Page 633.

"Plastic Deformation and Wear of Gramophone Records", D. A. Barlow, BSRA Journal, Feb 1958.

"Groove Deformation in Gramophone Records", D. A. Barlow, Wireless World, April 1964, Page 160.

"Metallised Paper Capacitors", Technical Staff of H. H. Hunt Ltd, Wireless World, Dec 1949, Supplement Page 10.

"Drawing Circuit Diagrams", L. Bainbridge-Bell, (Royal Naval Scientific Service), Wireless World, May 1949, (see also December 1948).

"Two Channel Stereophonic Sound Systems", Wireless World, May 1956, Page 206.

"Radiotron Designers Handbook" edited by Fritz Langford-Smith, published by Wireless Press for Amalgamated Wireless Valve Company Pty Ltd.

"High Quality Sound Reproduction", by James Moir, published by Chapman and Hall Ltd, 1961.

Some Abbreviations, Symbols and Units

Length:
1"= 1 inch = 25.4mm
1 mil = 0.001"

Pressure:
dB = deciBel; 94dB = 1 Pascal

Frequency:
Hz = Hertz = cycles per second = cps
1kHz = 1,000 Hz

Electrical Potential Difference
V = Volt; 1mV = 1milliVolt = 0.001V
HT = high tension (ie: high voltage)
LT = low tension (ie: low voltage)

Electrical Current
A = Ampere; 1A = 1,000mA
DC = direct current; AC = alternating current

Electrical Power
1 Watt = 1 Joule per second = 1 Volt.Amp (DC)

Electrical Resistance
Ω = Ohms; 1kΩ = 1000 Ohms; 1MΩ = 1,000kΩ

Electrical Capacitance
F = Farad; 1μF = 0.000001F = 1,000nF
1nF = 1,000pF

Other:
Pty Ltd = Proprietary Limited
LP = Long Play(ing)

Packing boxes for Leak equipment in the late 1950's and early 1960's featured this surprisingly "hip" design on the box-label.

This version of the Point One Stereo pre-amp with a matching Trough-Line II FM tuner made a desirable package for the American market.

Index

Symbols

D

E

F

G

H

I

J

K

L

M

N

O

P

U

V

W

Abused, unloved and then unused for decades, but still in operational condition when discovered. (Photo courtesy Patrick Cheffins, United Kingdom)

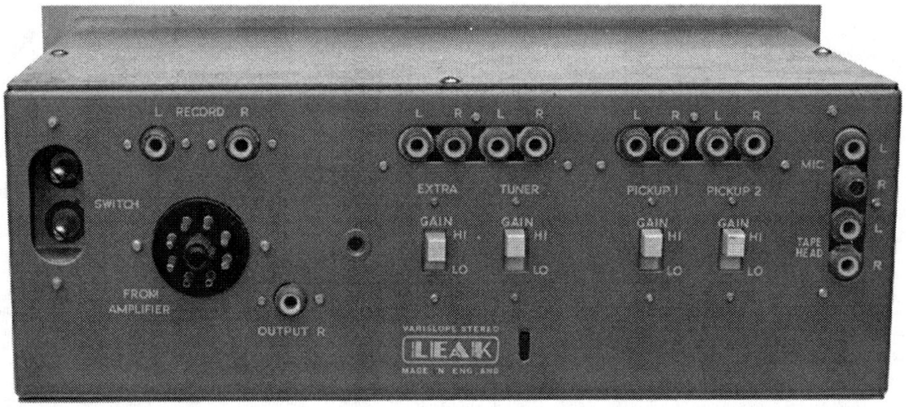

Rear panel of the Varislope Stereo pre-amp. Note that a single RCA plug for "Output Right" was provided so the pre-amp could be used with two mono power amps - one was connected via the normal multi-pin cable, and the other was fed audio from the "Output Right" connector.

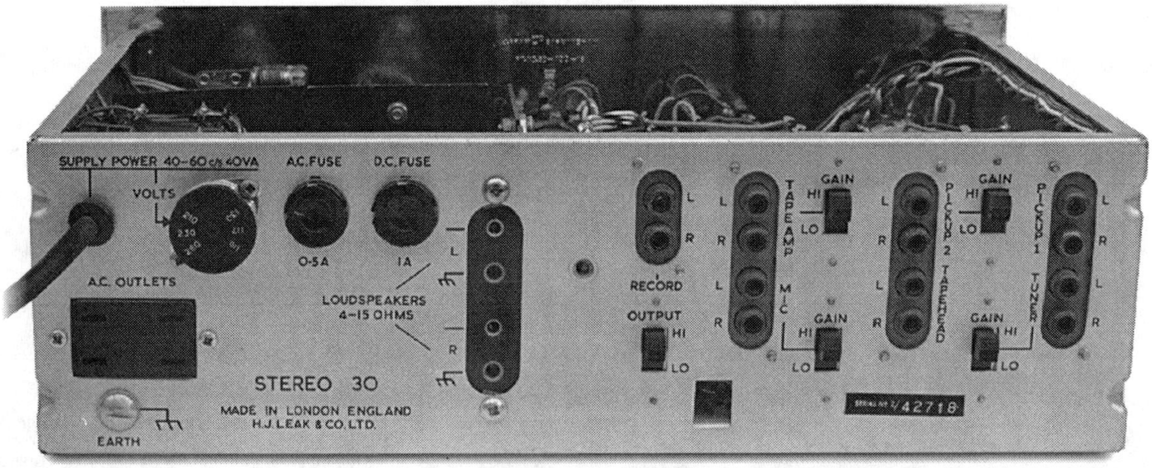

Rear panel of the Stereo 30, showing adjustable gain settings for each input.

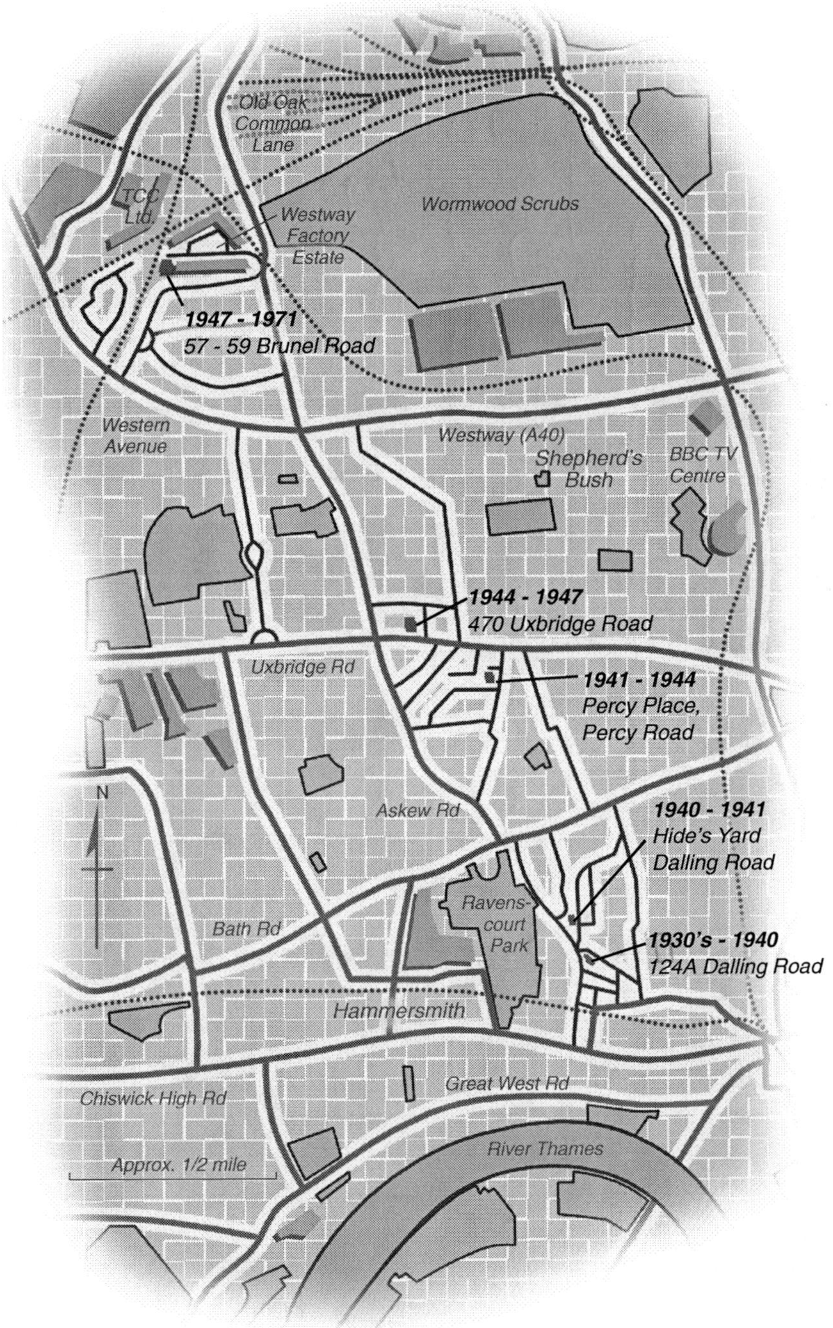

Above: A map of West London showing the five locations for H. J. Leak & Co. Ltd. from the 1930's to 1971.

Overleaf Below: The traditionally styled H. J. Leak & Co. Ltd. address and contact details used at the foot of many company publications.

About the Author

Stephen Spicer is an Australian with an interest in audio technology from the 1930's onwards. Nearing his teenage years in 1969, his audio hobby started with a Garrard record changer, a Ronette ceramic pickup, a single-ended 6V6 amplifier and a Rola loudspeaker mounted on an open-baffle.

After undertaking secondary schooling at Scotch College (Melbourne), he attended Monash University where he qualified with a Bachelor of Electrical Engineering (Honours), and has worked in the field of telecommunications research since graduating in 1981. Currently he leads project work in the area of multimedia and video-based Internet services.

He is a member of the Melbourne Audio Club and the Victorian 3D Society where, in the latter, he has developed skills in stereoscopic photography. These interests have been brought together in this book to create a detailed, well illustrated history of the Leak brand-name.

Harold Leak (left) running with his Chief Engineer, Ted Ashley, in a three-legged race.
The photo was taken during a sports-day at the school attended by John Leak in the 1950's.
(Photo courtesy E. H. Ashley.)

LONDON
ENGLAND
LEAK
RESEARCH · ENGINEERING · MANUFACTURING

LONDON
ENGLAND

LEAK

RESEARCH • ENGINEERING • MANUFACTURING

For advertisements inserted by
Audio Amateur Press
as required.

For advertisements inserted by
Audio Amateur Press
as required.

For advertisements inserted by
Audio Amateur Press
as required.

For advertisements inserted by
Audio Amateur Press
as required.

For advertisements inserted by
Audio Amateur Press
as required.